Metal Oxide Nanomaterials: From Fundamental to Applications

Metal Oxide Nanomaterials: From Fundamental to Applications

Editors

Yuanbing Mao
Santosh Gupta

MDPI • Basel • Beijing • Wuhan • Barcelona • Belgrade • Manchester • Tokyo • Cluj • Tianjin

Editors

Yuanbing Mao
Department of Chemistry
Illinois Institute of Technology
Chicago
United States

Santosh Gupta
Radiochemistry Division
Bhabha Atomic Research Centre
Mumbai
India

Editorial Office
MDPI
St. Alban-Anlage 66
4052 Basel, Switzerland

This is a reprint of articles from the Special Issue published online in the open access journal *Nanomaterials* (ISSN 2079-4991) (available at: www.mdpi.com/journal/nanomaterials/special_issues/metal_oxide_fundamental_applications).

For citation purposes, cite each article independently as indicated on the article page online and as indicated below:

LastName, A.A.; LastName, B.B.; LastName, C.C. Article Title. *Journal Name* **Year**, *Volume Number*, Page Range.

ISBN 978-3-0365-6210-0 (Hbk)
ISBN 978-3-0365-6209-4 (PDF)

© 2023 by the authors. Articles in this book are Open Access and distributed under the Creative Commons Attribution (CC BY) license, which allows users to download, copy and build upon published articles, as long as the author and publisher are properly credited, which ensures maximum dissemination and a wider impact of our publications.

The book as a whole is distributed by MDPI under the terms and conditions of the Creative Commons license CC BY-NC-ND.

Contents

About the Editors vii

Preface to "Metal Oxide Nanomaterials: From Fundamental to Applications" ix

Yuanbing Mao and Santosh K. Gupta
Metal Oxide Nanomaterials: From Fundamentals to Applications
Reprinted from: *Nanomaterials* 2022, 12, 4340, doi:10.3390/nano12234340 1

Santosh K. Gupta, Hisham Abdou, Carlo U. Segre and Yuanbing Mao
Excitation-Dependent Photoluminescence of BaZrO$_3$:Eu^{3+} Crystals
Reprinted from: *Nanomaterials* 2022, 12, 3028, doi:10.3390/nano12173028 5

Alexandru Enesca and Luminita Andronic
UV-Vis Activated Cu$_2$O/SnO$_2$/WO$_3$ Heterostructure for Photocatalytic Removal of Pesticides
Reprinted from: *Nanomaterials* 2022, 12, 2648, doi:10.3390/nano12152648 21

Yuan-Chang Liang and Tsun-Hsuan Li
Sputtering-Assisted Synthesis of Copper Oxide–Titanium Oxide Nanorods and Their Photoactive Performances
Reprinted from: *Nanomaterials* 2022, 12, 2634, doi:10.3390/nano12152634 37

Xiaodong Si, Mingliang Luo, Mingzhong Li, Yuben Ma, Yige Huang and Jingyang Pu
Experimental Study on the Stability of a Novel Nanocomposite-Enhanced Viscoelastic Surfactant Solution as a Fracturing Fluid under Unconventional Reservoir Stimulation
Reprinted from: *Nanomaterials* 2022, 12, 812, doi:10.3390/nano12050812 57

Xiaoyan Liu, Shaotong Feng, Caihua Wang, Dayun Yan, Lei Chen and Bao Wang
Wettability Improvement in Oil–Water Separation by Nano-Pillar ZnO Texturing
Reprinted from: *Nanomaterials* 2022, 12, 740, doi:10.3390/nano12050740 73

Maria V. Maevskaya, Aida V. Rudakova, Alexei V. Emeline and Detlef W. Bahnemann
Effect of Cu$_2$O Substrate on Photoinduced Hydrophilicity of TiO$_2$ and ZnO Nanocoatings
Reprinted from: *Nanomaterials* 2021, 11, 1526, doi:10.3390/nano11061526 87

Simeon Simeonov, Anna Szekeres, Dencho Spassov, Mihai Anastasescu, Ioana Stanculescu and Madalina Nicolescu et al.
Investigation of the Effects of Rapid Thermal Annealing on the Electron Transport Mechanism in Nitrogen-Doped ZnO Thin Films Grown by RF Magnetron Sputtering
Reprinted from: *Nanomaterials* 2021, 12, 19, doi:10.3390/nano12010019 99

Livia Naszályi Nagy, Evert Dhaene, Matthias Van Zele, Judith Mihály, Szilvia Klébert and Zoltán Varga et al.
Silica@zirconia Core@shell Nanoparticles for Nucleic Acid Building Block Sorption
Reprinted from: *Nanomaterials* 2021, 11, 2166, doi:10.3390/nano11092166 117

Tao Li, Wen Yin, Shouwu Gao, Yaning Sun, Peilong Xu and Shaohua Wu et al.
The Combination of Two-Dimensional Nanomaterials with Metal Oxide Nanoparticles for Gas Sensors: A Review
Reprinted from: *Nanomaterials* 2022, 12, 982, doi:10.3390/nano12060982 139

Jongbae Kim and Taejong Paik
Recent Advances in Fabrication of Flexible, Thermochromic Vanadium Dioxide Films for Smart Windows
Reprinted from: *Nanomaterials* 2021, 11, 2674, doi:10.3390/nano11102674 179

About the Editors

Yuanbing Mao

Dr. Yuanbing Mao is the Department Chair and a Professor of Chemistry at the Illinois Institute of Technology. He achieved his B.Sc. degree at Xiangtan University, his M.Sc. degree at the Institute of Chemistry, Chinese Academy of Sciences, and his Ph.D. degree from the State University of New York at Stony Brook. He has earned several awards, including the Department of Defense Young Investigator Award and the Outstanding Mentorship Award from the Council on Undergraduate Research, and is a recipient of the DOE Visiting Faculty Program. His research interests include nanomaterials, solid-state science, and nanoscience, with expertise in optoelectronics, energy storage, and conversion, as well as environmental remediation. To date, he has published more than 150 peer-reviewed journal articles, as well as some book chapters and patents.

Santosh Gupta

Dr. Santosh Kumar Gupta joined the 53rd batch of Bhabha Atomic Research Centre Training School and, subsequently, joined the Radiochemistry Division in 2010. Dr. Gupta is a graduate of Delhi University (B.Sc.), Indian Institute of Technology, Delhi (M.Sc.), and the Homi Bhabha National Institute at Mumbai, India (Ph.D.). Dr. Gupta is the recipient of several awards notable among them are Fulbright, Indo-US, and JSPS Fellowship, DAE Young Scientist and Group achievement award, Scientific India Young Scientist award, Society of Materials Chemistry and Indian association of Nuclear Chemists and allied scientists Bronze medal for year 2021. He has received the Fulbright Outreach award twice in 2018 and 2019 to deliver talks at Texas A&M and FICS University, USA. He has also been bestowed with membership of the Indian National Young Academy of Sciences (INSA-INYAS) and National Academy of Sciences (NASc) and young associate of Maharashtra Academy of Sciences (MASc) for the year 2021. He has received Royal Society of Chemistry certificate for among 10% most cited author and ICDD certificate for contribution of new XRD pattern. He was featured in Top 2% most influential scientists worldwide based on the Scopus publications impact consecutively for the year 2020 and 2021 prepared by Prof. John PA Loannidis of Stanford University and his team and published by Elsevier. As of today, he has published more than 205 journal articles with around 4800 citations and an h-index of 43 and i-10 index of 115. His area research focuses on designing light emitting materials for health, energy and environment etc.

Preface to "Metal Oxide Nanomaterials: From Fundamental to Applications"

The fundamental understanding and applications of metal oxide nanomaterials have been explored and expanded extensively over the past couple of decades, and continuously so, with creative and diligent work performed globally by researchers and scientists. Nanomaterials, more generally, have indeed revolutionized the world due to their unique properties and growing applications in all spheres of humankind, encompassing energy, health, and the environment. However, there are no recent collections of such diverse work, such as books, connecting their fundamental aspects and application perspectives in a concise fashion to give a broad view of the current status of this fascinating field. Therefore, we invited authors to contribute either comprehensive review articles or original research articles covering the most recent progress and new developments in the fundamental understanding of the synthesis and properties and the exploration of the utilization of metal oxide nanomaterials.

Specifically, in this Special Issue entitled "Metal Oxide Nanomaterials: From Fundamentals to Applications", we collected eight original research articles and two comprehensive review articles to highlight the development and understanding of different types of metal oxide nanoparticles and their use for applications in luminescence, photocatalysis, water–oil separation, optoelectronics, gas sensors, energy-saving smart windows, etc. The wide variety of applications covered by the ten articles published here is proof of the growing attention that the use of metal oxide nanomaterials has received in recent years.

This Special Issue presents just the tip of the iceberg of the broad, dynamic, and active fundamental research and applications in the developing field of metal oxide nanomaterials by collecting a few examples of the latest advancements. We hope that readers will enjoy reading these articles and find them useful for their research.

We appreciate all the contributors of this Special Issue for their wisdom and solid scientific work. We also thank our Section Managing Editors, Tracy Jin and Winston Yi, for their tireless support.

Yuanbing Mao and Santosh Gupta
Editors

Editorial

Metal Oxide Nanomaterials: From Fundamentals to Applications

Yuanbing Mao [1,*] and Santosh K. Gupta [2,3,*]

1 Department of Chemistry, Illinois Institute of Technology, 3101 South Dearborn Street, Chicago, IL 60616, USA
2 Radiochemistry Division, Bhabha Atomic Research Centre, Trombay, Mumbai 400085, India
3 Homi Bhabha National Institute, Anushaktinagar, Mumbai 400094, India
* Correspondence: ymao17@iit.edu (Y.M.); santoshg@barc.gov.in (S.K.G.);
Tel.: +1-312-567-3815 (Y.M.); +91-22-25590636 (S.K.G.)

This Special Issue of *Nanomaterials*, "Metal Oxide Nanomaterials: From Fundamentals to Applications", highlights the development and understanding of different types of metal oxide nanoparticles and their use for applications in luminescence, photocatalysis, water–oil separation, optoelectronics, gas sensors, energy-saving smart windows, etc. The wide variety of applications covered by the 10 articles published here is proof of the growing attention that the use of metal oxide nanomaterials has received in recent years. Here, nanomaterials are defined, based on the October 2011 European Commission's definition, as materials with one or more external dimension in the range of 1–100 nm. As the surface area per mass of a material increases, a greater proportion of the material can come into contact with the surrounding materials, thus affecting reactivity [1,2]. Nanomaterials have indeed revolutionized the world due to their unique properties and growing applications in all spheres of humankind, encompassing energy, health, and the environment.

Within this Special Issue, Gupta et al. designed and demonstrated excitation energy tunable light emission from barium zirconium oxide crystals as color-tunable phosphors [3]. The authors tuned the emission light from $BaZrO_3:Eu^{3+}$ crystals from orange to red based on the charge transfer and f-f transition excitation of an Eu^{3+} dopant, which is dictated by its magnetic and electric dipole transition probabilities. Enesca et al. showed the potential of $Cu_2O/SnO_2/WO_3$ heterostructure powder in the efficient removal of pesticides photocatalytically [4]. The photocatalytic mechanism corresponds to a charge transfer based on this three-component structure, where Cu_2O exhibited a reduction potential responsible for O_2 production and WO_3 had an oxidation potential responsible for OH· generation. Liang and his colleagues successfully decorated TiO_2 nanorods with a copper oxide layer through sputtering and post-annealing, which resulted in improved light absorption and photo-induced charge separations [5]. This led the composite nanorods to have enhanced photoactivity compared to the pristine TiO_2 nanorods.

Other fascinating properties of nanocomposite based on metal oxides nanoparticles, particularly from ZnO, were explored by several other research groups [6]. For example, Si et al. synthesized $Fe_3O_4@ZnO$ nanocomposites (NCs) to improve the stability of the viscoelastic surfactant (VES) fracturing fluid [7]. At a loading of 0.1 wt.%, this NC-VES nanocomposite showed superior stability at 95 °C or at a high shear rate and good sand-carrying performance and gel-breaking properties. Designing a surface with special wettability is an important approach to improving the separation efficiency of oil and water. Liu and co-authors demonstrated superhydrophobicity in both oil and water from their stainless-steel metal fibers coated with sol–gel-derived ZnO nano-pillars [8]. They found that their ZnO-coated stainless-steel metal fibers had a static underwater oil contact angle of 151.4° ± 0.8° and an underoil water contact angle of 152.7° ± 0.6° and was a highly promising candidate for both water-in-oil and oil-in-water separation in the industry. Maevskaya et al. studied the effect of Cu_2O on the photo-induced alteration of the hydrophilicity of TiO_2 and ZnO surfaces [9]. The Cu_2O/TiO_2 and Cu_2O/ZnO heterostructures showed photo-induced decay of the surface hydrophilicity caused by both

UV and visible light irradiation. Simeonov and his group carried out defect engineering of ZnO by nitrogen doping [10]. They demonstrated that nitrogen doping in ZnO led to an abundance of oxygen and zinc vacancies and interstitials and contributed to enhanced electron transport properties in ZnO:N films. As another example demonstrating advancements in the roles of structure and morphology in material properties, Naszályi Nagy et al. prepared silica NPs with a diameter of 50 nm and covered them with a monoclinic/cubic zirconia shell using a green, cheap, and up-scalable sol–gel method [11]. They confirmed that these silica@zirconia core@shell NPs bind as muchas 207 mg of deoxynucleoside monophosphates on 1 g of this nanocarrier at neutral physiological pH while maintaining good colloidal stability.

Two review articles are also included in this Special Issues. In the first review, Li et al. comprehensively discussed the development of two-dimensional (2D) nanomaterials with metal oxide nanoparticles for gas sensing applications [12]. They further emphasized recent advances in the fabrication of gas sensors based on metal oxides, 2D nanomaterials, and 2D material/metal oxide composites with highly sensitive and selective functions. In the second review article, Kim et al. presented recent advances in fabricating flexible thermochromic $VO_2(M)$ thin films using vacuum deposition methods and solution-based processes and discussed their optical properties for potential applications in energy-saving smart windows and several other emerging technologies [13].

In summary, this Special Issue presents just the tip of the iceberg of the broad, dynamic, and active fundamental research and applications in the developing field of metal oxide nanomaterials by collecting a few examples of the latest advancements. We hope that the readers enjoy reading these articles and find them useful for their research.

Author Contributions: Both authors have contributed equally. All authors have read and agreed to the published version of the manuscript.

Funding: Y.M. thanks the financial support by the IIT startup funds. S.K.G. thanks the United States-India Education Foundation (USIEF, India) and the Institute of International Education (IIE, USA) for his Fulbright Nehru Postdoctoral Fellowship (award #2268/FNPDR/2017).

Conflicts of Interest: The authors declare no conflict of interest.

References

1. Gupta, S.K.; Mao, Y. A review on molten salt synthesis of metal oxide nanomaterials: Status, opportunity, and challenge. *Prog. Mater. Sci.* **2021**, *117*, 100734. [CrossRef]
2. Gupta, S.K.; Mao, Y. Recent Developments on Molten Salt Synthesis of Inorganic Nanomaterials: A Review. *J. Phys. Chem. C* **2021**, *125*, 6508–6533. [CrossRef]
3. Gupta, S.K.; Abdou, H.; Segre, C.U.; Mao, Y. Excitation-Dependent Photoluminescence of $BaZrO_3$:Eu^{3+} Crystals. *Nanomaterials* **2022**, *12*, 3028. [CrossRef] [PubMed]
4. Enesca, A.; Andronic, L. UV-Vis Activated $Cu_2O/SnO_2/WO_3$ Heterostructure for Photocatalytic Removal of Pesticides. *Nanomaterials* **2022**, *12*, 2648. [CrossRef] [PubMed]
5. Liang, Y.-C.; Li, T.-H. Sputtering-assisted synthesis of copper oxide–titanium oxide nanorods and their photoactive performances. *Nanomaterials* **2022**, *12*, 2634. [CrossRef] [PubMed]
6. Gupta, S.K.; Mohan, S.; Valdez, M.; Lozano, K.; Mao, Y. Enhanced sensitivity of caterpillar-like ZnO nanostructure towards amine vapor sensing. *Mater. Res. Bull.* **2021**, *142*, 111419. [CrossRef]
7. Si, X.; Luo, M.; Li, M.; Ma, Y.; Huang, Y.; Pu, J. Experimental Study on the Stability of a Novel Nanocomposite-Enhanced Viscoelastic Surfactant Solution as a Fracturing Fluid under Unconventional Reservoir Stimulation. *Nanomaterials* **2022**, *12*, 812. [CrossRef]
8. Liu, X.; Feng, S.; Wang, C.; Yan, D.; Chen, L.; Wang, B. Wettability Improvement in Oil–Water Separation by Nano-Pillar ZnO Texturing. *Nanomaterials* **2022**, *12*, 740. [CrossRef] [PubMed]
9. Maevskaya, M.V.; Rudakova, A.V.; Emeline, A.V.; Bahnemann, D.W. Effect of Cu_2O substrate on photoinduced hydrophilicity of TiO_2 and ZnO nanocoatings. *Nanomaterials* **2021**, *11*, 1526. [CrossRef] [PubMed]
10. Simeonov, S.; Szekeres, A.; Spassov, D.; Anastasescu, M.; Stanculescu, I.; Nicolescu, M.; Aperathitis, E.; Modreanu, M.; Gartner, M. Investigation of the Effects of Rapid Thermal Annealing on the Electron Transport Mechanism in Nitrogen-Doped ZnO Thin Films Grown by RF Magnetron Sputtering. *Nanomaterials* **2021**, *12*, 19. [CrossRef]

11. Naszályi Nagy, L.; Dhaene, E.; Van Zele, M.; Mihály, J.; Klébert, S.; Varga, Z.; Kövér, K.E.; De Buysser, K.; Van Driessche, I.; Martins, J.C. Silica@zirconia Core@shell Nanoparticles for Nucleic Acid Building Block Sorption. *Nanomaterials* **2021**, *11*, 2166. [CrossRef]
12. Li, T.; Yin, W.; Gao, S.; Sun, Y.; Xu, P.; Wu, S.; Kong, H.; Yang, G.; Wei, G. The combination of two-dimensional nanomaterials with metal oxide nanoparticles for gas sensors: A review. *Nanomaterials* **2022**, *12*, 982. [CrossRef] [PubMed]
13. Kim, J.; Paik, T. Recent advances in fabrication of flexible, thermochromic vanadium dioxide films for smart windows. *Nanomaterials* **2021**, *11*, 2674. [CrossRef] [PubMed]

Article

Excitation-Dependent Photoluminescence of BaZrO$_3$:Eu^{3+} Crystals

Santosh K. Gupta [1,2], Hisham Abdou [3], Carlo U. Segre [4] and Yuanbing Mao [5,*]

[1] Radiochemistry Division, Bhabha Atomic Research Centre, Trombay, Mumbai 400085, India
[2] Homi Bhabha National Institute, Anushakti Nagar, Mumbai 400094, India
[3] Department of Chemistry, University of Texas Rio Grande Valley, 1201 West University Drive, Edinburg, TX 78539, USA
[4] Center for Synchrotron Radiation Research and Instrumentation and Department of Physics, Illinois Institute of Technology, Chicago, IL 60616, USA
[5] Department of Chemistry, Illinois Institute of Technology, 3105 South Dearborn Street, Chicago, IL 60616, USA
* Correspondence: ymao17@iit.edu; Tel.: +1-312-567-3815

Abstract: The elucidation of local structure, excitation-dependent spectroscopy, and defect engineering in lanthanide ion-doped phosphors was a focal point of research. In this work, we have studied Eu^{3+}-doped BaZrO$_3$ (BZOE) submicron crystals that were synthesized by a molten salt method. The BZOE crystals show orange–red emission tunability under the host and dopant excitations at 279 nm and 395 nm, respectively, and the difference is determined in terms of the asymmetry ratio, Stark splitting, and intensity of the uncommon $^5D_0 \rightarrow {}^7F_0$ transition. These distinct spectral features remain unaltered under different excitations for the BZOE crystals with Eu^{3+} concentrations of 0–10.0%. The 2.0% Eu^{3+}-doped BZOE crystals display the best optical performance in terms of excitation/emission intensity, lifetime, and quantum yield. The X-ray absorption near the edge structure spectral data suggest europium, barium, and zirconium ions to be stabilized in +3, +2, and +4 oxidation states, respectively. The extended X-ray absorption fine structure spectral analysis confirms that, below 2.0% doping, the Eu^{3+} ions occupy the six-coordinated Zr^{4+} sites. This work gives complete information about the BZOE phosphor in terms of the dopant oxidation state, the local structure, the excitation-dependent photoluminescence (PL), the concentration-dependent PL, and the origin of PL. Such a complete photophysical analysis opens up a new pathway in perovskite research in the area of phosphors and scintillators with tunable properties.

Keywords: BaZrO$_3$; europium; luminescence; EXAFS; defect

1. Introduction

The trivalent europium ion Eu^{3+} is considered to be one of the most sensitive lanthanide ions that displays environment- and symmetry-sensitive emissions owing to its pure magnetic dipole transition (MDT, $\Delta J = \pm 1$), hypersensitive electric dipole transition (HEDT, $\Delta J = \pm 2$), and neither magnetic nor electric $^5D_0 \rightarrow {}^7F_0$ ($\Delta J = 0$) transition [1–4]. When Eu^{3+} is localized at a highly symmetric site with a center of inversion (C_i), its MDT predominates over its EDT. If Eu^{3+} is situated at a highly asymmetric site, its emission is the other way around [5]. In addition, the Eu^{3+} ion is one of the most fascinating dopant ions for quality red phosphors with a high quantum yield (QY), a decent thermal stability, and a long luminescence lifetime [6,7].

Perovskites with a generic formula ABO$_3$ are in high demand as luminescence hosts due to their structural flexibility, wide band gap, ease of doping, and ability to accommodate lanthanide ions at both A and B sites [8–10]. Among them, BaZrO$_3$ (BZO) is a unique material due to its wide tunable band gap (5.6 eV) [11], high refractive index [12], high proton conductivity, and high chemical and mechanical stability [10,13–15]. It has various applications in the areas of luminescence [16], catalysis [16,17], proton-conducting solid oxide fuel cells [18,19], and many others. Eu^{3+} ion-doped ABO$_3$ perovskites have attracted a lot of attention due to their high thermal and chemical stability, low environmental toxicity, and various applications in photocatalysis, white light generation [20], light emitting diodes (LEDs) [21], and bioimaging [22].

One can probe the local sites of Eu^{3+} ions in ABO_3 perovskites based on the ratio, spectral splitting, and appearance of $^5D_0 \rightarrow {}^7F_J$ (J = 0–4) emissions.[1] This kind of study is crucial to make materials with optimum light emitting properties. For example, Kunti et al. recently observed MDT and HEDT emissions with $I_{MDT} >>> I_{HEDT}$ along with the host emission under the 275 nm excitation from their $BaZrO_3$:Eu samples synthesized by the solid state route [23]. There was a systematic host-to-dopant energy transfer with an increasing Eu^{3+} doping concentration. Based on the analysis of extended X-ray absorption fine structure (EXAFS) spectroscopic data, they concluded that Eu^{3+} ions were localized at Zr^{4+} sites [23]. Gupta, one of the co-authors of the current manuscript, and his coworkers observed spectral profiles with $I_{HEDT} >>> I_{MDT}$ under various excitations from gel combustion-synthesized BZO:Eu samples [12], which was exactly opposite to what was observed by Kunti et al. [23]. Gupta et al. also proposed that a large fraction of Eu^{3+} ions occupied Zr^{4+} sites based on population analysis of lifetime spectra [12]. Kanie et al. synthesized BZO samples with different sizes and shapes and investigated their effects on luminescence [24].

There were also reports on Eu^{3+}-doped perovskites of $SrZrO_3$, $SrSnO_3$, $BaTiO_3$, $BaSnO_3$, and $BaZr_xTi_{1-x}O_3$. For example, Basu et al. proposed that Eu^{3+} ions resided at Sr^{2+} sites at low dopant concentrations and were distributed at both Sr^{2+} and Sn^{4+} sites at high doping levels in their polyol-synthesized $SrSnO_3$ nanoparticles based on EXAFS measurements [25]. The same group further proposed that Eu^{3+} ions occupied the centrosymmetric Sr^{2+} sites up to 1.5% Eu^{3+} doping and, beyond that, the synthesized $SrSnO_3$ nanoparticles formed a separate europium oxide phase based on time resolved emission spectroscopy (TRES) and electron paramagnetic resonance (EPR) studies. Similarly, based on EXAFS studies, Rabufetti et al. found that Eu^{3+} ions resided at Ba^{2+} sites at low dopant concentrations (up to 4% Eu^{3+} doping) but were distributed at both Ba^{2+} and Ti^{4+} sites at high doping levels in their vapor-diffusion sol-gel-synthesized $BaTiO_3$ nanocrystals [26]. Canu et al. have tuned the photoluminescence properties of Eu^{3+}-doped $BaZr_xTi_{1-x}O_3$ perovskite by applying an electric field [27].

There are also reports that studied the effect of changing the A cation of $AZrO_3$:Eu on the luminescence emission intensities [28]. Katyayan et al. studied the impact of co-doping Tb^{3+} with Eu^{3+} on the optical and spectroscopic characteristics of BZO perovskite [29]. Another study investigated the effect of particle size and morphology on the fluorescence behaviors of these metal oxides [24].

Color tunability is achieved from samples with the same dopants and hosts by simply varying the excitation wavelength. For example, Gupta et al. showed different emission characteristics of $SrZrO_3$:Eu^{3+} nanoparticles in terms of the asymmetry ratio (A_{21}) under the excitations with host absorption, charge transfer, and the *f-f* band of Eu^{3+} [30]. Guo et al. synthesized a Bi^{3+} and Eu^{3+} ion co-doped $Ba_9Lu_2Si_6O_{24}$ single-phased phosphor via a conventional high-temperature solid-state reaction [31]. They demonstrated that the relative emission intensity of Bi^{3+} luminescent centers tightly depends on the incident excitation wavelength due to the complex energy transfer processes among these Bi^{3+} centers.

Furthermore, even though the induced electric dipole (ED) $^5D_0 \rightarrow {}^7F_0$ transition is strictly forbidden by the ΔJ selection rule of the Judd–Ofelt theory, there are reported occurrences of it as a well-known example of the breakdown of the selection rules of the Judd–Ofelt theory [1]. For example, Guzmán-Olguín et al. showed an unusual great intensity of the $^5D_0 \rightarrow {}^7F_0$ transition centered at 580 nm when they excited their Eu^{3+}-doped $BaHfO_3$ perovskite ceramic under UV radiation with the wavelength associated with the charge transfer band (272 nm), while this transition was very weak when the sample was excited at 396 or 466 nm wavelengths [32]. One of the co-authors of this manuscript, Gupta, with his co-workers, reported the presence of two Stark components in the $^5D_0 \rightarrow {}^7F_0$ transition from their $Nd_2Zr_2O_7$:Eu phosphor when excited at 256 nm [33].

It is clear that there is no systematic investigation of the luminescence of BZO:Eu nor studies on its $^5D_0 \rightarrow {}^7F_0$ transition under host and Eu^{3+} excitations. In this work, we have first synthesized BZO:Eu submicron crystals using an environmentally friendly molten salt synthesis (MSS) method based on the report by Zhou et al. using barium oxalate and zirconium oxide as precursors and a KOH/NaOH salt mixture as the reaction medium [34]. We studied tuning the red to orange emission ratio from the BZO:Eu crystals by modulating the excitation wavelength and deciphered the local site occupancy of Eu^{3+} ions in BZO with Eu, Ba, and Zr-edge EXAFS analysis. More importantly, other than the weak $^5D_0 \rightarrow {}^7F_3$ at 653 nm, we observed a strong $^5D_0 \rightarrow {}^7F_0$ transition, which is known to be strictly forbidden by both EDT and MDT of Eu^{3+} ions, as based on the Judd–Ofelt theory. This observation suggests the deviation of luminescence properties of the Eu^{3+} dopant in the BZO host from the Judd–Ofelt theory. In other words, it indicates that Eu^{3+} ions are localized in highly asymmetric environments, e.g., C_n, C_{nv}, and C_s point group symmetry, so that the selection rules are relaxed to some extent by the mixing of a low-energy charge transfer state with the $4f^6$ configuration [1]. Moreover, designing functional materials that display excitation wavelength-dependent color tunability and understanding structure–property correlation is invaluable to materials scientists.

2. Experimental

The synthesis and instrumentation characterization of the BZO and BZOE submicron crystals are described in detail in the electronic supplementary information as S1. Briefly, six $Ba_{1-x}ZrO_3$:x%Eu^{3+} (x = 0, 0.5, 1.0, 2.0, 5.0, 10.0) samples were synthesized using the MSS method following a procedure published previously with one of the co-authors of this manuscript. Based on the Eu^{3+} doping levels, the synthesized $Ba_{1-x}ZrO_3$:x%Eu^{3+} samples with x = 0, 0.5, 1.0, 2.0, 5.0, 10.0 are designated as BZO, BZOE-0.5, BZOE-1, BZOE-2, BZOE-5, and BZOE-10, respectively.

3. Results and Discussion

3.1. XRD Patterns

The XRD patterns of the BZO and BZO:Eu samples (Figure 1a) demonstrated that the diffraction peaks of all samples match with the cubic perovskite phase (Pm-$3m$) of BZO (JCPDS No. 74-1299) and no impurity peaks were observed. The substitution of Eu^{3+} for constituent ions is evidently aliovalent and may generate oxygen vacancies when resided at a Zr^{4+} site. In case if some fraction resides at a Ba^{2+} site, the charge compensation may invoke the creation of barium vacancies. As seen in Table 1, the cell parameter variation is complex, which means different defect complex generations at different doping levels.

Table 1. Lattice constants and crystallite sizes of the BZOE obtained from Rietveld refinement of the XRD data shown in Figure 1a.

%Eu	0.0	0.5	1.0	2.0	5.0	10.0
a (Å)	4.1947 (2)	4.1954 (3)	4.1944 (3)	4.1952 (3)	4.1976 (3)	4.1971 (2)
Size (nm)	156 (5)	127 (4)	111 (3)	127 (3)	102 (2)	162 (5)

3.2. FTIR and Raman Spectroscopy

To further confirm the formation of the perovskite phase and rule out the formation of other phases, FTIR spectra of the samples were collected (Figure 1b). The only observed peak around 570 cm^{-1} can be assigned to the anti-symmetric stretching Zr–O bond of the octahedral ZrO_6 unit of the $BaZrO_3$ lattice [35–37].

In the Raman spectra of the BZO and BZOE samples (Figure 1c), the peak around 600–900 cm^{-1} is attributed to the symmetric stretch (ν) of the Zr–O bonds in $BaZrO_3$ [23]. With an increasing Eu^{3+} doping level, two extra peaks around 283 and 338 cm^{-1} that correspond to symmetric A_g and degenerated F_g modes of the stretching vibrations of the C_2-octahedron (Eu_2-O) started to appear [38]. This means that Eu^{3+} ions stop going into the BZO lattice and precipitate as a separate phase of Eu_2O_3 at the doping concentration of 10.0%, which is similar to what Basu et al. observed from their polyol-synthesized $SrSnO_3$ nanoparticles based on EXAFS measurements [25]. The peaks between 100 and 230 cm^{-1} can be assigned to $BaCO_3$ impurity, which did not show up in the XRD patterns and FTIR spectra due to a low percentage. The carbonate phase probably resulted from the chemisorption of atmospheric CO_2 on the surface of the BZO crystals upon its exposure to air. It was reported that the existence of such an impurity phase has no effect on the luminescence properties of the BZO perovskite [8].

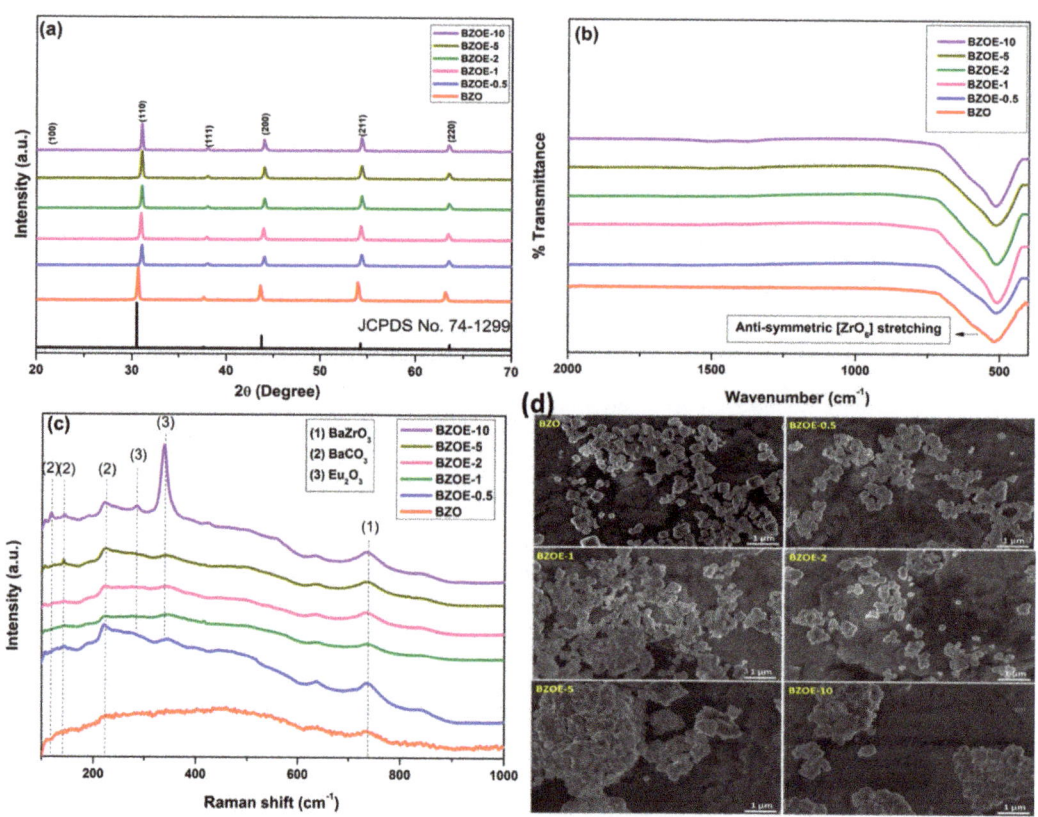

Figure 1. (**a**) XRD patterns, (**b**) FTIR spectra, (**c**) Raman spectra, and (**d**) SEM images of the BZO and BZOE samples.

3.3. SEM Images

The SEM images of the BZO and BZOE samples (Figure 1d) demonstrated that the particles were composed of a mixture of spheres and cubes with well-defined edges. In our earlier work, we found that cubic BZO microcrystals predominated when the synthesis was conducted at a higher annealing temperature. There was almost an equal number of spherical and cubical particles from these samples. No difference in the shape of the particles was noticed from these samples with different Eu^{3+} doping concentrations. However, the agglomeration of the particles increased with an increasing Eu^{3+} concentration. Based on the particle size distribution histograms of these samples obtained using the ImageJ software (Figure S1) and the crystallite sizes obtained from the XRD data (Table 1, Figure S2), no clear correlation between the average particle size and the Eu^{3+} doping concentration was established.

3.4. X-ray Absorption Spectroscopy

3.4.1. XANES

Figure 2a–c shows the normalized XANES spectra of three BZOE samples along with their standards (either the undoped BZO crystals or commercial Eu_2O_3 powder) at the Ba L_3 (Figure 2a), Zr K (Figure 2b), and Eu L_3 (Figure 2c) edges, respectively. The normalized XANES spectra at the Ba L_3-edge shown in Figure 2a are characterized by a sharp white line, which is the main absorption peak due to the transition $2p_{3/2} \rightarrow 5d$. There is no appreciable difference in this edge upon Eu^{3+} doping.

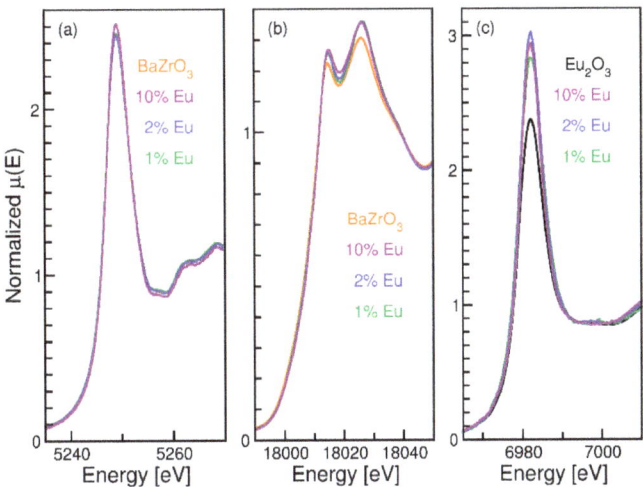

Figure 2. Normalized XANES spectra of the BZOE samples and standards at the (**a**) Ba L_3-edge, (**b**) Zr K-edge, and (**c**) Eu L_3-edge.

It can be seen from Figure 2b that the Zr absorption edges of the BZOE samples coincide with that of the BZO sample, confirming that the oxidation state of Zr is 4+ in the doped samples. Two peaks, A (18,010 eV) and B (18,021 eV), observed in the BZO and BZOE samples just above the Zr absorption edge, are similar to those obtained by Fassbender et al. [39] and Giannici et al. [40] and are due to the octahedral oxygen coordination of Zr^{4+} in the samples. The overall shape of the Zr XANES spectra remains nearly unchanged upon Eu^{3+} doping, and it is suggested that the octahedral symmetry of the Zr atom does not break with doping. The slight increase in the A and B peaks of the doped samples compared to BZO suggest that the Eu^{3+} solubility is very limited (no more than 1–2%).

The Eu L_3-edge XANES spectra of the samples (Figure 2c) show that the absorption edges coincide with that of the standard Eu_2O_3 sample, suggesting that the Eu dopant remains in the Eu^{3+} oxidation state in the BZOE samples. The increase in the white line at the edge indicates an increase in the empty Eu d-states at the Fermi level in the BZOE samples compared to Eu_2O_3, suggesting that the Eu^{3+} is at least partially in a different local environment than in Eu_2O_3.

3.4.2. EXAFS

Figure 3 presents the k^2-weighted Fourier transformed spectra, $|\chi(R)|$, of the BZO and BZOE samples (and Eu_2O_3 standard) at the Ba L_3 (Figure 3a), Zr K (Figure 3b), and Eu L_3 (Figure 3c) edges. For BZO in a cubic perovskite structure (ABO_3) with the space group Pm-$3m$, the Zr atoms are coordinated with 6 O atoms in a regular octahedral (6-fold (ZrO_6)) shape and the Ba atoms are coordinated with 12 O atoms in a cuboctahedral (12-fold (BaO_{12})) shape in the first coordination shells. Theoretical EXAFS spectra have been generated using the above structure for the Ba (Figure S3), Zr (Figure S4), and Eu (Figure S5) edges of the BZO and BZOE samples and fitted to the experimental data.

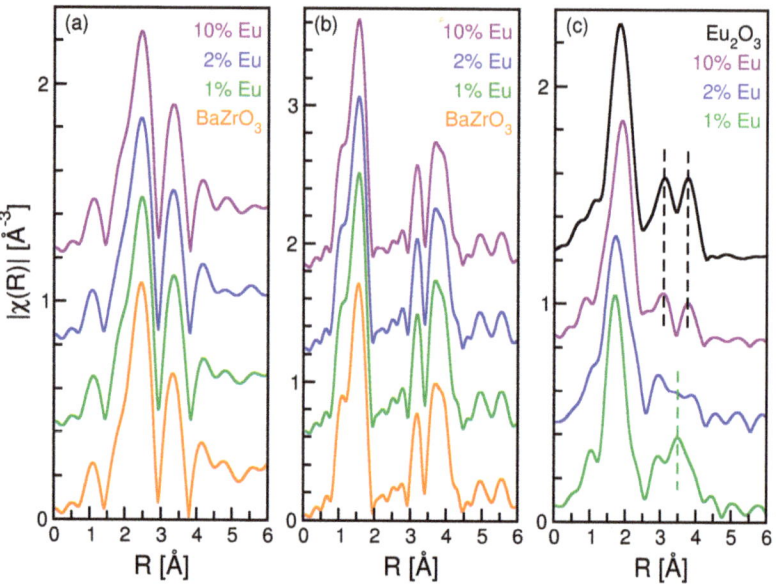

Figure 3. Fourier transformed spectra of the BZOE samples and standards at the (**a**) Ba L$_3$ edge, (**b**) Zr K edge, and (**c**) Eu L$_3$ edge. Spectra are shifted vertically for clarity.

Specifically, the Ba edge was fitted with a structural model including three paths, Ba–O (12-fold), Ba–Zr (8-fold), and Ba–Ba (6-fold). For the Ba edge fits, the path degeneracies were held constant and the σ^2 of the Ba–Zr and Ba–Ba paths were constrained to be identical. Windows of 2.0 Å^{-1} < k < 8.5 Å^{-1} with dk = 2.0 and 1.5 Å < R < 3.7 Å with dR = 0.2 were used for the Fourier transformation and fits, respectively. The fit results for the Ba–O paths of each sample are presented in Table 2 and they indicate that the Ba environment is unchanged by Eu doping. The results for the other paths are in Table S1.

Table 2. Values of the amplitude reduction factor (S_0^2) or path degeneracy (N), bond length, and disorder factor for the near neighbor paths obtained from EXAFS analysis of the BZO and BZOE samples at the Ba L$_3$, Zr K, and Eu L$_3$ edges.

Scattering Path	Parameter	BZO	BZOE-1	BZOE-2	BZOE-10
Ba–O N = 12	S_0^2	0.74 ± 0.16	0.78 ± 0.18	0.80 ± 0.17	0.78 ± 0.17
	R (Å)	2.91 ± 0.02	2.91 ± 0.02	2.91 ± 0.02	2.91 ± 0.02
	σ^2	0.011 ± 0.005	0.013 ± 0.005	0.013 ± 0.005	0.013 ± 0.005
Zr–O N = 6	S_0^2	0.90 ± 0.10	1.0 ± 0.1	1.0 ± 0.1	1.1 ± 0.1
	R (Å)	2.10 ± 0.01	2.10 ± 0.01	2.10 ± 0.01	2.11 ± 0.01
	σ^2	0.004 ± 0.001	0.005 ± 0.002	0.005 ± 0.002	0.006 ± 0.002
Scattering Path	Parameter	Eu$_2$O$_3$	BZOE-1	BZOE-2	BZOE-10
Eu–O S_0^2 = 0.86	N	7	6.7 ± 1.7	9.8 ± 1.9	7.9 ± 0.7
	R (Å)	2.35 ± 0.01	2.27 ± 0.03	2.33 ± 0.02	2.39 ± 0.01
	σ^2	0.012 ± 0.002	0.012 ± 0.002	0.012 ± 0.002	0.012 ± 0.002

The Zr edges were modeled out to 4 Å using three single scattering paths, Zr–O (6-fold), Zr–Ba (8-fold), and Zr–Zr (6-fold), plus the three high amplitude linear multiple scattering paths. The path degeneracies were held constant for all paths and all the multiple scattering paths are constrained to have the same ΔR as the Zr–Zr path and a common σ^2 parameter. Windows of 2.0 Å^{-1} < k < 13.0 Å^{-1} with dk = 2.0 and 1.0 Å < R < 4.3 Å with dR = 0.2 were used for the Fourier transformation and fits, respectively. Table 2 reports the fit results for the Zr–O single scattering path with the other paths reported in Table S2. For all samples, the Ba–Zr and Zr–Ba paths are distances and disorder parameters and are consistent as are the Ba–Ba and Zr–Zr paths. The quality of the Zr edge fits (Figure S4) are limited by the constraints applied but are consistent across all samples, suggesting only limited Eu^{3+} doping at the Zr site.

The Eu L$_3$ EXAFS presented in Figure 3c clearly show that the 10.0% doped sample exhibits two peaks at ~3.1 Å and ~3.7 Å, which are characteristic of Eu$_2$O$_3$. As the doping level is reduced, these two peaks vanish to be replaced by two small peaks at ~3.0 Å and ~3.5 Å, similar to those observed in the Zr edge EXAFS. The Eu edges were modeled only to the first shell, as the attempts to model the longer paths were unsuccessful. The amplitude reduction factor S_0^2 was held constant at the value determined by fitting the Eu$_2$O$_3$ data. The path length, degeneracy, and disorder of the single Eu–O path being modeled were allowed to vary. Windows of 2.0 Å^{-1} < k < 9.0 Å^{-1} with dk = 2.0 and 1.0 Å < R < 2.4 Å with dR = 0.2 were used for the Fourier transformation and fits, respectively (Figure S5). The resulting fit parameters are reported in Table 2 and it is clear that the Eu–O distance and the path degeneracy increase with the doping level. At 1.0% doping, the Eu–O distance is 2.25 Å, which is longer than the Zr–O distance but significantly shorter than both the Ba–O and the Eu–O distances in Eu$_2$O$_3$. As the doping level increases, the Eu–O distance increases and then exceeds that found in Eu$_2$O$_3$. Similarly, the path degeneracy for the 1.0% sample is close to six, as would be expected for doping on the Zr site and increases to a value greater than that in Eu$_2$O$_3$. These results strongly suggest that Eu^{3+} at low doping levels sits at the Zr site and has a solubility limit between 1.0–2.0%. At higher concentrations, Eu^{3+} ions are found in a Eu$_2$O$_3$-like local environment. This result is consistent with the change in the white line of the Eu XANES, which increases for 1.0% and 2.0% but decreases for 10%.

3.5. PL Spectra

The concentration-dependent excitation spectra (λ_{em} = 625 nm, Figure 4a) and emission spectra (λ_{ex} = 279 nm and 395 nm, Figure 4b,c) of the BZOE samples demonstrated characteristic PL features of the Eu^{3+} dopant in solid-state hosts [41,42]. In general, there is no change of the excitation and emission spectral profiles, Stark splitting, and relative intensity of excitation and emission peaks under the same excitation wavelength among the BZOE samples with the tested Eu^{3+} doping concentrations. The spectra also clearly show that the 2.0% Eu^{3+}-doped sample, BZOE-2, has the highest emission intensity among our samples using the MDT $^5D_0 \rightarrow {}^7F_1$ as an example (Figure 4d).

The excitation spectra of the BZOE samples with λ_{em} = 625 nm corresponding to the $^5D_0 \rightarrow {}^7F_2$ transition of Eu^{3+} ions consisted of two main features (Figure 4a): a broad band extending from 240–320 nm and several fine peaks in the range of 350–500 nm. The broad band peaking around 279 nm is attributed to the allowed charge transfer band (CTB) of electrons from the filled 2p orbital of O^{2-} to the vacant 4d-orbital of the Eu^{3+} ion. The fine peaks around 361, 375, 383, 387, 395, 405, 414, 456, 465, and 472 nm are attributed to the intra f-f transitions of Eu^{3+} ions. The main peaks located at 395 nm and 465 nm are attributed to $^7F_0 \rightarrow {}^5L_6$ and $^7F_0 \rightarrow {}^5D_2$, respectively.

Under the excitations of λ_{ex} = 279 and 395 nm, the emission spectra of the BZOE samples displayed the CTB and several fine peaks corresponding to $^5D_0 \rightarrow {}^7F_J$ (J = 0–4) transitions of Eu^{3+} in the spectral range of 550–750 nm (Figure 4b). Interestingly, several significant differences in terms of the appearance of $^5D_0 \rightarrow {}^7F_0$ transition, the asymmetry ratio (A_{21}), and Stark splitting were observed from the emission spectra of the BZOE samples in the spectral range of 550–750 nm under these two excitation wavelengths.

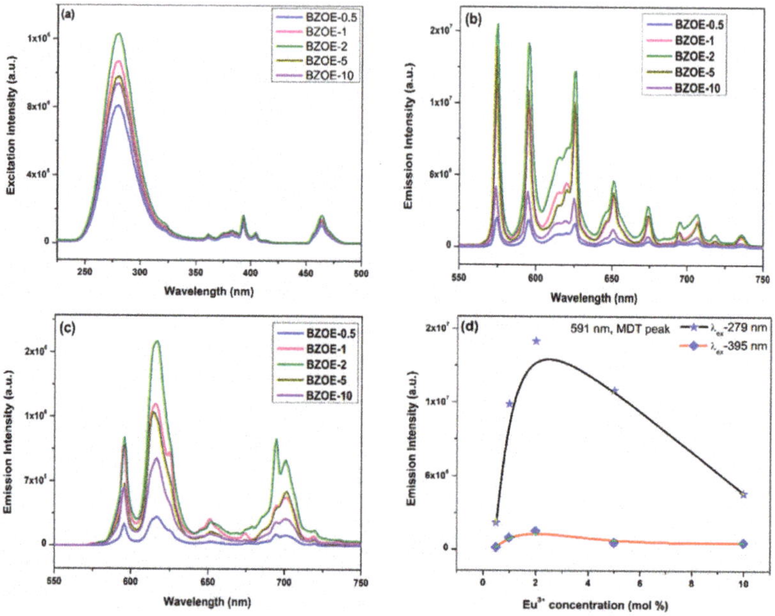

Figure 4. (a) Excitation spectra with λ_{em} = 625 nm, emission spectra under (b) λ_{ex} = 279 nm and (c) λ_{ex} = 395 nm of the BZOE samples. (d) Effects of Eu^{3+} doping concentration of the BZOE samples on integrated MDT emission intensity of $^5D_0 \rightarrow {}^7F_1$ transition.

Figure 5a shows a close look of the emission spectra using the BZOE-2 sample as an example. Specifically, under the 395 nm excitation, intense emission bands at 591 nm ($^5D_0 \rightarrow {}^7F_1$, MDT), 612 nm ($^5D_0 \rightarrow {}^7F_2$, HEDT), 701 nm ($^5D_0 \rightarrow {}^7F_4$), and 653 nm ($^5D_0 \rightarrow {}^7F_3$, weak peak) were observed [43]. There is no signature of $^5D_0 \rightarrow {}^7F_0$ transition. The integral intensity of the HEDT at 612 nm is stronger than that of the MDT at 591 nm. The $^5D_0 \rightarrow {}^7F_3$ transition is known to be allowed by neither MDT nor EDT. It is forbidden in nature according to the Judd–Ofelt (J–O) theory but could gain intensity via J-mixing. The $^5D_0 \rightarrow {}^7F_4$ transition is also considered as an ED transition [1].

Under the CTB excitation at 279 nm, we observed several interesting emission features compared to the emission spectrum recorded under 395 nm excitation (Figure 5a): (a) an unusually intense 575 nm peak corresponding to $^5D_0 \rightarrow {}^7F_0$ transition, which otherwise is forbidden by both ED and MD transitions [44], (b) increased Stark splitting, (c) enhanced intensity of the $^5D_0 \rightarrow {}^7F_3$ peak, and (d) a significant change of the A_{21} value. The possible reasons for these observations will be further discussed in the following sections.

3.6. PL Lifetime Spectra and QY

Figure 6a,b show the results of the luminescence lifetime measurements of the BZOE-2 sample under the excitations at 279 and 395 nm with three different emission wavelengths of 575, 591, and 612 nm corresponding to $^5D_0 \rightarrow {}^7F_0$, $^5D_0 \rightarrow {}^7F_1$, and $^5D_0 \rightarrow {}^7F_2$ transitions, respectively. For the BZOE-2 sample, the luminescence lifetime curves recorded under the 279 nm excitation (Figure 6a) demonstrated a biexponential behavior with two slopes and they can be approximated using the following equation:

$$I = A_0 + A_1 \exp(-t/\tau_1) + A_2 \exp(-t/\tau_2) \tag{1}$$

where A_1 and A_2 are the derived preexponential factors, and τ_1 and τ_2 are the lifetime values of the fast and slow decay components, respectively. The luminescence lifetime curves recorded under the 395 nm excitation (Figure 6b) could be fitted with monoexponential decay.

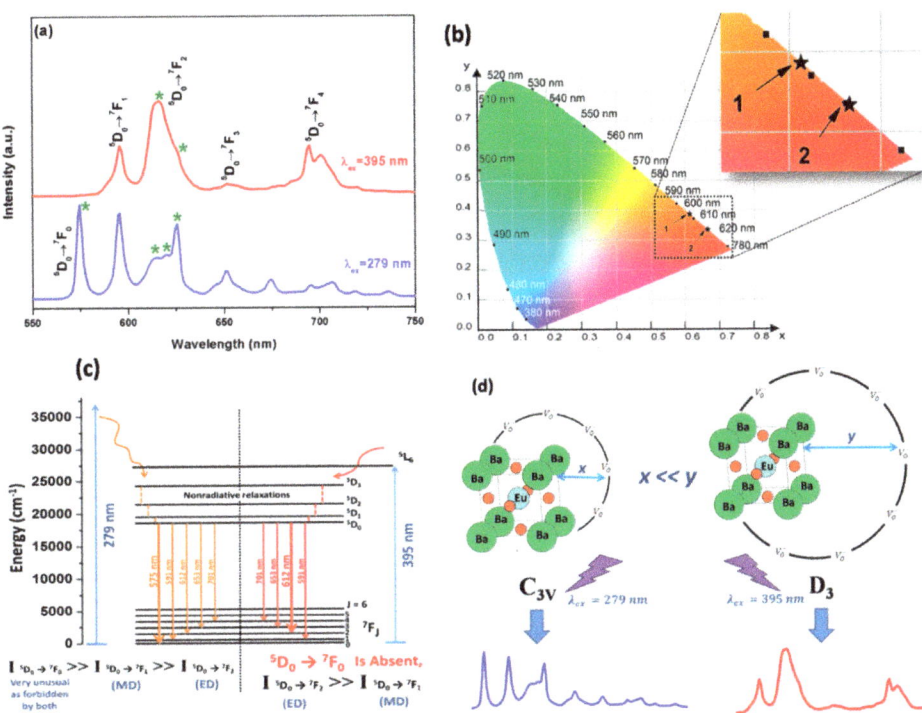

Figure 5. (**a**) Emission spectra of the BZOE-2 sample recorded at 77 K and (**b**) corresponding color coordinate diagram of the BZOE-2 sample under 279 and 395 nm excitations with * indicating the Stark components in (**a**) and arrows 1 and 2 pointing to the coordinates in (**b**), respectively. (**c**) Proposed photophysical processes happening under 279 and 395 nm excitations. (**d**) Schematic showing site selective excitations under 279 nm and 395 nm for the BZOE submicron crystals.

The population of Eu^{3+} ions with a particular lifetime is obtained by using the formula:

$$\% \text{ of species } n = (A_n * \tau_n)/(\Sigma_{n=1,2} A_n * \tau_n)] * 100 \qquad (2)$$

Under λ_{ex} = 279 nm, there were two lifetime values for all three emissions: short lifetime T_s (~360–460 μs, 15%) and long lifetime T_l (~1.0–1.5 ms, 85%). On the other hand, under λ_{ex} = 395 nm, only one short lifetime value was obtained as T_s (~370–580 μs).

The decay profiles of all other BZOE samples are mentioned in Figure S6. Under λ_{ex} = 279 nm and λ_{em} = 625 nm, the average lifetime values of the BZOE samples with Eu^{3+} doping levels of 0.5, 1.0, 2.0, 5.0, and 10.0% were 789, 820, 950, 853, and 813 μs, respectively. The effect of Eu^{3+} concentration on the average lifetime value of the BZOE samples (Figure 6c) indicated that the average lifetime value increased up to a 2.0% Eu^{3+} doping level. Beyond that doping concentration, there was a reduction due to concentration quenching, which is consistent with the phenomenon observed from the PL excitation and emission spectra shown in Figure 4.

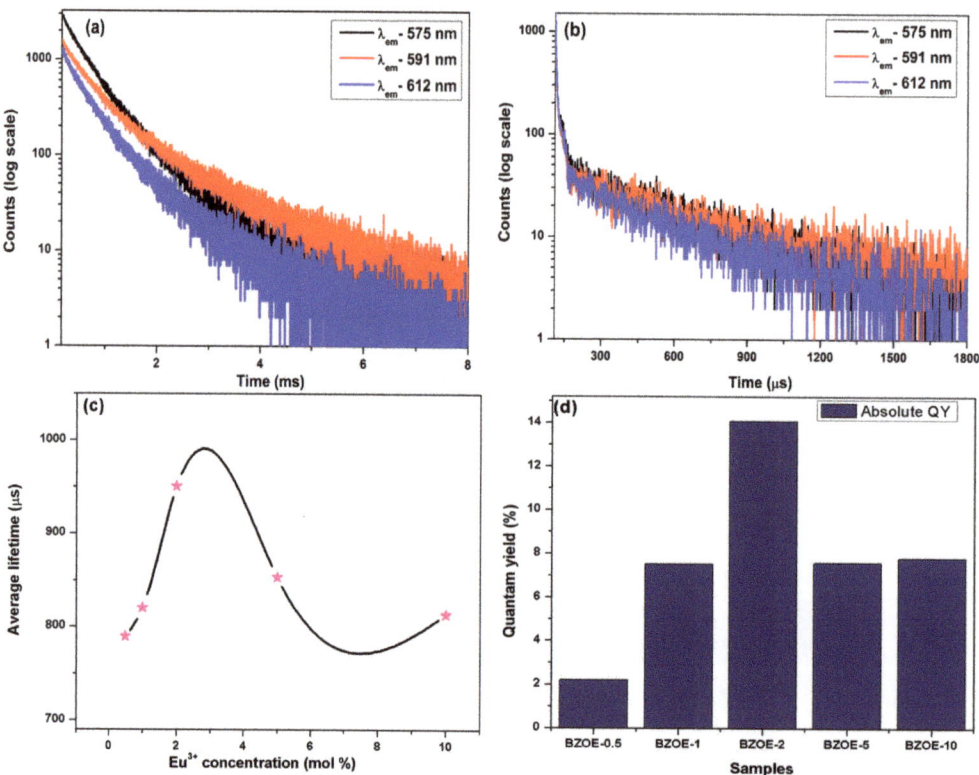

Figure 6. PL decay profiles of the BZOE-2 sample under excitation wavelengths of (**a**) 279 nm and (**b**) 395 nm at three different emission wavelengths of 575, 591, and 612 nm corresponding to $^5D_0 \rightarrow {}^7F_0$, $^5D_0 \rightarrow {}^7F_1$, and $^5D_0 \rightarrow {}^7F_2$ transitions of Eu^{3+} ions, respectively. (**c**) Effects of Eu^{3+} doping concentration of the BZOE samples on (**c**) average lifetime with * indicated the studied Eu^{3+} concentration (mol %) and (**d**) quantum yield (λ_{ex} = 279 nm and λ_{em} = 625 nm).

Quantum yield (QY) is an important parameter to evaluate the properties and application potentials of phosphors. We measured and calculated the QY of our BZOE samples using the following equation:

$$QY = \frac{\int F_S}{\int L_R + \int L_S} \qquad (3)$$

where F_S represents the emission spectrum of a sample, L_R is the excitation spectrum from an empty integrating sphere (without any sample), and L_S means the excitation spectrum of a sample. The effect of the Eu^{3+} concentration on the QY value of the BZOE samples (Figure 6d) indicated that the QY value of the BZOE samples increased from 2.2% to 14.0% as the Eu^{3+} doping level increased from 0.5% to 2.0%. After higher dopant concentrations, the QY value reduced to ~7.6–7.9%. This is again attributed to concentration quenching arising from non-radiative energy transfer among Eu^{3+} ions at high doping concentrations.

Concentration quenching is one of the most dominant phenomena that takes place at high dopant concentrations. It is attributed to increasing resonant energy transfer between Eu^{3+} ions at a high dopant concentration, which results in decreasing radiative emissions. To better understand the mechanism of the concentration quenching phenomenon of our

BZOE samples, the critical distance (r_c) between Eu^{3+} dopant ions and quenching sites was calculated using the following equation:

$$r_c = 2\left(\frac{3V}{4\pi X_c N}\right)^{\frac{1}{3}} \quad (4)$$

where V, X_c, and N are the volume of the unit cell, the critical concentration of Eu^{3+} and the number of cations per unit cell, respectively. The values of these three variables for our BZOE samples are 73.6575 Å^3, 0.02, and 8, respectively. Hence, the calculated critical distance r_c value was 9.58 Å. Since the Eu^{3+}–Eu^{3+} critical distance is more than 5 Å, multipolar interactions are responsible for the concentration quenching of our BZOE crystals. Therefore, various PL studies indicated that there is a close correlation between the doping concentration with the excitation and emission intensity, luminescence lifetime, and the QY of the BZOE crystals.

3.7. J–O Analysis

To explain the observed luminescence performance, J–O parameters were determined to provide empirical relations between the local site symmetry of Eu^{3+} ions in the BZO lattice, the crystal field strength of the BZO host lattice, and the Eu–O bond covalency and polarizability in the BZOE samples.[45] Based on various mathematical formulations, we have derived the radiative/non-radiative transition rates and the internal quantum efficiency of the BZOE-2 sample [45–47]. Various important optical parameters were calculated for the BZOE samples under the excitations at 279 and 395 nm (Table 3). The BZOE-2 sample had a higher internal quantum efficiency (IQE) under the 279 nm excitation compared to 395 nm excitation. Its non-radiative transition (A_{NR}, 787.4 s^{-1}) and radiative transition (A_R, 212.77 s^{-1}) values under 279 nm were lower compared to those under 395 nm excitation (A_{NR} = 2331 s^{-1} and A_R = 369 s^{-1}). When changing λ_{ex} from 279 nm to 395 nm, the increase in the A_{NR} value was higher than that of the A_R value.

Table 3. Calculated J–O parameters and radiative properties of the BZOE-2 sample (A_R = radiative Rate, A_{NR} = nonradiative rate, Ω_n = the Judd–Ofelt parameter, and β_n = branching ratio).

BZOE-2	A_R (s^{-1})	A_{NR} (s^{-1})	η(%)	Ω_2 (×10^{-20})	Ω_4 (×10^{-20})	β_1(%)	β_2(%)	β_4(%)	Ω_2/Ω_4
λ_{ex} = 279 nm	212.77	787.4	21.3	1.04	0.917	23.5	42.4	18.6	1.13
λ_{ex} = 395 nm	369	2331	13.7	2.27	2.78	13.6	53.7	32.5	0.82

For the J–O parameters, Ω_2 (the short range parameter) gives information related to the covalent character, local symmetry, and structural distortion in the vicinity of Eu^{3+} ions, whereas Ω_4 intensity parameters (the long range parameter) provides bulk information such as the viscosity and rigidity of the host lattice [48]. Under the 279 nm excitation, the observed trend of the J–O parameters ($\Omega_4 < \Omega_2$) suggested that excited Eu^{3+} ions were mostly localized in a highly asymmetric and distorted environment. On the other hand, under the 395 nm excitation, the J–O parameter trend reversed with $\Omega_4 > \Omega_2$, which confirmed that a large fraction of excited Eu^{3+} ions occupies relatively less distorted and asymmetric sites. The value of the J–O ratio (Ω_2/Ω_4) of lower than one suggests a high asymmetry of the Eu^{3+} environment where its value higher than one suggests a low asymmetry. The fractional distribution of branching ratios suggests that, under the 279 and 395 nm excitations, photon parts emitted via MDT are 23.5% and 13.6%, whereas those emitted via HEDT are 42.4% and 53.7%, respectively.

3.8. Discussion

As marked by numbers one and two on the color coordinated diagram (Figure 5b), the intense peaks around 575 nm and 612 nm impart orange emissions under λ_{ex} = 279 nm and red emissions under λ_{ex} = 395 nm, respectively. It demonstrated that one can achieve orange–red color tunability by selectively exciting the same material with dopant or host excitations. The different photophysical processes happening under these two excitations are schematically depicted in Figure 5d. The different spectral features of the BZOE samples observed under the 279 and 395 nm excitations suggest that the excited Eu^{3+} ions are relaxed through different channels to the ground states.

Some authors have proposed theoretical models for the observed $^5D_0 \rightarrow {}^7F_0$ transition, including the breakdown of the closure approximation in the Judd–Ofelt theory and third order perturbation theory [1,32,33]. The most obvious explanation assumes that this transition is due to J-mixing or to the mixing of low-lying charge-transfer states into the wave functions of the $4f^6$ configuration. Experimentally, the number of Stark components of the $^5D_0 \rightarrow {}^7F_0$ transition indicates the number of local sites of Eu^{3+} ions in host lattices. It is normally allowed when Eu^{3+} ions are situated at sites lacking inversion symmetry [1,49]. The presence of an unsplitted single band of the $^5D_0 \rightarrow {}^7F_0$ transition under λ_{ex} = 279 nm suggests that a large fraction of Eu^{3+} ions are located at the non-inversion symmetric sites in the BZOE submicron crystals. This hypothesis is further supported by the appearance of forbidden $^5D_0 \rightarrow {}^7F_3$ peaks with large Stark splitting [30,44]. On the other hand, under λ_{ex} = 395 nm, the observed phenomena, including the absence of $^5D_0 \rightarrow {}^7F_0$ transition, weak $^5D_0 \rightarrow {}^7F_3$ transition, and a low extent of Stark splitting of $^5D_0 \rightarrow {}^7F_1$ and $^5D_0 \rightarrow {}^7F_2$ transitions, suggest that a large fraction of Eu^{3+} ions at doping sites, which are less asymmetric or distorted, are selectively excited.

Based on the selection rules of point group symmetry, the $^5D_0 \rightarrow {}^7F_0$ transition appears when Eu^{3+} dopants are located at sites lacking an inversion center with 10 designated non-cubic point groups, including C_{6v}, C_6, C_{3v}, C_3, C_{4v}, C_4, C_{2v}, C_2, C_s, and C_1 [50]. The $^5D_0 \rightarrow {}^7F_0$ transition is not allowed in cubic groups with inversion symmetry such as T, T_d, and O or non-cubic point groups without inversion symmetry such as D_2, D_3, D_{3h}, C_{3h}, D_3, D_4, S_4, D_{2d}, D_{4d}, and D_6 [49].

The ideal BZO is a perfect cubic perovskite with O_h point group symmetry (space group: Pm-$3m$), which has 12-coordinated Ba^{2+} sites and 6-coordinated Zr^{4+} sites in cuboctahedra and octahedral geometries, respectively [51]. The observed emission spectra are in line with Eu^{3+} ions occupying the Zr^{4+} sites in the BZOE samples even with the following ionic radii values of Ba^{2+} (r_{ion} = 161 pm @ CN = 12), Zr^{4+} (r_{ion} = 72 pm @ CN = 6), and Eu^{3+} ions (r_{ion} = 95 pm @ CN = 6). Substituting Eu^{3+} ions at the Zr^{4+} sites distorts the symmetric ideal perovskite structure of BZO and invokes charge compensation by oxygen vacancies, which reduce the point group symmetry from O_h to further lower symmetry. This is consistent with our EXAFS analysis (Figure 3), especially at a low Eu^{3+} doping level before the low amount of Eu_2O_3 phase forms.

It has been reported that the emission of Eu^{3+} dopant in a cubic structure with the O_h point group should only have a single unsplitted $^5D_0 \rightarrow {}^7F_1$ transition peak [52]. By considering the most sensitive peaks for $^5D_0 \rightarrow {}^7F_0$ and $^5D_0 \rightarrow {}^7F_2$ transitions, there are 0 and 2 Stark components under λ_{ex} = 395 nm and 1 and 3 Stark components under λ_{ex} = 279 nm, respectively (Figure 5a). This observation suggested D_3 and C_{3V} point group symmetry around Eu^{3+} ions in our BZOE samples [52].

The Kroger–Vink notation for the substitution, wherein trivalent Eu^{3+} ions occupy tetravalent Zr^{4+} sites, is formulated below [53]:

$$2Eu^{\cdots} + Zr_{Zr}^{\cdots} \leftrightarrow Eu'_{Zr} + V_{\ddot{O}} \tag{5}$$

Defects such as $V_{\ddot{O}}$ and Eu'_{Zr} in the BZOE crystals provide additional pathways for non-radiative relaxation. They tend to quench PL by absorbing emitted photon energy from Eu^{3+} ion centers ($Eu_{Ba}^{\cdot}$) [47]. Hence, although Eu^{3+} ions occupy Zr^{4+} sites (Eu'_{Zr}), we assume that there are enough oxygen vacancies surrounding them with random distribution. There would be two scenarios: one with enough Eu'_{Zr} surrounded by oxygen vacancies in a close vicinity (x), designated as xEu'_{Zr}, and another with Eu'_{Zr} surrounded by oxygen vacancies at a much farther-off distance (y), designated as yEu'_{Zr}, such as y >> x. The point group symmetry of xEu'_{Zr}, as discussed above, is C_{3v}, and that of yEu'_{Zr} is D_3. As schematically shown in Figure 5d, upon the excitation with the Eu^{3+} f-f band at 395 nm, the prevalent excited species is yEu'_{Zr}, whereas upon excitation with the host CTB selectively, a large fraction excited species is xEu'_{Zr}.

4. Conclusions

In this work, BZOE submicron crystals with varied Eu^{3+} doping concentrations were synthesized using the molten salt method. XANES and EXAFS spectroscopies confirm that Eu is stabilized in a +3 oxidation state at Zr^{4+} s at a low doping concentration, while a separate Eu_2O_3 phase forms at the highest 10% doping level. Based on the PL measurement, it was established that europium is localized at Zr^{4+} sites in two different environments: one close to zirconium vacancies with C_{3v} symmetry and one far off from zirconium vacancies with D_3 symmetry. Interestingly, when excited at the charge transfer band of the BZO host at 279 nm, a large fraction of Eu^{3+} ions at non-symmetric C_{3v} sites were excited to give a highly intense $^5D_0 \rightarrow {}^7F_0$ transition, large spectral splitting, and intense MDT peaks compared to HEDT peaks. On the other hand, when excited at a dopant transition wavelength of 395 nm, a relatively large fraction of Eu^{3+} dopants, which are far off from zirconium vacancies with D_3 symmetry, were excited to give no $^5D_0 \rightarrow {}^7F_0$ transition, highly intense HEDT peaks compared to MDT peaks, and fewer Stark components. This excitation wavelength dependence induces emission light tunability of orange light at λ_{ex} = 279 nm and red light at λ_{ex} = 395 nm from the BZOE samples. This observation is further justified by the trend of the J–O parameters, especially with $\Omega_4 < \Omega_2$ at λ_{ex} = 279 nm and $\Omega_4 > \Omega_2$ at λ_{ex} = 395 nm. This work demonstrates the role of local dopant sites, defects, excitation wavelengths, and doping concentrations on optimizing the optical properties of lanthanide-doped perovskite phosphors for efficient optoelectronics and scintillator applications.

Supplementary Materials: The following are available online at https://www.mdpi.com/article/10.3390/nano12173028/s1, Experimental details. Figure S1. Size distribution plot for BZOE for different europium concentration derived using ImageJ software; Figure S2: Rietveld refinements for the BZOE samples; Figure S3: Ba L_3 edge fits for the three Eu-doped samples; Table S1: Fit parameters for Ba L_3 edge fits; Figure S4: Zr K edge fits for the three Eu-doped samples; Table S2: Fit parameters for Zr K edge fits; Figure S5: Eu L_3 edge fits for the three Eu-doped samples; Figure S6: PL decay profiles of the various BZOE sample under excitation wavelengths of 279 nm and emission wavelengths of 625 nm corresponding $^5D_0 \rightarrow {}^7F_2$ transitions of Eu^{3+} ions. References [34,54–57] are cited in the Supplementary Materials.

Author Contributions: Conceptualization, Y.M.; methodology, Y.M., S.K.G. and C.U.S.; validation, S.K.G., H.A. and C.U.S.; formal analysis, S.K.G. and C.U.S.; investigation, S.K.G. and C.U.S.; resources, Y.M.; data curation, S.K.G., H.A. and C.U.S.; writing—original draft preparation, S.K.G. and C.U.S.; writing—review and editing, Y.M.; visualization, S.K.G. and C.U.S.; supervision, Y.M.; project administration, Y.M.; funding acquisition, Y.M. All authors have read and agreed to the published version of the manuscript.

Funding: YM thanks the financial support by the IIT startup funds. SKG thanks the United States-India Education Foundation (USIEF, India) and the Institute of International Education (IIE, USA) for his Fulbright Nehru Postdoctoral Fellowship (Award #2268/FNPDR/2017). MRCAT operations are supported by the Department of Energy and the MRCAT member institutions. This research used resources of the Advanced Photon Source, a U.S. Department of Energy (DOE) Office of Science User Facility operated for the DOE Office of Science by Argonne National Laboratory under Contract No. DE-AC02-06CH11357.

Institutional Review Board Statement: Not applicable.

Informed Consent Statement: Not applicable.

Data Availability Statement: Data is contained within the article or supplementary materials.

Conflicts of Interest: The authors declare no conflict of interest.

References

1. Binnemans, K. Interpretation of europium(III) spectra. *Coord. Chem. Rev.* **2015**, *295*, 1–45. [CrossRef]
2. Gupta, S.K.; Rajeshwari, B.; Achary, S.N.; Patwe, S.J.; Tyagi, A.K.; Natarajan, V.; Kadam, R.M. Europium Luminescence as a Structural Probe: Structure-Dependent Changes in Eu^{3+}-Substituted $Th(C_2O_4)_2 \cdot xH_2O$ (x = 6, 2, and 0). *Eur. J. Inorg. Chem.* **2015**, *2015*, 4429–4436. [CrossRef]
3. Atuchin, V.; Aleksandrovsky, A.; Chimitova, O.; Gavrilova, T.; Krylov, A.; Molokeev, M.; Oreshonkov, A.; Bazarov, B.; Bazarova, J. Synthesis and spectroscopic properties of monoclinic α-$Eu_2(MoO_4)_3$. *J. Phys. Chem. C* **2014**, *118*, 15404–15411. [CrossRef]
4. Atuchin, V.; Subanakov, A.; Aleksandrovsky, A.; Bazarov, B.; Bazarova, J.; Gavrilova, T.; Krylov, A.; Molokeev, M.; Oreshonkov, A.; Stefanovich, S.Y. Structural and spectroscopic properties of new noncentrosymmetric self-activated borate $Rb_3EuB_6O_{12}$ with B_5O_{10} units. *Mater. Des.* **2018**, *140*, 488–494. [CrossRef]
5. Kitagawa, Y.; Ueda, J.; Fujii, K.; Yashima, M.; Funahashi, S.; Nakanishi, T.; Takeda, T.; Hirosaki, N.; Hongo, K.; Maezono, R.; et al. Site-Selective Eu^{3+} Luminescence in the Monoclinic Phase of $YSiO_2N$. *Chem. Mater.* **2021**, *33*, 8873–8885. [CrossRef]
6. Gupta, S.K.; Zuniga, J.P.; Ghosh, P.S.; Abdou, M.; Mao, Y. Correlating Structure and Luminescence Properties of Undoped and Eu^{3+}-Doped $La_2Hf_2O_7$ Nanoparticles Prepared with Different Coprecipitating pH Values through Experimental and Theoretical Studies. *Inorg. Chem.* **2018**, *57*, 11815–11830. [CrossRef]
7. Zhang, Y.; Xu, J.; Cui, Q.; Yang, B. Eu^{3+}-doped $Bi_4Si_3O_{12}$ red phosphor for solid state lighting: Microwave synthesis, characterization, photoluminescence properties and thermal quenching mechanisms. *Sci. Rep.* **2017**, *7*, 42464. [CrossRef]
8. Gupta, S.K.; Ghosh, P.S.; Yadav, A.K.; Pathak, N.; Arya, A.; Jha, S.N.; Bhattacharyya, D.; Kadam, R.M. Luminescence properties of $SrZrO_3/Tb^{3+}$ perovskite: Host-dopant energy-transfer dynamics and local structure of Tb^{3+}. *Inorg. Chem.* **2016**, *55*, 1728–1740. [CrossRef]
9. Kunkel, N.; Meijerink, A.; Springborg, M.; Kohlmann, H. Eu(ii) luminescence in the perovskite host lattices $KMgH_3$, $NaMgH_3$ and mixed crystals $LiBa_xSr_{1-x}H_3$. *J. Mater. Chem. C* **2014**, *2*, 4799–4804. [CrossRef]
10. Orvis, T.; Surendran, M.; Liu, Y.; Niu, S.; Muramoto, S.; Grutter, A.J.; Ravichandran, J. Electron Doping $BaZrO_3$ via Topochemical Reduction. *ACS Appl. Mater. Interfaces* **2019**, *11*, 21720–21726. [CrossRef] [PubMed]
11. Leonidov, I.I.; Tsidilkovski, V.I.; Tropin, E.S.; Vlasov, M.I.; Putilov, L.P. Acceptor doping, hydration and band-gap engineering of $BaZrO_3$. *Mater. Lett.* **2018**, *212*, 336–338. [CrossRef]
12. Gupta, S.K.; Pathak, N.; Kadam, R. An efficient gel-combustion synthesis of visible light emitting barium zirconate perovskite nanoceramics: Probing the photoluminescence of Sm^{3+} and Eu^{3+} doped $BaZrO_3$. *J. Lumin.* **2016**, *169*, 106–114. [CrossRef]
13. Charoonsuk, T.; Vittayakorn, N. Soft-mechanochemical synthesis of monodispersed $BaZrO_3$ sub-microspheres: Phase formation and growth mechanism. *Mater. Des.* **2017**, *118*, 44–52. [CrossRef]
14. Guo, L.; Zhong, C.; Wang, X.; Li, L. Synthesis and photoluminescence properties of Er^{3+} doped $BaZrO_3$ nanotube arrays. *J. Alloys Compd.* **2012**, *530*, 22–25. [CrossRef]
15. Vøllestad, E.; Strandbakke, R.; Tarach, M.; Catalán-Martínez, D.; Fontaine, M.-L.; Beeaff, D.; Clark, D.R.; Serra, J.M.; Norby, T. Mixed proton and electron conducting double perovskite anodes for stable and efficient tubular proton ceramic electrolysers. *Nat. Mater.* **2019**, *18*, 752–759. [CrossRef]
16. Ma, X.; Zhang, J.; Li, H.; Duan, B.; Guo, L.; Que, M.; Wang, Y. Violet blue long-lasting phosphorescence properties of Mg-doped $BaZrO_3$ and its ability to assist photocatalysis. *J. Alloys Compd.* **2013**, *580*, 564–569. [CrossRef]
17. Foo, G.S.; Polo-Garzon, F.; Fung, V.; Jiang, D.-E.; Overbury, S.H.; Wu, Z. Acid–Base Reactivity of Perovskite Catalysts Probed via Conversion of 2-Propanol over Titanates and Zirconates. *ACS Catal.* **2017**, *7*, 4423–4434. [CrossRef]
18. Ding, J.; Balachandran, J.; Sang, X.; Guo, W.; Veith, G.M.; Bridges, C.A.; Rouleau, C.M.; Poplawsky, J.D.; Bassiri-Gharb, N.; Ganesh, P. Influence of Nonstoichiometry on Proton Conductivity in Thin-Film Yttrium-Doped Barium Zirconate. *ACS Appl. Mater. Interfaces* **2018**, *10*, 4816–4823. [CrossRef]
19. Polfus, J.M.; Yildiz, B.; Tuller, H.L.; Bredesen, R. Adsorption of CO_2 and Facile Carbonate Formation on $BaZrO_3$ Surfaces. *J. Phys. Chem. C* **2018**, *122*, 307–314. [CrossRef]
20. Qi, S.; Wei, D.; Huang, Y.; Kim, S.I.; Yu, Y.M.; Seo, H.J. Microstructure of Eu^{3+}-Doped Perovskites-Type Niobate Ceramic $La_3Mg_2NbO_9$. *J. Am. Ceram. Soc.* **2014**, *97*, 501–506. [CrossRef]
21. Xie, J.; Shi, Y.; Zhang, F.; Li, G. $CaSnO_3:Tb^{3+},Eu^{3+}$: A distorted-perovskite structure phosphor with tunable photoluminescence properties. *J. Mater. Sci.* **2016**, *51*, 7471–7479. [CrossRef]
22. Pazik, R.; Tekoriute, R.; Håkansson, S.; Wiglusz, R.; Strek, W.; Seisenbaeva, G.A.; Gun'ko, Y.K.; Kessler, V.G. Precursor and solvent effects in the nonhydrolytic synthesis of complex oxide nanoparticles for bioimaging applications by the ether elimination (Bradley) reaction. *Chem. Eur. J.* **2009**, *15*, 6820–6826. [CrossRef] [PubMed]
23. Kunti, A.K.; Patra, N.; Harris, R.A.; Sharma, S.K.; Bhattacharyya, D.; Jha, S.N.; Swart, H.C. Local Structure and Spectroscopic Properties of Eu^{3+}-Doped $BaZrO_3$. *Inorg. Chem.* **2019**, *58*, 3073–3089. [CrossRef] [PubMed]
24. Kanie, K.; Seino, Y.; Matsubara, M.; Nakaya, M.; Muramatsu, A. Hydrothermal synthesis of $BaZrO_3$ fine particles controlled in size and shape and fluorescence behavior by europium doping. *New J. Chem.* **2014**, *38*, 3548–3555. [CrossRef]
25. Basu, S.; Patel, D.K.; Nuwad, J.; Sudarsan, V.; Jha, S.N.; Bhattacharyya, D.; Vatsa, R.K.; Kulshreshtha, S.K. Probing local environments in Eu^{3+} doped $SrSnO_3$ nano-rods by luminescence and Sr K-edge EXAFS techniques. *Chem. Phys. Lett.* **2013**, *561–562*, 82–86. [CrossRef]

26. Rabuffetti, F.A.; Culver, S.P.; Lee, J.S.; Brutchey, R.L. Local structural investigation of Eu^{3+}-doped $BaTiO_3$ nanocrystals. *Nanoscale* 2014, *6*, 2909–2914. [CrossRef] [PubMed]
27. Canu, G.; Bottaro, G.; Buscaglia, M.T.; Costa, C.; Condurache, O.; Curecheriu, L.; Mitoseriu, L.; Buscaglia, V.; Armelao, L. Ferroelectric order driven Eu^{3+} photoluminescence in $BaZr_xTi_{1-x}O_3$ perovskite. *Sci. Reports* 2019, *9*, 6441. [CrossRef]
28. Drag-Jarząbek, A.; John, Ł.; Petrus, R.; Kosińska-Klähn, M.; Sobota, P. Alkaline Earth Metal Zirconate Perovskites $MZrO_3$ (M = Ba^{2+}, Sr^{2+}, Ca^{2+}) Derived from Molecular Precursors and Doped with Eu^{3+} Ions. *Chem. Eur. J.* 2016, *22*, 4780–4788. [CrossRef]
29. Katyayan, S.; Agrawal, S. Effect of rare earth doping on optical and spectroscopic characteristics of $BaZrO_3:Eu^{3+},Tb^{3+}$ perovskites. *Methods Appl. Fluoresc.* 2018, *6*, 035002. [CrossRef]
30. Gupta, S.K.; Mohapatra, M.; Natarajan, V.; Godbole, S.V. Site-specific luminescence of Eu^{3+} in gel-combustion-derived strontium zirconate perovskite nanophosphors. *J. Mater. Sci.* 2012, *47*, 3504–3515. [CrossRef]
31. Guo, Y.; Park, S.H.; Choi, B.C.; Jeong, J.H.; Kim, J.H. Dual-Mode Manipulating Multicenter Photoluminescence in a Single-Phased $Ba_9Lu_2Si_6O_{24}:Bi^{3+},Eu^{3+}$ Phosphor to Realize White Light/Tunable Emissions. *Sci. Rep.* 2017, *7*, 15884. [CrossRef]
32. Guzmán-Olguín, J.; Esquivel, R.L.; Jasso, G.T.; Guzmán-Mendoza, J.; Montalvo, T.R.; García-Hipólito, M.; Falcony, C. Luminescent behavior of Eu^{3+} doped $BaHfO_3$ perovskite ceramic under UV radiation. *Appl. Radiat. Isot.* 2019, *153*, 108815. [CrossRef]
33. Gupta, S.K.; Reghukumar, C.; Kadam, R. Eu^{3+} local site analysis and emission characteristics of novel $Nd_2Zr_2O_7$:Eu phosphor: Insight into the effect of europium concentration on its photoluminescence properties. *RSC Adv.* 2016, *6*, 53614–53624. [CrossRef]
34. Zhou, H.; Mao, Y.; Wong, S.S. Shape control and spectroscopy of crystalline $BaZrO_3$ perovskite particles. *J. Mater. Chem.* 2007, *17*, 1707–1713. [CrossRef]
35. Charoonsuk, T.; Vittayakorn, W.; Vittayakorn, N.; Seeharaj, P.; Maensiri, S. Sonochemical synthesis of monodispersed perovskite barium zirconate ($BaZrO_3$) by using an ethanol–water mixed solvent. *Ceram. Int.* 2015, *41*, S87–S94. [CrossRef]
36. Li, C.-C.; Chang, S.-J.; Lee, J.-T.; Liao, W.-S. Efficient hydroxylation of $BaTiO_3$ nanoparticles by using hydrogen peroxide. *Colloids Surf. A Physicochem. Eng. Asp.* 2010, *361*, 143–149. [CrossRef]
37. Wirunchit, S.; Charoonsuk, T.; Vittayakorn, N. Facile sonochemical synthesis of near spherical barium zirconate titanate ($BaZr_{1-y}Ti_yO_3$; BZT); perovskite stability and formation mechanism. *RSC Adv.* 2015, *5*, 38061–38074. [CrossRef]
38. Zeng, C.-H.; Zheng, K.; Lou, K.-L.; Meng, X.-T.; Yan, Z.-Q.; Ye, Z.-N.; Su, R.-R.; Zhong, S. Synthesis of porous europium oxide particles for photoelectrochemical water splitting. *Electrochim. Acta* 2015, *165*, 396–401. [CrossRef]
39. Fassbender, R.U.; Lilge, T.S.; Cava, S.; Andrés, J.; da Silva, L.F.; Mastelaro, V.R.; Longo, E.; Moreira, M.L. Fingerprints of short-range and long-range structure in $BaZr_{1-x}Hf_xO_3$ solid solutions: An experimental and theoretical study. *Phys. Chem. Chem. Phys.* 2015, *17*, 11341–11349. [CrossRef]
40. Giannici, F.; Longo, A.; Balerna, A.; Kreuer, K.-D.; Martorana, A. Proton Dynamics in In:$BaZrO_3$: Insights on the Atomic and Electronic Structure from X-ray Absorption Spectroscopy. *Chem. Mater.* 2009, *21*, 2641–2649. [CrossRef]
41. Shi, P.; Xia, Z.; Molokeev, M.S.; Atuchin, V.V. Crystal chemistry and luminescence properties of red-emitting $CsGd_{1-x}Eu_x(MoO_4)_2$ solid-solution phosphors. *Dalton Trans.* 2014, *43*, 9669–9676. [CrossRef] [PubMed]
42. Denisenko, Y.G.; Molokeev, M.S.; Oreshonkov, A.S.; Krylov, A.S.; Aleksandrovsky, A.S.; Azarapin, N.O.; Andreev, O.V.; Razumkova, I.A.; Atuchin, V.V. Crystal structure, vibrational, spectroscopic and thermochemical properties of double sulfate crystalline hydrate [$CsEu(H_2O)_3(SO_4)_2$]·H_2O and its thermal dehydration product $CsEu(SO_4)_2$. *Crystals* 2021, *11*, 1027. [CrossRef]
43. Gupta, S.K.; Ghosh, P.S.; Yadav, A.K.; Jha, S.N.; Bhattacharyya, D.; Kadam, R.M. Origin of Blue-Green Emission in α-$Zn_2P_2O_7$ and Local Structure of Ln^{3+} Ion in α-$Zn_2P_2O_7$:Ln^{3+} (Ln = Sm, Eu): Time-Resolved Photoluminescence, EXAFS, and DFT Measurements. *Inorg. Chem.* 2017, *56*, 167–178. [CrossRef] [PubMed]
44. Gupta, S.K.; Mohapatra, M.; Kaity, S.; Natarajan, V.; Godbole, S.V. Structure and site selective luminescence of sol–gel derived Eu:Sr_2SiO_4. *J. Lumin.* 2012, *132*, 1329–1338. [CrossRef]
45. Jain, N.; Paroha, R.; Singh, R.K.; Mishra, V.K.; Chaurasiya, S.K.; Singh, R.A.; Singh, J. Synthesis and Rational design of Europium and Lithium Doped Sodium Zinc Molybdate with Red Emission for Optical Imaging. *Sci. Rep.* 2019, *9*, 2472. [CrossRef]
46. Gupta, S.K.; Mohapatra, M.; Godbole, S.V.; Natarajan, V. On the unusual photoluminescence of Eu^{3+} in α-$Zn_2P_2O_7$: A time resolved emission spectrometric and Judd–Ofelt study. *RSC Adv.* 2013, *3*, 20046–20053. [CrossRef]
47. Gupta, S.K.; Sudarshan, K.; Ghosh, P.S.; Srivastava, A.P.; Bevara, S.; Pujari, P.K.; Kadam, R.M. Role of various defects in the photoluminescence characteristics of nanocrystalline $Nd_2Zr_2O_7$: An investigation through spectroscopic and DFT calculations. *J. Mater. Chem. C* 2016, *4*, 4988–5000. [CrossRef]
48. Vats, B.G.; Gupta, S.K.; Keskar, M.; Phatak, R.; Mukherjee, S.; Kannan, S. The effect of vanadium substitution on photoluminescent properties of $KSrLa(PO_4)_x(VO_4)_{2-x}$:$Eu^{3+}$ phosphors, a new variant of phosphovanadates. *New J. Chem.* 2016, *40*, 1799–1806. [CrossRef]
49. Tanner, P.A. Some misconceptions concerning the electronic spectra of tri-positive europium and cerium. *Chem. Soc. Rev.* 2013, *42*, 5090–5101. [CrossRef]
50. Chen, X.; Liu, G. The standard and anomalous crystal-field spectra of Eu^{3+}. *J. Solid State Chem.* 2005, *178*, 419–428. [CrossRef]
51. Manju, P.; Ajith, M.R.; Jaiswal-Nagar, D. Synthesis and characterization of $BaZrO_3$ nanoparticles by citrate-nitrate sol-gel auto-combustion technique: Systematic study for the formation of dense $BaZrO_3$ ceramics. *J. Eur. Ceram. Soc.* 2019, *39*, 3756–3767. [CrossRef]

52. Ju, Q.; Liu, Y.; Li, R.; Liu, L.; Luo, W.; Chen, X. Optical Spectroscopy of Eu^{3+}-Doped BaFCl Nanocrystals. *J. Phys. Chem. C* **2009**, *113*, 2309–2315. [CrossRef]
53. Gupta, S.K.; Sudarshan, K.; Yadav, A.K.; Gupta, R.; Bhattacharyya, D.; Jha, S.N.; Kadam, R.M. Deciphering the Role of Charge Compensator in Optical Properties of $SrWO_4$:Eu^{3+}:A (A = Li^+, Na^+, K^+): Spectroscopic Insight Using Photoluminescence, Positron Annihilation, and X-ray Absorption. *Inorg. Chem.* **2018**, *57*, 821–832. [CrossRef]
54. Toby, B.H.; Von Dreele, R.B. GSAS-II: The genesis of a modern open-source all purpose crystallography software package. *J. Appl. Crystallogr.* **2013**, *46*, 544–549. [CrossRef]
55. Kropf, A.; Katsoudas, J.; Chattopadhyay, S.; Shibata, T.; Lang, E.; Zyryanov, V.; Ravel, B.; McIvor, K.; Kemner, K.; Scheckel, K. The new MRCAT (Sector 10) bending magnet beamline at the advanced photon source. *AIP Conf. Proc.* **2010**, *1234*, 299–302.
56. Newville, M. IFEFFIT: Interactive XAFS analysis and FEFF fitting. *J. Synchrotron Radiat.* **2001**, *8*, 322–324. [CrossRef]
57. Ravel, B.; Newville, M. ATHENA, ARTEMIS, HEPHAESTUS: Data analysis for X-ray absorption spectroscopy using IFEFFIT. *J. Synchrotron Radiat.* **2005**, *12*, 537–541. [CrossRef]

Article

UV-Vis Activated $Cu_2O/SnO_2/WO_3$ Heterostructure for Photocatalytic Removal of Pesticides

Alexandru Enesca * and Luminita Andronic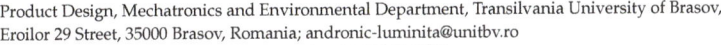

Product Design, Mechatronics and Environmental Department, Transilvania University of Brasov, Eroilor 29 Street, 35000 Brasov, Romania; andronic-luminita@unitbv.ro
* Correspondence: aenesca@unitbv.ro; Tel.: +40-732-71-2472

Abstract: A three-steps sol–gel method was used to obtain a $Cu_2O/SnO_2/WO_3$ heterostructure powder, deposited as film by spray pyrolysis. The porous morphology of the final heterostructure was constructed starting with fiber-like WO_3 acting as substrate for SnO_2 development. The SnO_2/WO_3 sample provide nucleation and grew sites for Cu_2O formation. Diffraction evaluation indicated that all samples contained crystalline structures with crystallite size varying from 42.4 Å (Cu_2O) to 81.8 Å (WO_3). Elemental analysis confirmed that the samples were homogeneous in composition and had an oxygen excess due to the annealing treatments. Photocatalytic properties were tested in the presence of three pesticides—pirimicarb, S-metolachlor (S-MCh), and metalaxyl (MET)—chosen based on their resilience and toxicity. The photocatalytic activity of the $Cu_2O/SnO_2/WO_3$ heterostructure was compared with WO_3, SnO_2, Cu_2O, Cu_2O/SnO_2, Cu_2O/WO_3, and SnO_2/WO_3 samples. The results indicated that the three-component heterostructure had the highest photocatalytic efficiency toward all pesticides. The highest photocatalytic efficiency was obtained toward S-MCh (86%) using a $Cu_2O/SnO_2/WO_3$ sample and the lowest correspond to MET (8.2%) removal using a Cu_2O mono-component sample. TOC analysis indicated that not all the removal efficiency could be attributed to mineralization, and by-product formation is possible. $Cu_2O/SnO_2/WO_3$ is able to induce 81.3% mineralization of S-MCh, while Cu_2O exhibited 5.7% mineralization of S-MCh. The three-run cyclic tests showed that $Cu_2O/SnO_2/WO_3$, WO_3, and SnO_2/WO_3 exhibited good photocatalytic stability without requiring additional procedures. The photocatalytic mechanism corresponds to a Z-scheme charge transfer based on a three-component structure, where Cu_2O exhibits reduction potential responsible for O_2 production and WO_3 has oxidation potential responsible for HO· generation.

Keywords: pesticides; wastewater; photocatalysis; metal oxides; semiconductors

Citation: Enesca, A.; Andronic, L. UV-Vis Activated $Cu_2O/SnO_2/WO_3$ Heterostructure for Photocatalytic Removal of Pesticides. *Nanomaterials* **2022**, *12*, 2648. https://doi.org/10.3390/nano12152648

Academic Editors: Diego Cazorla-Amorós, Yuanbing Mao and Santosh Kumar Gupta

Received: 2 June 2022
Accepted: 28 July 2022
Published: 1 August 2022

Publisher's Note: MDPI stays neutral with regard to jurisdictional claims in published maps and institutional affiliations.

Copyright: © 2022 by the authors. Licensee MDPI, Basel, Switzerland. This article is an open access article distributed under the terms and conditions of the Creative Commons Attribution (CC BY) license (https://creativecommons.org/licenses/by/4.0/).

1. Introduction

The agricultural sector is considered one of the larger water consumers. However, due to population growth and climate change, the water shortage represents an important challenge to overcome [1,2]. The recycled water used in crop irrigation usually contains high quantities of pesticides, due to inefficient wastewater-treatment plant technologies. The risk is even higher when the pesticides interfere with the food chain as a result of transfer from water to soil and from soil to plants through roots [3–5].

Pesticide consumption in agricultural areas has significantly increased in the last few years. Only in 2014 was the pesticide consumption reported by FAOSTAT (North America was not included) above 3,000,000 tons [6]. Epidemiological studies have shown the carcinogenic potential of certain pesticides able to produce birth defects, endocrine deficiencies, cardiovascular diseases, or even disorders of the reproductive organs [7–9]. Traditional methods used to reduce pesticides form wastewater, such as ozonation or chloride dioxide insertion, are effective ways to address the problem, but raise economic and environmental issues [10,11]. Chlorine dioxide (ClO_2) is considered an alternative to NaClO products able to work as a disinfection agent for the wastewater reused in the agricultural sector. However, the ClO_2^- and ClO_3^- biproducts formed during the water treatment can damage the thyroid gland and cause anemia [12,13].

Photocatalysis stands out as a modern and sustainable technology based on advanced oxidation processes (AOPs). In terms of function of the photocatalytic materials involved in the process, the wavelength may vary from UVA to the entire visible spectrum [14,15]. Under irradiation, the catalysts generate oxidative and superoxidative species that promote a complex series of oxidation reactions, aiming to induce the complete mineralization of the organic pollutants [16]. Monocomponent catalysts, such as TiO_2 [17], CuO [18], SnO_2 [19], WO_3 [20], ZnO [21], and Cu_2S [22], exhibit a series of disadvantages, such as limited light absorption, low chemical stability, and fast charge carrier recombination. In order to overcome these issues, heterostructures and composite materials have been developed. Similar systems based on metal oxide heterostructures like $TiO_2/WO_3/ZnO$ [23], BiOI/BiOCl [24], $BaTiO_3/KNbO$ [25], TiO_2/CuO [26], and CuO/Bi_2WO_6 [27] exhibit lower charge recombination rates and high photocatalytic efficiency toward dye molecules. The presence of sulfur compounds as part of heterostructure matrices, such as $SnS_2/ZnInS_4$ [28], $CoTiO_3/ZnCdS_2$ [29], CdS/CeO_2 [30], $CoWO_4/CoS$ [31], and $ZnCdS_2/MoS$ [32], extend the light absorbance spectrum without solving the chemical stability issues in acid or alkaline environments. Composite materials based on $g-C_3N_4$ [33], $g-C_4N_6$ [34], and r-GO [35] exhibit high surface-active materials able to efficiently eliminate the pollutants using both absorption and photocatalytic processes.

In this work, the photocatalytic activity of $Cu_2O/SnO_2/WO_3$ was evaluated toward three pesticides—pirimicarb (PIR; insecticide), S-metolachlor (S-MCh; herbicide) and metalaxyl (MET; fungicide)—used in the agricultural sector. The pollutants were selected based on their toxicity and long-term persistence in soil and water. Pesticide diffusion into plants is an important source of food contamination that then expands in the human body with negative impacts on long-term health. A heterostructure powder was obtained by a three-step sol–gel synthesis followed by spray deposition as film. Each component was chosen based on its spectral absorbance and suitable position of the energy bands, able to reduce the charge carrier's recombination rate and improve long-term photocatalytic activity. The synergic action of high Cu_2O visible absorption and facile charge migration through the heterostructure provides a good alternative for oxidative species generation required for pollutant mineralization. Additionally, the study focused on free-titania photocatalysts as an alternative to traditional materials. Crystalline composition influence on surface composition and film morphology were studied. Photocatalytic efficiency considers the mechanism of oxidative species generation and includes UV-vis and TOC/TN evaluation. The results indicate that the $Cu_2O/SnO_2/WO_3$ heterostructure is suitable for pesticide removal and can be considered a sustainable future alternative.

2. Experimental

2.1. Material Synthesis and Heterostructure Deposition

2.1.1. Heterostructure Powder Synthesis

The heterostructure synthesis consisted of a three-step sol–gel (Figure 1) method followed by thermal treatments to remove all the solvent content.

Step 1. WO_3 powder was obtained from a precursor composed of tungsten hexachloride (99.8%, WCl_6, AcrosOrganics, Gell, Germany) dissolved in a mixed solvent of absolute ethanol (100%, C_2H_6O, Sigma Aldrich, Munich, Germany) and 2-propanol (100%, C_3H_8O, Sigma Aldrich, Munich, Germany). The precursor was magnetically stirred at 140 rpm for 240 min in the dark until a yellow homogeneous solution formed. Gel development took place after the drop-by-drop addition of 0.24 mol sodium hydroxide (99.98%, NaOH, Honeywell, Charlotte, NC, USA). The gel was kept in the dark for 12 h and the final precipitate was centrifuged. The yellow powder (Figure 1, step1) was thermally treated at 410 °C for 4 h.

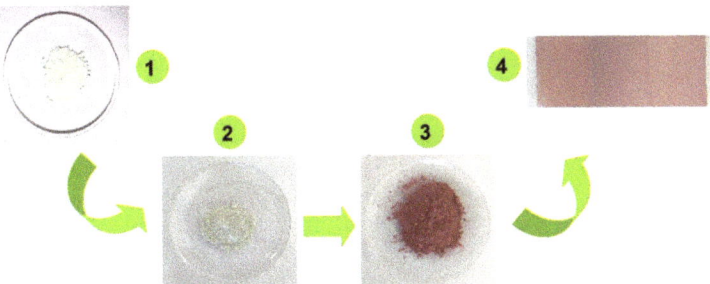

Figure 1. Steps for heterostructure development.

Step 2. SnO$_2$/WO$_3$ powder was prepared by inserting the WO$_3$ powder previously obtained into a precursor based on tin tetrachloride (99.7%, SnCl$_4$, Sigma Aldrich, Munich, Germany) dissolved in absolute methanol (100%, CH$_4$O, Sigma Aldrich, Munich, Germany). The mixture was stirred at 140 rpm for 220 min in dark to ensure the uniform dispersion of WO$_3$ powder into the SnO$_2$ precursor. By slowly adding 0.18 mol sodium hydroxide (99.98%, NaOH, Honeywell, Charlotte, NC, USA) a yellowish white precipitate (Figure 1, step 2) formed, which was centrifuged and annealed at 410 °C for 6 h.

Step 3. the Cu$_2$O/SnO$_2$/WO$_3$ heterostructure was developed by adding the SnO$_2$/WO$_3$ powder previously obtained into a precursor composed of 0.03 mol copper acetate (Cu(CH$_3$COO)$_2$, 97%, Sigma Aldrich, Munich, Germany) and deionized water. After 120 min of magnetic stirring at 140 rpm in dark, a mixture of NaOH, glucose (C$_6$H$_{12}$O$_6$, 99.8%, Sigma Aldrich, Munich, Germany), and deionized water was slowly added until the precipitate had formed. Due to the copper tendency to change from the +1 to +2 oxidation state, after centrifugation the reddish powder (Figure 1, step 3) was dried for 12 h at room temperature.

2.1.2. Film Deposition

The heterostructure films (Figure 1, step 4) were obtained by cold spray deposition. The deposition precursor was obtained by dispersing 50 mg Cu$_2$O/SnO$_2$/WO$_3$ powder in a 50 mL mixture of ethanol and 2-propanol by ultrasound bath. A small quantity (0.1 mL) of Triton X was added to the precursor and magnetically mixed for 30 min. The 2 × 2 cm^2 microscope glass substrate was firstly degreased and then cleaned by successive immersion in acetone and ethanol. The clean substrate was preheated at 45 °C for 120 min and then the precursor was sprayed at 0.5 bars. Breaks of 12 min were kept between each deposition sequence in order to allow solvent evaporation.

2.2. Photocatalytic Experiments

The photocatalytic experiments were done using a rectangular photoreactor containing 4 UVA light sources (18 W, black tubes, T8, 320–370 nm, λ_{max} = 360 nm, 3Lx flux intensity, Philips) and 4 Vis sources (18 W, cold tubes, TL-D 80/865, 400–700 nm, λ_{max} = 560 nm, 28 Lx flux intensity), a ventilation engine to ensure stable humidity, and temperature sensors required to keep a stable 25 °C during the experiments. The light sources were placed in radial position to provide maximum light radiation and ensure uniform light distribution. The energy of the incident light on the sample was 12.63 mW/cm^2.

Pirimicarb (PIR; insecticide, C$_{11}$H$_{18}$N$_4$O$_2$, 98%, Sigma Aldrich, Munich, Germany), S-metolachlor (S-MCh; herbicide, C$_{15}$H$_{22}$ClNO$_2$, 98.5%, LGC, Augsburg, Germany), and metalaxyl (MET; fungicide, C$_{15}$H$_{21}$NO$_4$, 98%, Sigma Aldrich, Munich, Germany) were used to prepare 50 mg/L aqueous solution. In 40 mL pesticide solution was inserted one 2 × 2 cm^2 piece of microscopic glass covered with Cu$_2$O/SnO$_2$/WO$_3$. Quartz recipients were used during the experiments. The samples were kept in the dark for 120 min to reach the k5 absorption–desorption equilibrium. Light exposure duration was 10 h and photocatalytic evaluation was done hourly.

2.3. Characterization

The crystalline composition was evaluated with an X-ray diffractometer (model D8 Discover, Bruker, Karlsruhe, Germany) using the locked-couple option. The scan step was 0.002 degree with a rate of 0.020 s/step from a 20 to 60 degree theta angle. The samples' morphology was studied in a high-vacuum regime with scanning electron microscopy (SEM; S–3400 N type 121 II, Hitachi, Tokyo, Japan) at an accelerated voltage of 10 kV. Field-emission scanning electron microscopy (FESEM SU8010, Fukuoka, Japan) was also used for morphological characterization, where the samples were covered with gold coating due to the low surface conductivity. The changes in the specific absorbance for each pollutant were registered by UV-vis spectrometry (Lambda 950, PerkinElmer, Waltham, MA, USA). The optical band gap energy of the single-component samples was estimated by the Wood and Tauc model. Pollutant mineralization was measured using a total organic carbon analyzer (TOC-L, model CPN, Shimadzu, Kyoto, Japan) and a total nitrogen analyzer (TNM-L, model RHOS, Shimadzu, Kyoto, Japan). The system was set to make three consecutive injections and to provide a mediated value. This evaluation provides essential information regarding the possible formation of by-products that are not identified during the UV-vis analysis.

The photocatalytic decolorization was evaluated by the UV-vis method. The changes in pollutant concentration were measured by UV-vis spectrometry based on the calibration curve corresponding to each pollutant. The absorption wavelength characteristics for each pollutant were PIR 315 nm, S-MCh 274 nm, and MET 267 nm. The photocatalytic efficiency was calculated based on the initial (C_0) and the final concentrations (C) using the following equation:

$$\eta = \left[\frac{(C_0 - C)}{C_0}\right] \cdot 100 \tag{1}$$

3. Results and Discussion

3.1. Composition and Morphology

The crystalline composition of each sample was evaluated by X-ray diffraction, and the results are presented in Figure 2. All the analyses were done on films obtained from the same quantity of precursor without additional thermal treatment. The monocomponent sample contained WO_3 with a monoclinic structure (ICCD 83-0951). The bicomponent sample contained monoclinic WO_3 and tetragonal SnO_2 (ICCD 41-1445). After the last synthesis step, the heterostructure exhibits all three metal oxides components: monoclinic WO_3, tetragonal SnO_2, and cubic Cu_2O (ICCD 71-3645). The synthesis procedure allows WO_3 and SnO_2 to act as nucleation sites for the following component, which facilitates the formation of crystalline structure. There was no indication of mixed metal oxide formation, but the presence of amorphous compounds cannot be excluded [36,37]. Due to the synthesis procedure, no preferential orientation was observed. In order to avoid the formation of mixed copper oxides, the heterostructure powder was dried at room temperature for 12 h and the films deposited at 45 °C. Without annealing treatment, the persistence of carbonaceous species originating from the sol–gel synthesis is possible and may influence the photocatalytic activity.

The changes in crystallite sizes were calculated with the Scherrer formula in Equation (2), using the previous diffraction analysis [38]:

$$D = \frac{0.9\lambda}{\beta \cos\theta} \tag{2}$$

where λ is the X-ray wavelength value (1.5406 Å for $CuK_{\alpha 1}$), β is the angular width measured at half-maximum intensity (FWHM) of the most representative peak, and θ is represents the Bragg's angle. The instrumental broadening was considered during the calculations. The crystallite sizes are included in Table 1 for the WO_3 and SnO_2 components annealed at the same temperature, but for different periods. Using WO_3 as a substrate for SnO_2 development will have a negligible impact on the crystalline size of the first component. Both metal oxides exhibit similar crystallite size, which may be related to the annealing process taking place at the same temperature. The annealing period was higher for WO$_3$, which shows the highest crystallite sizes. The same results are obtained when SnO$_2$/WO$_3$ is used as substrate for Cu$_2$O development. However, the absence of annealing on the last synthesis step will induce the formation of lower crystallite size values corresponding to the Cu$_2$O component. This behavior was also reported in other works [39,40], showing that if higher temperature is used the copper will pass in both oxidation states.

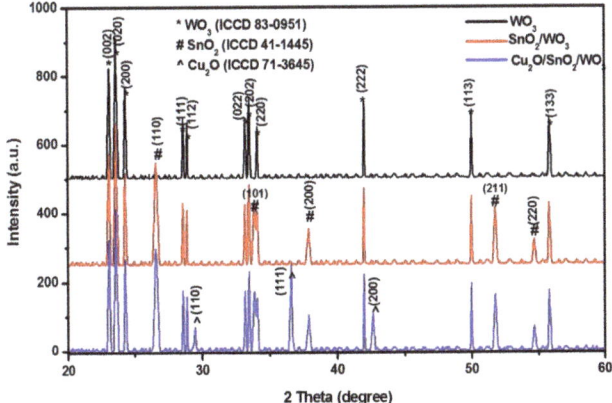

Figure 2. XRD patterns of the photocatalysts.

Table 1. Photocatalyst crystallite size values evaluated based on Scherrer formula.

Photocatalyst	Crystallite Size (Å)		
	WO$_3$	SnO$_2$	Cu$_2$O
WO$_3$	80.5	-	-
SnO$_2$/WO$_3$	81.8	77.1	-
Cu$_2$O/SnO$_2$/WO$_3$	81.6	77.3	42.4

Sample morphology was evaluated by SEM analysis, and the results are presented in Figure 3. The monocomponent sample exhibits fiber-like morphology (Figure 3a) corresponding to WO$_3$, which serves as substrate for SnO$_2$ development. The fibers have a random distribution in terms of diameter and length. The addition of the second component (SnO$_2$) will completely cover the WO$_3$ fibers and the morphology resembles overlapping pallets (Figure 3b). The final heterostructure containing all three components presents a porous morphology (Figure 3c), with Cu$_2$O uniformly distributed on the sample surface. Due to the low surface conductivity and the presence of aggregates, it was difficult to properly evaluate the grain size corresponding to the Cu$_2$O/SnO$_2$/WO$_3$ sample. The WO$_3$ and SnO$_2$ provide preferential nucleation sites for the following component and preserve a close interface contact between the components [41].

EDX analysis was used to evaluate the elemental composition of the samples and to calculate the component ratio. The analyses were done in three different areas of the samples and the results were similar (<2% abatement), which is an indicator of sample-component homogeneous distribution. The results were compared with the theoretical values calculated based on the stoichiometric compounds, and the results are presented in Table 2. In all three samples, oxygen excess was identified due to the WO$_3$ and SnO$_2$/WO$_3$ thermal treatment at 410 °C in oxygen-rich atmosphere. However, the heterostructure oxygen excess decreases due to the Cu$_2$O addition, which was not subjected to annealing [42,43]. During the annealing treatment, the number of oxygen vacancies is diminished due to the oxygen diffusion. The copper ratio compared with the other metal ions is similar which confirms that the weight–ratio relationship between the components was maintained during the synthesis and film deposition.

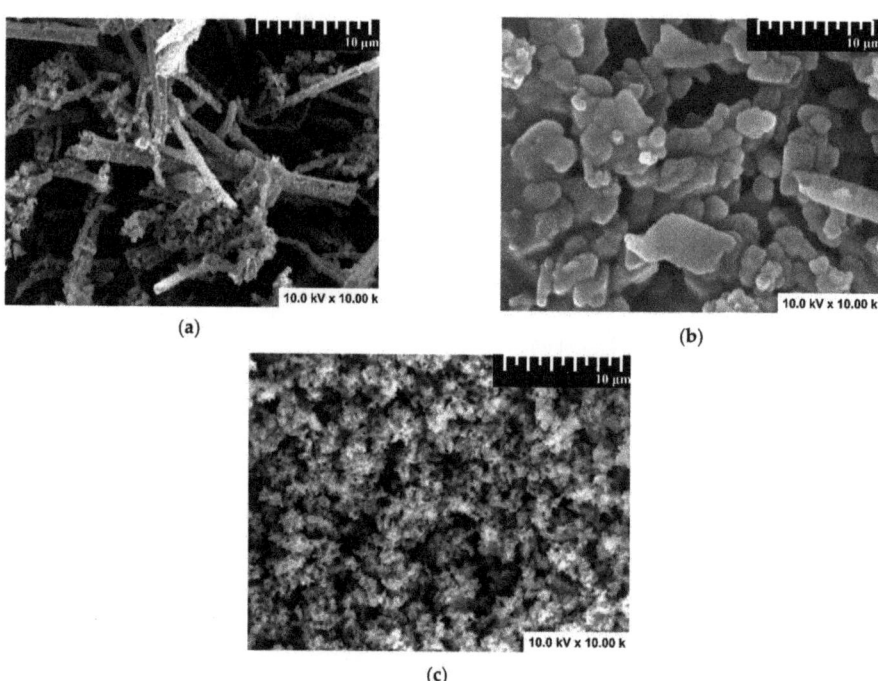

Figure 3. SEM images of (a) WO$_3$, (b) SnO$_2$/WO$_3$, and (c) Cu$_2$O/SnO$_2$/WO$_3$ samples.

Table 2. Photocatalyst elemental composition by EDX.

Samples	Elemental Composition [at.%]				
	W	Sn	Cu	O	O$_{th}$ [1]
WO$_3$	23.8	-	-	76.2	71.4
SnO$_2$/WO$_3$	12.4	14.7	-	72.9	68.9
Cu$_2$O/SnO$_2$/WO$_3$	10.6	11.9	13.3	64.2	62.2

[1] Theoretic content calculated based on stoichiometry.

3.2. Photocatalytic Activity

Three pesticide molecules were chosen based on their toxicity and residual persistence in water and soil. Pirimicarb (PIR) is an insecticide that can cause skin irritation, damage to brain functions, and is suspected of causing cancer [44,45]. S-metolachlor (S-MCh) is a herbicide inducing allergic skin irritation, anemia, and has a very toxic potential to aquatic life with long-lasting effects [46,47]. Metalaxyl (MET) is a fungicide with direct impact on the respiratory system and has long persistence in the aquatic environment [48]. This study integrated the large spectrum of pesticide to compare the influence of the photocatalysts on the pollutant molecule. It must be underlined that all the experimental conditions in terms of pollutant concentration, photocatalyst dosage, radiation type, and exposure periods were the same.

3.2.1. Degradation Efficiency and Kinetics

The lowest photocatalytic efficiencies (Figure 4a,c,e) correspond to bare Cu_2O sample, which exhibit 10.5%, 11.7%, and 8.2% toward PIR, S-MCh, and MET, respectively. Similar photocatalytic activities were obtained for SnO_2 and Cu_2O/WO_3, where coupling Cu_2O with WO_3 induced the presence of a high potential gap between the valence energy of Cu_2O and that corresponding to WO_3. The WO_3 monocomponent sample was able to remove 29.8% of PIR, 32.5% of S-MCh, and 12% of MET. Due to this band-gap energy (3.31 eV), the WO_3 photocatalyst used only a small fraction of the light radiation found at the border between UV and Vis spectra. When the second component is added, the coupled SnO_2/WO_3 benefits from the synergic effect of simultaneous charge carrier photogeneration in both semiconductors [49]. The photocatalytic removal efficiencies corresponding to SnO_2/WO_3 sample increase at 41.5% for PIR, 53.1% for S-MCh, and 26.8% for MET. The higher photocatalytic efficiencies correspond to the $Cu_2O/SnO_2/WO_3$ heterostructure, which uses the extended UV and Vis spectra due to the Cu_2O insertion with a band gap of 2.14 eV. In this case, the photocatalytic removal efficiency was 79.7% for PIR, 86.5% for S-MCh, and 50.6% for MET. The heterostructure photocatalytic removal efficiency is not a sum of the monocomponent sample efficiencies, as the contact between components induces a shift of the energy bands with consequences on the charge carrier's mobility. The lower MET photocatalytic removal efficiency values compared with the other pesticides indicate that the oxidation process is influenced by the pollutant molecule structure and the chemistry compatibility with the photocatalyst surface [50].

The kinetic evaluation was done using the simplified Langmuir–Hinshelwood mathematical Equation (3) and the results are presented in Figure 4b,d,f.

$$\ln \frac{C}{C_0} = -kt \tag{3}$$

The results indicate superior constant rates for S-MCh removal compared with PIR and MET pollutants. Additionally, for all three pollutants the photocatalytic activity of the $Cu_2O/SnO_2/WO_3$ heterostructure is 2× faster than that of SnO_2/WO_3 and 3× faster than that of the WO_3 monocomponent sample. As predicted, the lowest constant rates correspond to bare Cu_2O, which exhibits limited photocatalytic activity. However, when Cu_2O is put in contact with SnO_2 the constant rates increase 3×, which is a clear indicator of the heterostructure's ability to decrease the recombination rates and efficiently use the charge carries for oxidative species generation. These values indicate the contribution of each heterostructure component on the overall photocatalytic efficiency toward the pollutant's molecules. The heterostructure synthesis method has allowed the formation of stable interfaces that maintain high photocatalytic activity during the light irradiation [51].

In order to differentiate the mineralization induced by the photogenerated oxidative species from the partial oxidation products, TOC and TN were investigated (Figure 5a,c,e), as well as the corresponding kinetics (Figure 5b,d,f and Table 3). The TOC measurements for PIR and S-MCh indicate small differences on the photocatalytic activity. The lowest TOC reduction was registered for bare Cu_2O, followed by Cu_2O/SnO_2 and bare SnO_2. The TOC reduction corresponding to the WO_3 sample was 27.3% for PIR and 27.9% for S-MCh. Similarly, SnO_2/WO_3 (39.3% for PIR and 46.7% for S-MCh) and $Cu_2O/SnO_2/WO_3$ (73.4% for PIR and 81.3% for S-MCh) recorded slightly TOC reduction efficiency, indicating that not all the carbon components were mineralized. However, the TN evaluation confirms the photocatalytic activity presented in Figure 4, meaning that the nitrogen removed during the irradiation was completely mineralized. An interesting finding was observed for MET pollutant, where both TOC and TN confirm that the photocatalytic removal efficiencies can be attributed to mineralization. It must be outlined that more work must be done to increase the photocatalytic efficiency above 86% for S-MCh, 80% for PIR, and 50% for MET in order to have a significant impact on the long-term effects. The mineralization mechanism of pesticides considers the following steps: (i) charge carriers' development during the irradiation (Equation (4)); (ii) oxidative species (HO·) production during the reaction between the photogenerated holes and water molecules (Equation (5)); (iii) superoxidative species ($\cdot O_2^-$) development during the reaction between photogenerated electrons and dissolved oxygen (Equation (6)), and (iv) interaction between the (super)oxidative species and pesticide molecules (Equation (7)).

$$\text{Heterostructure} + h\nu \rightarrow e^- + h^+ \tag{4}$$

$$h^+ \text{ (Heterostructure)} + H_2O \rightarrow HO\cdot + H^+ \tag{5}$$

$$O_2 + e^- \text{ (Heterostructure)} \rightarrow \cdot O^-_2 \tag{6}$$

$$\text{Pesticides} + HO\cdot + \cdot O^-_2 \rightarrow xCO_2 + yH_2O + NO_2 \tag{7}$$

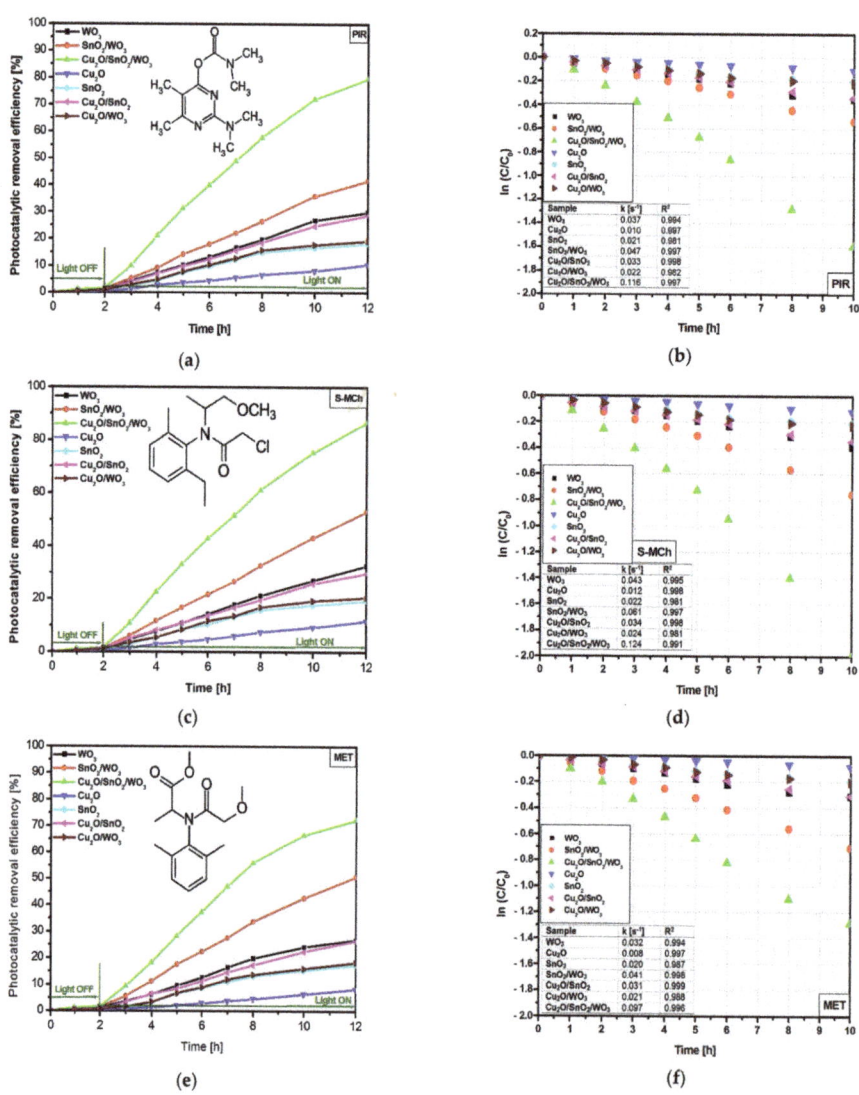

Figure 4. Photocatalytic removal efficiency and the corresponding kinetics toward PIR (**a**,**b**), S-MCh (**c**,**d**), and MET (**e**,**f**).

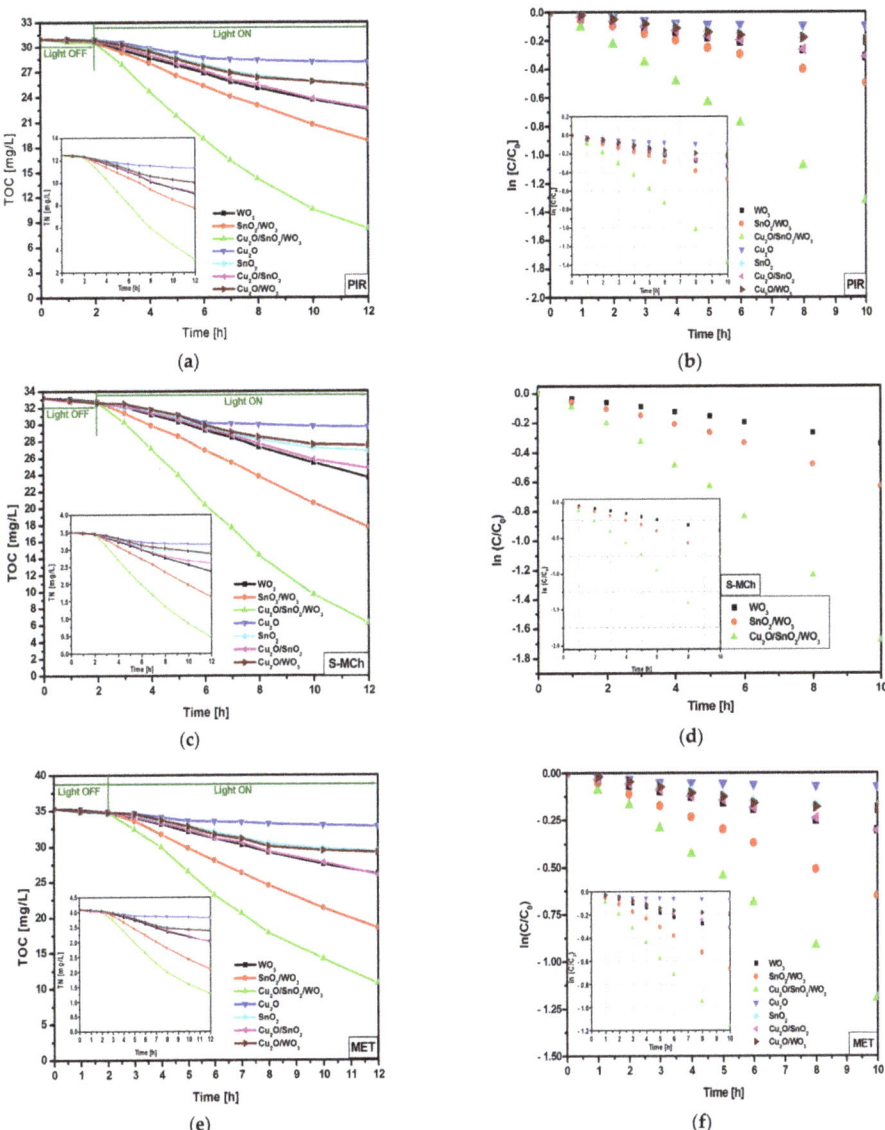

Figure 5. TOC, TN (inset), and the corresponding kinetics toward PIR (**a**,**b**), S-MCh (**c**,**d**), and MET (**e**,**f**).

The kinetic evaluation indicates that the WO_3 monocomponent sample has small and similar constant rates for all three pollutant molecules, which is a consequence of reduced light absorbance and high charge carrier recombination. Larger differences in the constant rates were obtained for SnO_2/WO_3 and $Cu_2O/SnO_2/WO_3$ samples, which confirm the photocatalytic removal efficiency results. The insertion of an additional semiconductor component will increase the photocatalytic efficiency where each partner will play a significant role on the charge carrier photogeneration and mobility. These characteristics are essential in order to produce oxidative (·OH) and superoxidative (·O_2^-) species able to induce pollutant mineralization [52–54].

Table 3. Kinetic parameters for TOC and TN evaluation.

Kinetic Parameters		PIR		S-MCh		MET	
		TOC	TN	TOC	TN	TOC	TN
WO$_3$	k (s^{-1})	0.031	0.033	0.033	0.038	0.030	0.032
	R^2	0.997	0.998	0.998	0.999	0.998	0.994
Cu$_2$O	k (s^{-1})	0.010	0.009	0.011	0.009	0.006	0.005
	R^2	0.911	0.947	0.919	0.928	0.920	0.866
SnO$_2$	k (s^{-1})	0.020	0.022	0.022	0.019	0.019	0.020
	R^2	0.978	0.989	0.982	0.965	0.981	0.969
SnO$_2$/WO$_3$	k (s^{-1})	0.049	0.047	0.062	0.074	0.045	0.043
	R^2	0.999	0.998	0.995	0.994	0.998	0.998
Cu$_2$O/SnO$_2$	k (s^{-1})	0.031	0.032	0.030	0.030	0.030	0.031
	R^2	0.998	0.998	0.998	0.989	0.999	0.999
Cu$_2$O/WO$_3$	k (s^{-1})	0.021	0.022	0.020	0.019	0.020	0.020
	R^2	0.972	0.987	0.975	0.978	0.973	0.963
Cu$_2$O/SnO$_2$/WO$_3$	k (s^{-1})	0.135	0.137	0.168	0.195	0.120	0.119
	R^2	0.998	0.994	0.990	0.986	0.997	0.998

3.2.2. Reusability and Mechanism of Charge Carrier Generation

The mechanism of charge carrier generation through the heterostructure can explain the role of each component in enhancing the photocatalytic efficiency toward the pesticide molecules. The band energy diagram (Figure 6a) was constructed considering the experimental band gap corresponding to each heterostructure component based on the Wood and Tauc model, as presented in Figure 6b–d. The methodology has been already presented in another paper [55] and is in good agreement with the literature [56,57]. The band gap may submit minor shift during the heterostructure development. The methodology takes into consideration the band-gap changes during the heterostructure's internal energy field developed during the irradiation. The diagram includes the energy band position based on Equations (8)–(11), which consider several key parameters: E_e, representing the free electron energy vs. hydrogen, χ_{cation} (eV), representing the absolute cationic electronegativity, χ_{cation} (P.u.) is the cationic specific electronegativity, where P.u. corresponds to the Pauling units, E_g is the band-gap energy, and $\chi_{semiconductor}$ represents the electronegativity of each semiconductor.

$$E_{VB} = \chi_{semiconductor} - E_e + 0.5E_g \quad (8)$$

$$E_{CB} = E_{VB} - E_g \quad (9)$$

$$\chi_{semiconductor}\ (eV) = 0.45 \times \chi_{cation}\ (eV) + 3.36 \quad (10)$$

$$\chi_{cation}(eV) = \frac{\chi_{cation\ (P.u.)} + 0.206}{0.336} \quad (11)$$

During the semiconductors' interface development, the band gap may shift due to the internal energy field. The diagram corresponds to a Z-scheme charge transfer mechanism where the electrons generated during the light irradiation from the Cu$_2$O conduction band (−0.35 eV) will transit the SnO$_2$ conduction band on the way to the WO$_3$ conduction band (+0.14 eV). The photogenerated electrons from SnO$_2$ and WO$_3$ conduction bands and the photogenerated holes from the Cu$_2$O valence band (+1.79 eV) are not involved in oxidative species development, owing to their potential. In this case, some of the charge carriers will recombine. However, the useful photogenerated electrons from the Cu$_2$O conduction band and the photogenerated holes from the SnO$_2$ (+2.45 eV) and WO$_3$ (3.44 eV) valence bands possess stronger redox ability and cannot recombine due to the electric field development in the charged separation region. The charge carrier's mobility is sustained by the combined drift and diffusion effect. Consequently, the synergic effect provided by the heterostructure semiconductor components enhances the production of oxidative and superoxidative radicals responsible for pollutant mineralization. The heterostructure requires further improvement in order to reduce the charge recombination and to increase the overall photocatalytic efficiency toward pesticides.

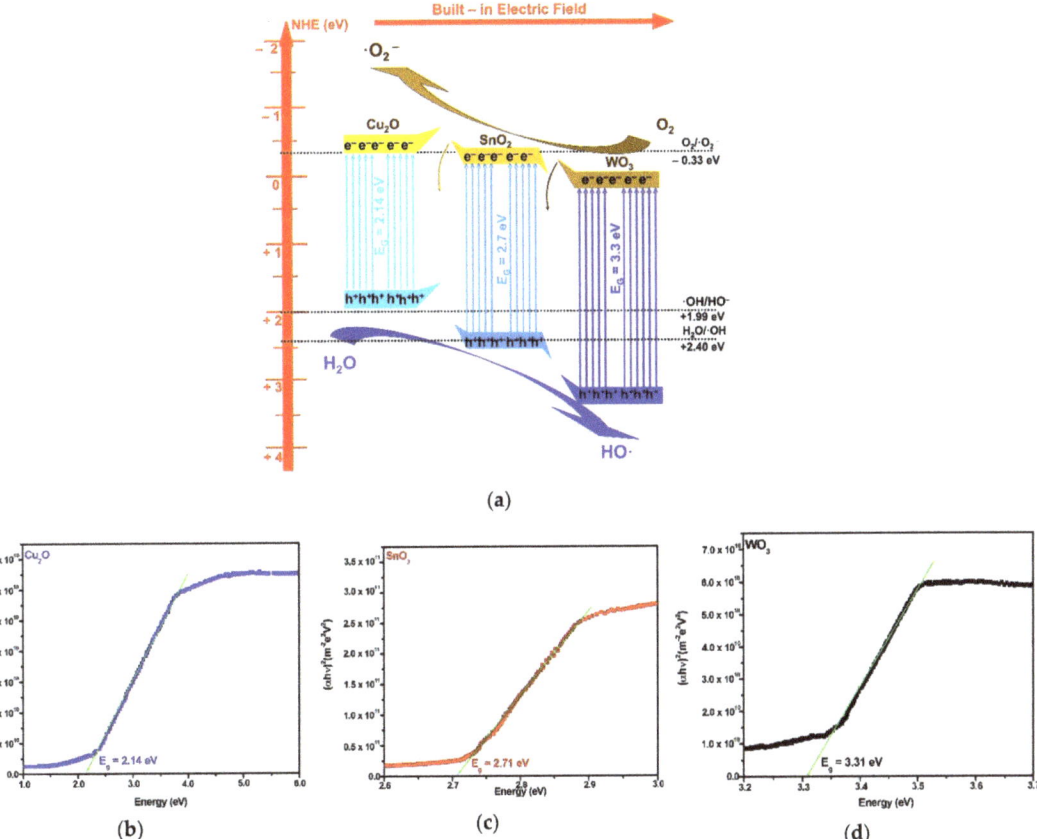

Figure 6. Heterostructure mechanism (a) and component band-gap values (b–d).

The reusability evaluation (Figure 7) was performed on the samples with the highest photocatalytic efficiency using a three-cycle assessment. The results indicate that for all three pollutant molecules, the $Cu_2O/SnO_2/WO_3$ exhibits good stability with negligible photocatalytic changes between cycles. The WO_3 and SnO_2/WO_3 show small abatement of less then 5%, mostly on the second cycle. The changes can be induced by pollutant molecule adsorption at the photocatalyst surface active centers, which may require a longer desorption/degradation period. The results indicate that the $Cu_2O/SnO_2/WO_3$ heterostructure can be used for multiple cycle assessment without influencing the photocatalytic activity toward pesticide molecules.

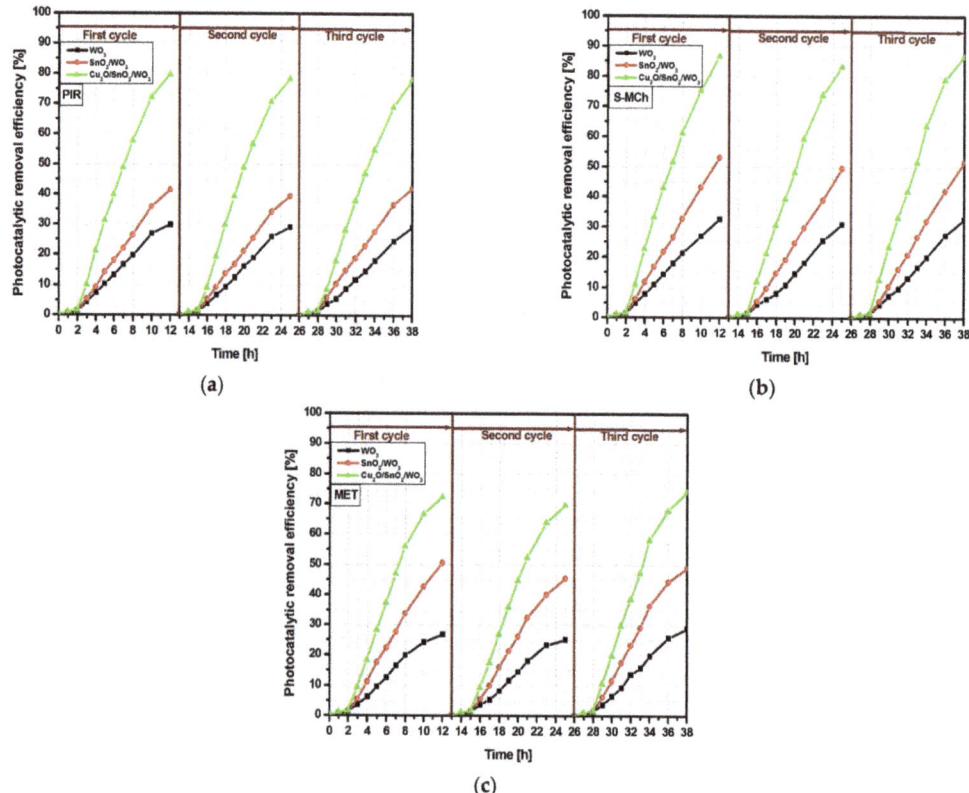

Figure 7. Photocatalytic activity during the 3-cycle sample reusability for (**a**) PIR, (**b**) S-MCh and (**c**) MET pesticides.

4. Conclusions

A heterostructure based on $Cu_2O/SnO_2/WO_3$ was synthesized using a three-step sol–gel method and deposited as films on glass substrate by spray deposition. The sample contains tetragonal SnO_2 and monoclinic WO_3 with similar crystallite sizes (≈ 80 Å), while cubic Cu_2O exhibits significantly lower crystallite sizes (≈ 30 Å). The porous morphology of the final heterostructure was constructed starting with fiber-like WO_3, which serves as substrate for SnO_2 development, and SnO_2/WO_3 provides the nucleation and growing sites for Cu_2O formation. The elemental analysis confirms that the samples are homogeneous in composition and exhibit oxygen excess due to the annealing treatments. The EDX measurements were done on three different areas, indicating that the final heterostructure preserves the component:weight ratio used during the synthesis.

Three pesticide molecules (pirimicarb, S-metolachlor, and metalaxyl) were used as pollutants reference and were chosen based on their resilience and toxicity. The photocatalytic activity of the $Cu_2O/SnO_2/WO_3$ heterostructure is superior to that of bare oxides or tandem systems. The lowest photocatalytic activity ($\approx 10\%$) corresponds to bare Cu_2O, followed by SnO_2 and Cu_2O/WO_3. The highest photocatalytic efficiency was obtained toward S-MCh (86%) using the $Cu_2O/SnO_2/WO_3$ sample and the lowest corresponds to MET (12%) removal using the WO_3 monocomponent sample. However, the TOC analysis indicates lower mineralization efficiencies that can be attributed to secondary product formation. The photocatalytic mechanism corresponds to a Z-scheme charge transfer where Cu_2O exhibits reduction potential and WO_3 oxidation potential. The charge carrier's mobility sustained by the combined drift and diffusion effect provides a synergic effect where the heterostructure semiconductor components enhance the (super)oxidative radicals' production. The reusability tests indicate that $Cu_2O/SnO_2/WO_3$ exhibits similar photocatalytic efficiency after a three-cycle assessment, which recommends this material for further experiments.

Author Contributions: Conceptualization, A.E.; methodology, A.E.; software, L.A.; validation, A.E. and L.A.; formal analysis, L.A.; investigation, L.A.; resources, A.E.; data curation, L.A.; writing—original draft preparation, A.E.; writing—review and editing, A.E.; visualization, L.A.; supervision, A.E.; project administration, A.E.; funding acquisition, A.E. All authors have read and agreed to the published version of the manuscript.

Funding: This work was supported by a grant of the Ministry of Research, Innovation, and Digitization, CNCS-UEFISCDI, project number PN-III-P4-PCE-2021-1020 (PCE87), within PNCDI III.

Institutional Review Board Statement: Not applicable.

Data Availability Statement: Data presented in this study are available upon request from the corresponding author.

Acknowledgments: The authors acknowledge the experimental investigation support given by the Tokyo University of Science and the advice given by L. Isac during the experimental work.

Conflicts of Interest: The authors declare no conflict of interest.

References

1. Alexandrino, D.A.M.; Almeida, C.M.R.; Carvalho, M.F. Revisiting pesticide pollution: The case of fluorinated pesticides. *Environ. Pollut.* **2022**, *292*, 118315. [CrossRef] [PubMed]
2. Fu, H.; Tan, P.; Wu, Z. Advances in organophosphorus pesticides pollution: Current status and challenges in ecotoxicological, sustainable agriculture, and degradation strategies. *J. Hazard. Mater.* **2022**, *424*, 127494. [CrossRef] [PubMed]
3. Quaglia, G.; Joris, I.; Seuntjens, P. A spatial approach to identify priority areas for pesticide pollution mitigation. *J. Environ. Manag.* **2019**, *246*, 583–593. [CrossRef] [PubMed]
4. An, X.; Wu, Z.; Yu, B. Biochar for simultaneously enhancing the slow-release performance of fertilizers and minimizing the pollution of pesticides. *J. Hazard. Mater.* **2021**, *407*, 124865. [CrossRef]
5. Tsaboula, A.; Menexes, G.; Papadopoulou-Mourkidou, E. Assessment and management of pesticide pollution at a river basin level part II: Optimization of pesticide monitoring networks on surface aquatic ecosystems by data analysis methods. *Sci. Total Environ.* **2019**, *653*, 1612–1622. [CrossRef]
6. Dudita, M.; Bogatu, C.; Enesca, A.; Duta, A. The influence of the additives composition and concentration on the properties of SnOx thin films used in photocatalysis. *Mater. Lett.* **2011**, *65*, 2185–2189. [CrossRef]
7. Kaushal, J.; Khatri, M.; Arya, S.K. A treatise on Organophosphate pesticide pollution: Current strategies and advancements in their environmental degradation and elimination. *Ecotox. Environ. Saf.* **2021**, *207*, 111483. [CrossRef]
8. Gong, W.; Barrett, H.; Qu, H. Application of biochar: An approach to attenuate the pollution of the chiral pesticide fipronil and its metabolites in leachate from activated sludge. *Process Saf. Environ.* **2021**, *149*, 936–945. [CrossRef]
9. Sahin, C.; Ekrem Karpuzcu, M. Mitigation of organophosphate pesticide pollution in agricultural watersheds. Mitigation of organophosphate pesticide pollution in agricultural watersheds. *Sci. Total Environ.* **2020**, *710*, 136261. [CrossRef]
10. Wołejko, E.; Jabłońska-Trypuć, A.; Łozowicka, B. Soil biological activity as an indicator of soil pollution with pesticides—A review. *Appl. Soil Ecol.* **2020**, *147*, 103356. [CrossRef]
11. Parada, J.; Rubilar, O.; Tortella, G.R. Combined pollution of copper nanoparticles and atrazine in soil: Effects on dissipation of the pesticide and on microbiological community profiles. *J. Hazard. Mater.* **2019**, *361*, 228–236. [CrossRef]
12. Solé, M.; Bonsignore, M.; Freitas, R. Exploring alternative biomarkers of pesticide pollution in clams. *Mar. Pollut. Bull.* **2018**, *136*, 61–67. [CrossRef]
13. Wan, N.F.; Ji, X.Y.; Li, B. An eco-engineering assessment index for chemical pesticide pollution management strategies to complex agro-ecosystems. *Ecol. Eng.* **2013**, *52*, 203–210. [CrossRef]
14. Enesca, A.; Duta, A.; Schoonman, J. Influence of tantalum dopant ions (Ta^{5+}) on the efficiency of the tungsten trioxide photoelectrode. *Phys. Status Solidi A* **2008**, *205*, 2038–2041. [CrossRef]
15. Horne, G.P.; Zalupski, P.R.; Mincher, B.J. Radiolytic degradation of formic acid and formate in aqueous solution: Modeling the final stages of organic mineralization under advanced oxidation process conditions. *Water Res.* **2020**, *186*, 116314. [CrossRef]
16. Mukherjee, A.; Mullick, A.; Moulik, S. Performance and energetic analysis of hydrodynamic cavitation and potential integration with existing advanced oxidation processes: A case study for real life greywater treatment. *Ultrason. Sonochem.* **2020**, *66*, 105116. [CrossRef]

17. Chen, C.J.; Wu, C.C.; Hsieh, L.T.; Chen, K.C. Treatment of Trichloroethylene with Photocatalyst-Coated Optical Fiber. *Water* **2019**, *11*, 2391. [CrossRef]
18. Zhu, X.; Zhao, Y.; Pu, M.; Zhang, Y.; Zhang, H.; Cheng, C. The Effect of Sintering Oxygen Partial Pressure on a $SmBiO_3$ Buffer Layer for Coated Conductors via Chemical Solution Deposition. *Coatings* **2016**, *6*, 50. [CrossRef]
19. Wu, S.; Dai, W. Microwave-Hydrothermal Synthesis of SnO_2-CNTs Hybrid Nanocomposites with Visible Light Photocatalytic Activity. *Nanomaterials* **2017**, *7*, 54. [CrossRef]
20. Antoniadou, M.; Arfanis, M.K.; Ibrahim, I.; Falaras, P. Bifunctional $g-C_3N_4/WO_3$ Thin Films for Photocatalytic Water Purification. *Water* **2019**, *11*, 2439. [CrossRef]
21. Dong, H.; Virtanen, S. Anodic ZnO Microsheet Coating on Zn with Sub-Surface Microtrenched Zn Layer Reduces Risk of Localized Corrosion and Improves Bioactivity of Pure Zn. *Coatings* **2021**, *11*, 486. [CrossRef]
22. Liao, Y.K.; Liu, Y.T.; Hsieh, D.H.; Shen, T.L.; Hsieh, M.Y.; Tzou, A.J.; Chen, S.C.; Tsai, Y.L.; Lin, W.S.; Chan, S.W.; et al. Breakthrough to Non-Vacuum Deposition of Single-Crystal, Ultra-Thin, Homogeneous Nanoparticle Layers: A Better Alternative to Chemical Bath Deposition and Atomic Layer Deposition. *Nanomaterials* **2017**, *7*, 78. [CrossRef]
23. Ramesh, K.; Gnanavel, B.; Shkir, M. Enhanced visible light photocatalytic degradation of bisphenol A (BPA) by reduced graphene oxide (RGO)–metal oxide (TiO_2, ZnO and WO_3) based nanocomposites. *Diam. Relat. Mater.* **2021**, *118*, 108514. [CrossRef]
24. Zhang, P.; Liang, H.; Li, C. A novel Z-scheme BiOI/BiOCl nanofibers photocatalyst prepared by one-pot solvothermal with efficient visible-light-driven photocatalytic activity. *Mater. Chem. Phys.* **2021**, *272*, 125031. [CrossRef]
25. Zhang, Y.; Shen, G.; Fan, W. The effect of piezo-photocatalysis on enhancing the charge carrier separation in $BaTiO_3/KNbO_3$ heterostructure photocatalyst. *Appl. Surf. Sci.* **2021**, *562*, 150164. [CrossRef]
26. Baneto, M.; Enesca, A.; Mihoreanu, C.; Lare, Y.; Jondo, K.; Napo, K.; Duta, A. Effects of the growth temperature on the properties of spray deposited $CuInS_2$ thin films for photovoltaic applications. *Ceram. Int.* **2015**, *41*, 4742–4749. [CrossRef]
27. Koutavarapu, R.; Syed, K.; Shim, J. An effective CuO/Bi_2WO_6 heterostructured photocatalyst: Analyzing a charge-transfer mechanism for the enhanced visible-light-driven photocatalytic degradation of tetracycline and organic pollutants. *Chemosphere* **2022**, *287*, 132015. [CrossRef]
28. Sridharan, M.; Maiyalagan, T. Recent progress in Tungsten disulphide based Photocatalyst for Hydrogen Production and Environmental Remediation. *Chem. Eng. J.* **2021**, *424*, 130393. [CrossRef]
29. Liu, T.; Yang, K.; Jin, Z. Visible-light driven S-scheme $Mn_{0.2}Cd_{0.8}S/CoTiO_3$ heterojunction for photocatalytic hydrogen evolution. *Renew. Energ.* **2021**, *173*, 389–400. [CrossRef]
30. Cui, Z.; Zhang, D. Caihong Fang, CdS/CeO_2 heterostructures as visible-light photocatalysts for the reduction of nitro to amine organics. *J. Alloys Compd.* **2021**, *885*, 160961. [CrossRef]
31. Flihh, S.M.; Ammar, S.H. Fabrication and photocatalytic degradation activity of core/shell $ZIF-67@CoWO_4@CoS$ heterostructure photocatalysts under visible light. *Environ. Nanotech. Monit. Manag.* **2021**, *16*, 100595. [CrossRef]
32. Zhai, D.; Lv, Z.; Lu, J. Pre-deposition growth of interfacial SiO_2 layer by low-oxygen-partial-pressure oxidation in the $Al_2O_3/4H-SiC$ MOS structure. *Microelectron. Eng.* **2021**, *244–246*, 111574. [CrossRef]
33. Semenov, M.Y.; Semenov, Y.M.; Silaev, A.V.; Begunova, L.A. Assessing the Self-Purification Capacity of Surface Waters in Lake Baikal Watershed. *Water* **2019**, *11*, 1505. [CrossRef]
34. Li, J.-Q.; Liu, L.; Fu, X.; Tang, D.; Wang, Y.; Hu, S.; Yan, Q.-L. Transformation of Combustion Nanocatalysts inside Solid Rocket Motor under Various Pressures. *Nanomaterials* **2019**, *9*, 381. [CrossRef]
35. Domene-López, D.; Sarabia-Riquelme, R.; García-Quesada, J.C.; Martin-Gullon, I. Custom-Made Chemically Modified Graphene Oxide to Improve the Anti-Scratch Resistance of Urethane-Acrylate Transparent Coatings. *Coatings* **2019**, *9*, 408. [CrossRef]
36. Ma, D.; Sun, D.; Shi, J.W. The synergy between electronic anchoring effect and internal electric field in CdS quantum dots decorated dandelion-like $Fe-CeO_2$ nanoflowers for improved photocatalytic hydrogen evolution. *J. Colloid Interface Sci.* **2019**, *549*, 179–188. [CrossRef]
37. Enesca, A.; Duta, A. Tailoring WO_3 thin layers using spray pyrolysis technique. *Phys. Status Solidi C* **2008**, *5*, 3499–3502. [CrossRef]
38. Güneri, E.; Göde, F.; Saatçi, B. The effect of Cu_2O layer on characteristic properties of $n-CdS/p-Cu_2O$ heterojunction. *J. Mol. Struct.* **2021**, *1241*, 130679. [CrossRef]
39. Li, X.; Raza, S.; Liu, C. Directly electrospinning synthesized Z-scheme heterojunction $TiO_2@Ag@Cu_2O$ nanofibers with enhanced photocatalytic degradation activity under solar light irradiation. *J. Environ. Chem. Eng.* **2021**, *9*, 106133. [CrossRef]
40. Fentahun, D.A.; Tyagi, A.; Kar, K.K. Numerically investigating the AZO/Cu_2O heterojunction solar cell using ZnO/CdS buffer layer. *Optik* **2021**, *228*, 166228. [CrossRef]
41. Moghanlou, A.O.; Bezaatpour, A.; Salimi, F. Cu_2O/rGO as an efficient photocatalyst for transferring of nitro group to amine group under visible light irradiation. *Mater. Sci. Semicon. Proc.* **2021**, *130*, 105838. [CrossRef]
42. Zhou, M.; Guo, Z.; Liu, Z. FeOOH as hole transfer layer to retard the photocorrosion of Cu_2O for enhanced photoelctrochemical performance. *Appl. Catal. B* **2020**, *260*, 118213. [CrossRef]
43. Zhu, Z.; Chu, H.; Jiang, L. Anti-microbial corrosion performance of concrete treated by Cu_2O electrodeposition: Influence of different treatment parameters. *Cem. Concr. Comp.* **2021**, *123*, 104195. [CrossRef]
44. Jepson, P.C.; Murray, K.; Neumeister, L. Selection of pesticides to reduce human and environmental health risks: A global guideline and minimum pesticides list. *Lancet Planet. Health* **2020**, *4*, e56–e63. [CrossRef]

45. Bhandari, G.; Atreya, K.; Geissen, V. Concentration and distribution of pesticide residues in soil: Non-dietary human health risk assessment. *Chemosphere* **2020**, *253*, 126594. [CrossRef]
46. Jiang, M.; Zhang, W.; Gong, W. Assessing transfer of pesticide residues from chrysanthemum flowers into tea solution and associated health risks. *Ecotox. Environ. Saf.* **2020**, *187*, 109859. [CrossRef]
47. Mandić-Rajčević, S.; Rubino, F.R.; Colosio, C. Establishing health-based biological exposure limits for pesticides: A proof of principle study using mancozeb. *Regul. Toxicol. Pharm.* **2020**, *115*, 104689. [CrossRef]
48. Gamboa, L.C.; Diaz, K.S.; van Wendel de Joode, B. Passive monitoring techniques to evaluate environmental pesticide exposure: Results from the Infant's Environmental Health study (ISA). *Environ. Res.* **2020**, *184*, 109243. [CrossRef]
49. Mouchaal, Y.; Enesca, A.; Mihoreanu, C.; Khelil, A.; Duta, A. Tuning the opto-electrical properties of SnO_2 thin films by Ag^{+1} and In^{+3} co-doping. *Mater. Sci. Eng. B* **2015**, *199*, 22–29. [CrossRef]
50. Zhang, L.; Li, Y.; Wang, H. Well-dispersed Pt nanocrystals on the heterostructured TiO_2/SnO_2 nanofibers and the enhanced photocatalytic properties. *Appl. Surf. Sci.* **2014**, *319*, 21–28. [CrossRef]
51. Hamadanian, M.; Jabbari, V.; Mutlay, I. Preparation of novel hetero-nanostructures and high efficient visible light-active photocatalyst using incorporation of CNT as an electron-transfer channel into the support TiO_2 and PbS. *J. Taiwan Inst. Chem. Eng.* **2013**, *44*, 748–757. [CrossRef]
52. Wang, Z.; Liang, X.; Zhu, R. Ag and Cu_2O modified 3D flower-like ZnO nanocomposites and evaluated by photocatalysis oxidation activity regulation. *Ceram. Int.* **2019**, *45*, 23310–23319. [CrossRef]
53. Tobajas, M.; Belver, C.; Rodríguez, J.J. Degradation of emerging pollutants in water under solar irradiation using novel TiO_2-ZnO/clay nanoarchitectures. *Chem. Eng. J.* **2017**, *309*, 596–606. [CrossRef]
54. Mohammadiyan, E.; Ghafuri, H.; Kakanejadifard, A. Synthesis and characterization of a magnetic $Fe_3O_4@CeO_2$ nanocomposite decorated with Ag nanoparticle and investigation of synergistic effects of Ag on photocatalytic activity. *Optik* **2018**, *166*, 39–48. [CrossRef]
55. Enesca, A.; Andronic, L.; Duta, A. The influence of surfactants on the crystalline structure, electrical and photocatalytic properties of hybrid multi-structured (SnO_2, TiO_2 and WO_3) thin films. *Appl. Surf. Sci.* **2012**, *258*, 4339–4346. [CrossRef]
56. Gao, C.; Li, J.; Shan, Z.; Huang, F.; Shen, H. Preparation and visible-light photocatalytic activity of In2S3/TiO2 composite. *Mater. Chem. Phys.* **2010**, *122*, 183–187. [CrossRef]
57. Mise, T.; Nakada, T. Low temperature growth and properties of Cu–In–Te based 433 thin films for narrow bandgap solar cells. *Thin Solid Films* **2010**, *518*, 5604–5609. [CrossRef]

Article

Sputtering-Assisted Synthesis of Copper Oxide–Titanium Oxide Nanorods and Their Photoactive Performances

Yuan-Chang Liang * and Tsun-Hsuan Li

Department of Optoelectronics and Materials Technology, National Taiwan Ocean University,
Keelung 20224, Taiwan; chris6080624@gmail.com
* Correspondence: yuanvictory@gmail.com or yuan@mail.ntou.edu.tw

Abstract: A TiO_2 nanorod template was successfully decorated with a copper oxide layer with various crystallographic phases using sputtering and postannealing procedures. The crystallographic phase of the layer attached to the TiO_2 was adjusted from a single Cu_2O phase or dual Cu_2O–CuO phase to a single CuO phase by changing the postannealing temperature from 200 °C to 400 °C. The decoration of the TiO_2 (TC) with a copper oxide layer improved the light absorption and photoinduced charge separation abilities. These factors resulted in the composite nanorods demonstrating enhanced photoactivity compared to that of the pristine TiO_2. The ternary phase composition of TC350 allowed it to achieve superior photoactive performance compared to the other composite nanorods. The possible Z-scheme carrier movement mechanism and the larger granular size of the attached layer of TC350 under irradiation accounted for the superior photocatalytic activity in the degradation of RhB dyes.

Keywords: microstructure; composites; photoactivity

Citation: Liang, Y.-C.; Li, T.-H. Sputtering-Assisted Synthesis of Copper Oxide–Titanium Oxide Nanorods and Their Photoactive Performances. *Nanomaterials* **2022**, *12*, 2634. https://doi.org/10.3390/nano12152634

Academic Editor: Francesc Viñes Solana

Received: 8 July 2022
Accepted: 28 July 2022
Published: 30 July 2022

Publisher's Note: MDPI stays neutral with regard to jurisdictional claims in published maps and institutional affiliations.

Copyright: © 2022 by the authors. Licensee MDPI, Basel, Switzerland. This article is an open access article distributed under the terms and conditions of the Creative Commons Attribution (CC BY) license (https://creativecommons.org/licenses/by/4.0/).

1. Introduction

TiO_2 nanorods are widely used as template for fabrication of photoexcited devices [1]. However, the main drawback of the intrinsic properties of TiO_2 is its large energy gap, which means that it only absorbs light in the ultraviolet region. Recent progress on coupling the heterogeneous structure of TiO_2 with visible light sensitizers has been demonstrated as a promising approach to substantially improve the light harvesting ability of the TiO_2 template. Several binary visible-light sensitizers, such as Bi_2O_3, Cu_2O, CuO, Fe_2O_3, CdS, and Bi_2S_3, have been adopted for coupling with TiO_2 templates to achieve improved photoactive performance [2–8]. Among these visible-light sensitizers, binary oxides provide a better, more suitable process and chemical compatibility for integration with TiO_2 templates in comparison with most sulfides. Notably, in comparison with n-n heterostructures, the construction of p-n heterostructures is a more promising approach for the enhancement of the photoactivity of TiO_2-based composites. The p-n junction generates an internal electric field that can effectively suppress the recombination of photogenerated carriers in the composite system [5,9]. In addition, in terms of charge transport mode, the Z scheme often appears in organic degradation, CO_2 reduction and photoelectric catalytic water splitting in heterostructured systems [10].

Among the various p-type visible-light sensitizers, copper oxides are distinguished by having diverse crystallographic phases and tunable band energy. Copper oxides are non-toxic and low cost materials rich in earth elements. Due to their low energy gap values, they have high optical absorption properties, resulting in excellent photoelectrochemical (PEC) performance and high energy conversion efficiency [11,12]. Recent work on the attachment of copper oxides onto TiO_2 to enhance photoactive performance has attributed this improvement to the formation of a p-n junction. For example, electrodeposition of p-type Cu_2O onto TiO_2 nanoarrays improved the light absorption capacity and enhanced the photocatalytic activity [13]. Furthermore, p-type CuO nanoparticles attached onto TiO_2

nanosheets effectively enhanced the photocatalytic activity for the oxidation of methanol to methyl formate [14]. CuO–Cu$_2$O co-coupled TiO$_2$ nanomaterials synthesized through chemical reduction and hydrolysis presented better charge separation rates and photocatalytic activity than those of pristine TiO$_2$ [15]. The above examples show that attachment of single CuO or Cu$_2$O or dual phase CuO–Cu$_2$O onto parent TiO$_2$ induces the formation of a p-n heterojunction between the copper oxide and the TiO$_2$, resulting in the composites possessing an internal electric field and suppressing the recombination of photogenerated carriers. These phenomena can effectively increase the photocatalytic ability of the pristine TiO$_2$. However, most investigations of the photoactivity of copper oxide–TiO$_2$ composite systems are based on a fixed decorated oxide phase (one of the following: CuO, Cu$_2$O, or CuO–Cu$_2$O); this is attributed to the fact that precise manipulation of the crystallographic phase of copper oxide is still highly challenging using most chemical or physical synthesis routes. Systematic investigations of the effects of phase evolution on the photoactivity of copper oxide–TiO$_2$ nanocomposite rods are still limited in number, and such information is an important reference for the design and tuning of the photoactive performance of copper oxide–TiO$_2$ nanocomposites.

Thin copper oxide films can be synthesized via diverse chemical and physical routes [16–18]. Physical deposition of thin copper oxide films with adjustable crystallographic phases is a promising approach to design copper oxide–TiO$_2$ nanocomposites with desirable photoactive performance for photoexcited device applications. It has been shown that the formation temperature of the crystalline copper oxide has profound effects on the crystallographic phases of the as-synthesized copper oxides [18,19]. However, such temperature-dependent copper oxide phase evolutions are not always similar between different studies because of the different copper oxide precursors initially formed and the different process parameters or routes used [20,21]. For example, a copper film was transformed into the Cu$_2$O phase after annealing at 250 °C under an atmospheric environment for 1 h. Moreover, a mixed phase of Cu$_2$O–CuO appeared when the annealing temperature was set between 250–350 °C. Finally, the CuO phase could be obtained with an annealing temperature above 350 °C [22]. In this study, a thin metallic copper film was sputter-coated onto a TiO$_2$ nanorod template. The crystallographic phase of the copper oxide layer formed by postannealing the pre-deposited copper film was tuned to manipulate the photoactive performance of the copper oxide-decorated TiO$_2$ nanorod composites. The approach used by this work to produce copper oxide-decorated TiO$_2$ composite nanorods differs from previous reference works [13–15]. Most copper oxide-decorated TiO$_2$ is synthesized through chemical routes. It is difficult to manipulate the copper oxide crystalline phase using these routes. Only one copper oxide phase is attached onto the TiO$_2$ template. In contrast, by combining a sputtering process and postannealing procedures in this work, we could easily design different copper oxide crystal phases on the TiO$_2$ templates. The correlation between the composition phase, microstructure, and photoactivity of the copper oxide layer attached onto the TiO$_2$ nanorod template was systematically investigated. The results presented herein are important references for the design of copper oxide–TiO$_2$ composite systems with desirable photoactivity for photoexcited device applications.

2. Experiments

The preparation of TiO$_2$ composite nanorods decorated with a copper oxide layer can be divided into two steps. The first step was to prepare TiO$_2$ nanorod arrays on F-doped SnO$_2$ glass substrates. The detailed preparation procedures have been described elsewhere [23]. The second step included modification of the surfaces of TiO$_2$ nanorods with a copper oxide layer by sputtering. A metallic copper disc with a size of 2 inches was used as the target. The metallic copper film was sputter-coated onto the surfaces of TiO$_2$ nanorods at room temperature under a pure argon atmosphere. The working pressure was 20 mtorr, and the sputtering power was fixed at 30 W. The sputtering duration was 12 min. The as-synthesized metallic copper layers on the TiO$_2$ nanorods were further subjected to an atmospheric annealing treatment for 1 h. The annealing temperature was varied between 200, 300, 350, and 400 °C to induce the formation of copper oxide from the metallic copper layer. The sample codes for the composite nanorods formed after 200, 300, 350, and 400 °C annealing were TC200, TC300, TC350, and TC400, respectively.

The crystallographic structures of the various samples were characterized with grazing incidence angle X-ray diffraction (GID; BRUKER D8 SSS, Karlsruhe, Germany) using monochromatic Cu-Kα radiation. A field emission scanning electron microscope (SEM; JSM-7900F, JEOL, Tokyo, Japan) equipped with an energy-dispersive X-ray spectrometer (EDS) was used for further investigations into the morphology and elemental distribution of the samples. A high-resolution transmission electron microscope equipped with EDS (HRTEM; Philips Tecnai F20 G2) was used to investigate the detailed structure and composition of the composite nanorods. An X-ray photoelectron spectroscopy (XPS ULVAC-PHI, PHI 5000 VersaProbe, Chigasaki, Japan) with Al Kα X-rays was used to detect the element binding states of the samples. The optical absorption spectral information for the samples was obtained with a UV-vis spectrophotometer (Jasco V750, Tokyo, Japan). Photoelectrochemical (PEC) performance and electrochemical impedance (EIS) were measured using a potentiostat (SP150, BioLogic, Seyssinet-Pariset, France). In the photoelectrochemical system, the effective area of the working electrode was 1.0 cm^2. The reference and counter electrodes were Ag/AgCl (in saturated KCl) and platinum wire, respectively. A 0.5 M aqueous Na_2SO_4 solution was used as the electrolyte in the measurement system. During the photoexcitation experiments, a 100 W xenon lamp was used as the light source. Rhodamine B (RhB) solution (10^{-5} M) was used as the target pollutant for photodegradation experiments, and residual RhB concentrations after different degradation durations were estimated using a UV-vis spectrophotometer.

3. Results and Discussion

Figure 1a shows SEM micrographs of TiO_2 nanorod templates. The TiO_2 nanorods have rectangular cross-section morphologies and smooth sidewalls. Figure 1b shows an SEM micrograph of a TiO_2 nanorod template coated with thin Cu films and postannealed at 200 °C. In comparison with the diameter of pristine TiO_2 nanorods, it can be seen that the diameter of the TiO_2 nanorods increased after copper coating and annealing at 200 °C. Furthermore, the morphology of the decorated layer wrapped in the outer layer of the TiO_2 was film-like, and the sidewalls of the TC200 became rough. When the annealing temperature increased to 300 °C, the morphology of TC300 differed from that of TC200. The continuous film-like decorated layer of the composite nanorods formed with the lower annealing temperature of 200 °C transformed into a layer consisting of numerous tiny particles for the composite nanorods annealed at 300 °C (Figure 1c). As the annealing temperature was further increased to 350 °C, a clearer granular surface morphology was observed in the decorated layer for TC350 (Figure 1d). The surface morphology of the decorated layer transformed from having small particle features initially into larger granular features with the increase in temperature from 300 to 350 °C. Notably, when the annealing temperature reached 400 °C, the surface granular crystals of TC400 were further coarsened and aggregated, as revealed in Figure 1e. In a high temperature environment, the rapid formation of crystal nuclei leads to nucleus aggregation between the crystal nuclei and the coalescence of the crystal nuclei might occur. Furthermore, from a thermodynamic point of view, the aggregation of surface particles and the growth of crystallites decrease the surface energy to a stable condition. These factors account for the coarser surface granular features in TC400 [24,25]. The corresponding SEM-EDS mapping images of the TC composite nanorods are presented in Figure 1f–i. The Cu and O compositional distribution, which presented the appearance of a column shape, is visibly displayed in all SEM-EDS mapping images, preliminarily revealing the copper oxide layer homogeneously decorated onto the TiO_2 nanorods after the copper film coating and postannealing procedures. In contrast, the Ti signal is distributed over a large area in the elemental mapping image and is not in a distinguishable column shape; this might be associated with the underlying effect of the

TiO$_2$ nanorod template on the TC composite nanorods. The EDS analysis demonstrated that the Cu/Ti atomic ratio of the representative sample (TC350) was 0.22.

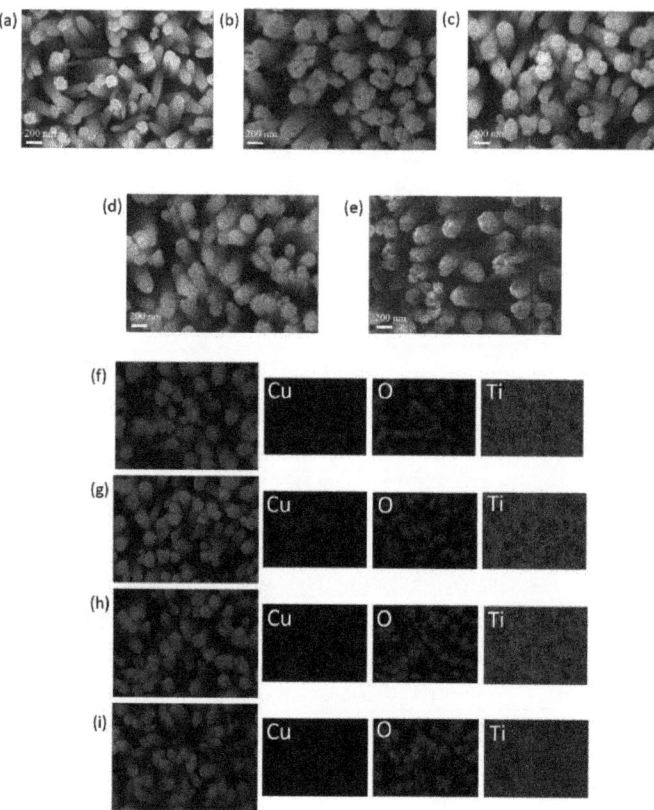

Figure 1. SEM images: (**a**) TiO$_2$, (**b**) TC200, (**c**) TC300, (**d**) TC350, and (**e**) TC400. Corresponding Cu, O, and Ti mapping images of the composite nanorods: (**f**) TC200, (**g**) TC300, (**h**) TC350, and (**i**) TC400.

Figure 2 shows the XRD patterns of various TC composite nanorods. In Figure 2, in addition to the Bragg reflection from the FTO substrate, several strong Bragg reflection peaks can be seen stably distributed at approximately 27.45°, 36.08°, 41.22°, 54.32°, and 56.64°, and they can be attributed to the (110), (101), (111), (211), and (220) crystal planes of the rutile TiO$_2$ phase, respectively (JCPDS 0211276). Figure 2a shows the XRD pattern of the TC200. Three Bragg reflections centered at approximately 29.55°, 36.41°, and 42.29° can be observed. These Bragg reflection peaks can be attributed to the (110), (111), and (200) planes of cuprite Cu$_2$O (JCPDS 05-0667), respectively. This confirms that the thin metallic copper film coated on the surfaces of the TiO$_2$ nanorods was thermally oxidized to form cuprite Cu$_2$O after annealing at 200 °C. This result is consistent with previous work on the full transformation of Cu thin films into cuprite Cu$_2$O after a 200 °C atmospheric annealing procedure [21]. The high crystallinity of the Cu$_2$O phase that appeared after the 200 °C atmospheric annealing procedure was a result of the easy binding of the copper atoms to oxygen atoms above 150 °C, which was mediated in accordance with Equation (1) [26]:

$$2Cu + 0.5O_2 \rightarrow Cu_2O \tag{1}$$

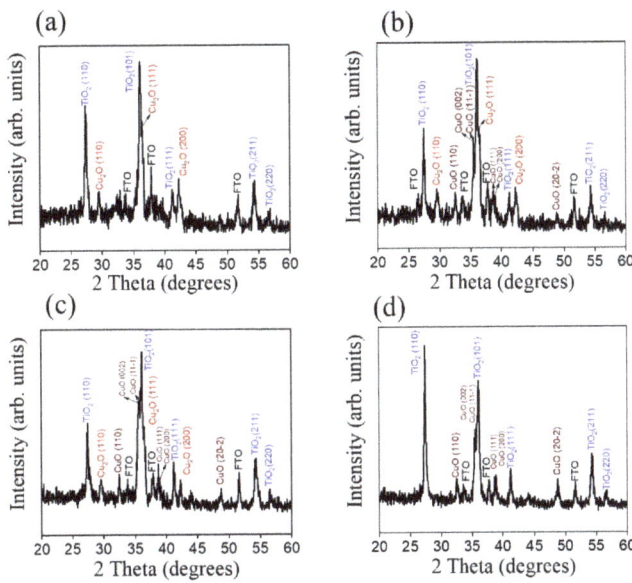

Figure 2. XRD patterns: (**a**) TC200, (**b**) TC300, (**c**) TC350, and (**d**) TC400.

Figure 2b,c show the XRD patterns of the TC300 and TC350. Compared with Figure 2a, six additional Bragg reflections can be observed in Figure 2b,c. These Bragg reflections are centered at approximately 32.50°, 35.41°, 35.54°, 38.70°, 38.90°, and 48.71°. These definite peaks match the characteristic peaks of tenorite CuO (JCPDS 48-1548) and correspond to (110), (002), (11-1), (111), (200), and (20-2), respectively. The characteristic peaks of Cu_2O and CuO coexist in Figure 2b,c, which proves that Cu_2O was partially converted into CuO when the sample was annealed above 300 °C. This result is very similar to that obtained by Sh. R. Adilov et al. In their work, a CuO oxide phase began to form when metallic copper films were annealed at 280 °C; furthermore, when the temperature was raised to 350 °C, a more obvious mixed phase of Cu_2O and CuO was obtained in their thin-film samples [27]. Comparatively, as the temperature was increased from 300 °C to 350 °C, the characteristic peaks of CuO became more intense, revealing improved CuO crystalline content and crystalline quality. Notably, the CuO layer initially formed on the thin-film samples would decline the further oxidation rate was increased due to the thickening of the oxide layer and the increased distance that ions have to diffuse. In order to keep the oxidation rate stable and control the copper oxide phase, the annealing temperature was further increased to 400 °C in this study. In Figure 2d, it can be clearly seen that a single, pure CuO phase replaced the coexisting Cu_2O and CuO phases in the films when the annealing temperature was raised to above 400 °C. This is associated with the fact that the initially formed Cu_2O phase is converted into a CuO phase at higher temperatures according to Equation (2) [28]:

$$Cu_2O + 0.5O_2 \rightarrow 2CuO \qquad (2)$$

The evolution of the copper oxide phase above 400 °C described herein has also been observed in previous work on the annealing temperature-dependent phase transformation of chemically deposited copper oxide films [29].

Figure 3a shows a low-magnification TEM image of a single TC200 nanorod. The entire TiO_2 nanorod was uniformly covered by a continuous Cu_2O film. Rough and irregular surface features can be observed on the sidewalls of the nanorod. The decorated copper oxide layer thickness was estimated to be approximately 32 nm. The feature that appeared corresponded to the previous SEM observations. High-resolution (HR) TEM images of different regions of the TC200 nanorod are shown in Figure 3b,c. However, due to the repeated stacking of TiO_2 and Cu_2O, the lattice fringe arrangements in the inner region of the images cannot be easily distinguished. In contrast, clear lattice fringe arrangements can be observed in the outer regions of the HRTEM images, indicating the crystalline features of the decorated Cu_2O layer. The spacing between these lattice fringes was measured to be approximately 0.24 nm and 0.3 nm in different orientations, and these lattice spacings corresponded to the interplanar spacings of the (111) and (110) planes of cuprite Cu_2O, respectively [30]. Figure 3d shows selected area electron diffraction (SAED) patterns of several TC200 composite nanorods. It shows several diffraction spots arranged in concentric circles with different radii. These concentric circles correspond to rutile TiO_2 ((110), (101), and (200)) and cuprite Cu_2O ((111), (211), (110), and (200)). This confirms the formation of a crystalline Cu_2O layer on the TiO_2 nanorod. Figure 3e shows the cross-sectional EDS line-scan profiling spectra, in which the signal of Ti is distributed across the inner region of the nanorod, the signal of O is uniformly distributed over the entire nanorod region, and the signal of Cu is concentrated in the outer region of the nanorod. This indicates that the main core of the nanorod was TiO_2 and the surface was covered with a layer of copper oxide. A TiO_2 composite nanorod well-decorated with a Cu_2O layer is visibly displayed.

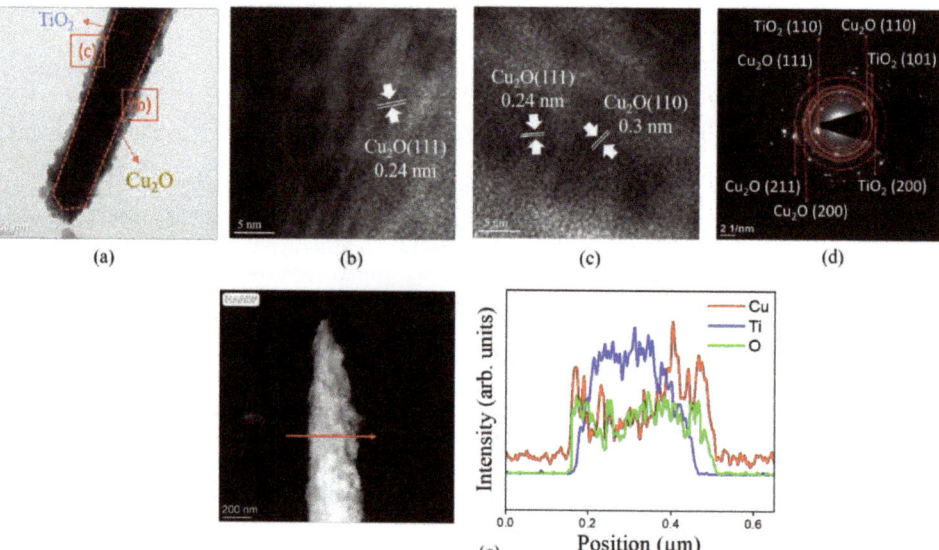

Figure 3. TEM analysis of TC200: (**a**) low-magnification TEM image; (**b**,**c**) HRTEM images of various regions of the composite rod as marked in (**a**); (**d**) SAED patterns of several TC200 nanorods; (**e**) EDS line scanning profiles across the composite rod.

Figure 4a shows a low magnification TEM image of a single TC350 nanorod. Compared with the TEM image of TC200 (Figure 3a), the copper film originally coated onto the surface of the TiO_2 nanorod was transformed into a discontinuously decorated layer after the postannealing procedure. The discontinuous decorated layer consisted of numerous granular crystallites with a particle size of approximately 37 nm. Figure 4b,c present HRTEM images of different peripheral regions from Figure 4a. The decorated particles were further analyzed using HRTEM. The lattice fringes arranged with spacings of 0.25 nm, 0.27 nm, and 0.24 nm could be measured in the different orientations. They corresponded to the (002) and (110) crystal planes of the CuO phase and the (111) crystal plane of the Cu_2O phase, respectively. The HRTEM images showed that the crystalline features of the Cu_2O

and CuO phases coexisted in the decorated discontinuous layer. This further confirmed the results for the previous XRD patterns. When the annealing temperature was raised above 300 °C, the original pure copper film was transformed into two different oxides, which coexisted on the surface of the TiO$_2$ nanorods. Figure 4d shows the SAED patterns of multiple TC350 composite nanorods. The contributions of the crystallographic planes of cuprite Cu$_2$O, tenorite CuO, and rutile TiO$_2$ are visibly exhibited, proving that the ternary phases of Cu$_2$O, CuO, and TiO$_2$ coexisted in the TC350 nanorods. Figure 4e presents the cross-sectional EDS line-scan profiling spectra of the TC350 nanorods. The Cu signal was very strong in the outer region, and the Ti signal was mainly distributed in the inner region of the composite nanorod. The O signal was evenly distributed over the composite nanorod. A TiO$_2$ composite nanorod well-shelled with copper oxide is demonstrated here, and the EDS analysis revealed that the Cu/Ti had an atomic ratio of 0.24. When the annealing temperature was further increased to 400 °C, as the low-magnification TEM image (Figure 5a) of the TC400 nanorod shows, the size of the particles wrapped over the sidewall surface of the TiO$_2$ nanorod changed significantly compared to TC350 (Figure 4a). The size of the particles wrapped over the surface of the TiO$_2$ nanorod was further increased to 55–70 nm. These seriously agglomerated particles on the TiO$_2$ with relative large sizes can be attributed to the marked increase in the annealing temperature, which led to a substantially increased rate of nucleation and accelerated crystal size growth under the given annealing condition. During particle coalescence, the initially formed copper oxide particles could migrate to the TiO$_2$ nanorod template surface and coalesce if motion yielded a reduction in overall system energy. Evidence for such a thermal annealing-induced Ostwald ripening process has been provided in other heterogeneous catalyst systems [31].

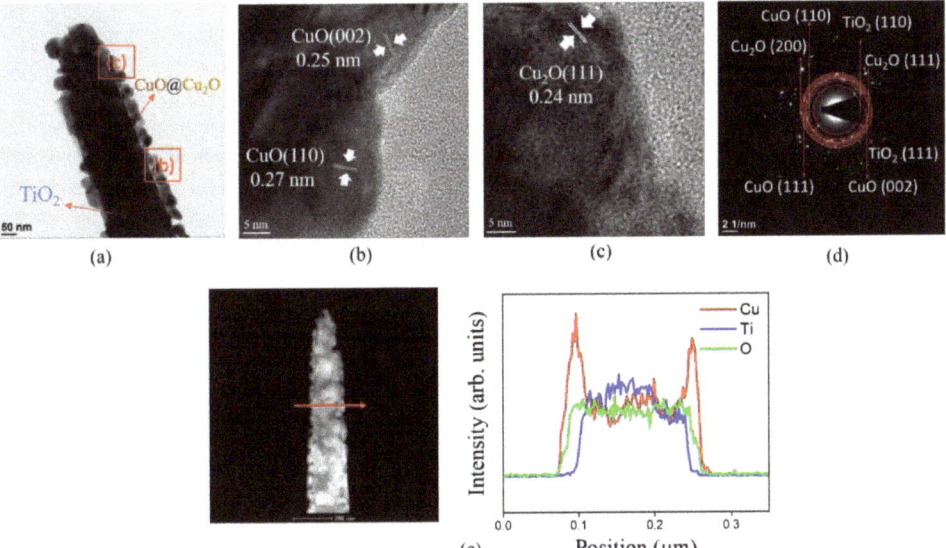

Figure 4. TEM analysis of TC350: (**a**) low-magnification TEM image; (**b**,**c**) HRTEM images of various regions of the composite rod as marked in (**a**); (**d**) SAED patterns of several TC350 nanorods; (**e**) EDS line scanning profiles across the composite rod.

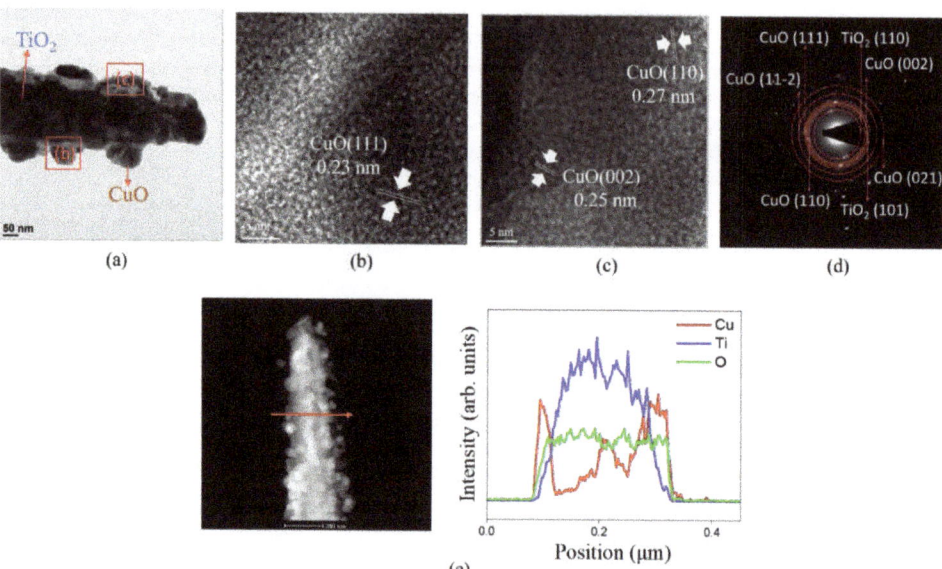

Figure 5. TEM analysis of TC400: (**a**) low-magnification TEM image; (**b,c**) HRTEM images of various regions of the composite rod as marked in (**a**); (**d**) SAED patterns of several TC400 nanorods; (**e**) EDS line scanning profiles across the composite rod.

When the annealing temperature was further increased to 400 °C, as the low-magnification TEM image (Figure 5a) of TC400 nanorod shows, the size of the particles wrapped over the sidewall surface of the TiO_2 nanorod changed significantly compared to TC350 (Figure 4a). The size of the particles wrapped over the surface of TiO_2 nanorod was further increased to 55–70 nm. These seriously agglomerated particles with a relative large size on the TiO_2 can be attributed to the marked increase in the annealing temperature, which led to a substantially increased rate of nucleation and accelerated crystal size growth under the given annealing condition. During particle coalescence, the initially formed copper oxide particles can migrate to the TiO_2 nanorod template surface and coalesce if motion yields an overall system-energy reduction. Evidence for such a thermal annealing-induced Ostwald ripening process has been found in other heterogeneous catalyst systems [31]. Figure 5b,c show HRTEM images of the periphery of TC400 nanorod. Lattice spacings of 0.23 nm, 0.25 nm, and 0.27 nm in different orientations can be measured in Figure 5b,c, which corresponded to the interplanar distances of CuO (111), (002), and (110), respectively. These results confirm that the large-sized particles attached to the surface of TiO_2 nanorod after annealing at 400 °C were CuO crystallites. Figure 5d shows the SAED patterns obtained from multiple TC400 composite nanorods. Obvious diffraction spots are arranged in concentric circles with different radii. Several crystallographic planes of CuO (111), (002), (11-2), (110), and (021) can be indexed in the SAED pattern. No other copper oxide phases were identified, indicating a TiO_2-CuO composite structure for the TC400 nanorods. The TEM structural analyses showed the same results as revealed in XRD patterns. In addition, the cross-sectional elemental profiling spectra shown in Figure 5e also demonstrated a good compositional distribution for the copper oxide-decorated TiO_2 nanorod composite structure. The TEM analysis results demonstrate that the annealing temperature effectively dominated the copper oxide phase and crystallite size on the TiO_2 nanorod template.

Figure 6a displays the high-resolution Cu 2p XPS spectra for TC200. The distinct peaks centered at approximately 932.4 eV and 952.4 eV can be attributed to Cu $2p_{3/2}$ and Cu $2p_{1/2}$, respectively. The Cu binding energies matched the Cu^{+1} binding state in the Cu_2O phase, and this was consistent with the results from a report on the XPS analysis of a sol-gel-derived thin Cu_2O film [32]. Figure 6b shows the high-resolution Cu 2p XPS spectra for TC 350. The appearance of the XPS spectra is similar to that observed in a study on $CuO@Cu_2O$ heterostructures derived using the solvothermal method [33]. In contrast to the Cu 2p spectra for TC200, oscillating satellite peaks could be detected for TC350 at the binding energies of approximately 942.2 eV and 961.8 eV, which further indicated the existence of a CuO phase in TC350. This has also been demonstrated in the Cu 2p spectra analysis of pristine Cu_2O and CuO thin films, in which pure Cu_2O and CuO could easily be observed without and with the appearance of satellite peaks from the XPS spectra, respectively [32]. The spectra detected herein were further separated into several contributions. The intense fitted peaks located at 933.5 eV and 953.4 eV (blue line) were attributed to Cu $2p_{3/2}$ and Cu $2p_{1/2}$ of the CuO phase, respectively. There was a difference of approximately 20 eV between the Cu $2p_{3/2}$ and Cu $2p_{1/2}$ peaks of CuO, which matches well with the reported results for hydrothermally derived CuO nanoflowers [34]. In contrast, two relatively weak peaks (green line) appeared at 932.6 eV and 952.2 eV, corresponding to Cu $2p_{3/2}$ and Cu $2p_{1/2}$ of the Cu_2O phase, respectively [35]. These results verify the coexistence of Cu_2O and CuO phases in the decorated copper oxide layer in TC350. Figure 6c presents the high-resolution XPS spectra for Cu 2p in TC400. The distinct appearance of the satellite peaks (located at 942.2 and 961.8 eV) was observed (Figure 6c). The characteristic peaks centered at the binding energies of 933.3 eV and 953.3 eV corresponded to Cu $2p_{3/2}$ and Cu $2p_{1/2}$ of the CuO phase, respectively. These XPS results demonstrate that an adjustable copper oxide phase was obtained in the decorated copper film layer by varying the annealing temperature.

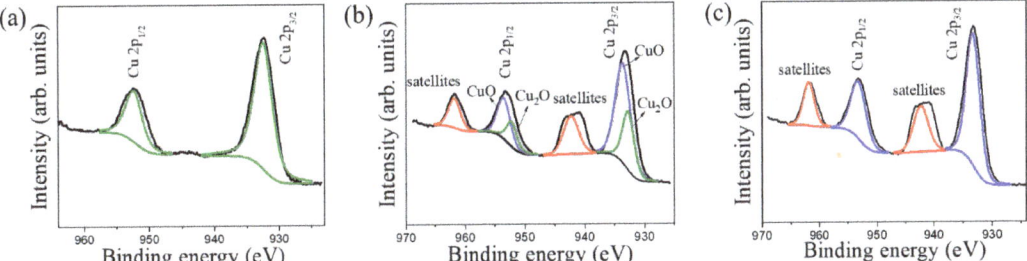

Figure 6. High-resolution XPS Cu 2p spectra: (**a**) TC200, (**b**) TC350, and (**c**) TC400.

Figure 7a shows the referenced Ti 2p core-level doublet spectra for the TiO_2 nanorod template. The high-resolution XPS spectra were deconvoluted into four subpeaks. The more intense subpeaks at 458.3 eV and 463.9 eV corresponded to Ti $2p_{3/2}$ and Ti $2p_{1/2}$ for the Ti^{4+} valence state in TiO_2, respectively. Furthermore, the subpeaks with weaker intensities and smaller binding energies of 457.2 eV and 462.9 eV corresponded to Ti $2p_{3/2}$ and Ti $2p_{1/2}$ in the Ti^{+3} valence state [36,37]. The presence of the mixed Ti^{4+}/Ti^{3+} valance state indicates the possible presence of oxygen vacancies on the surfaces of the TiO_2 nanorod template. Figure 7b shows a comparison of the Ti 2p core-level doublet spectra for TiO_2, TC200, TC350, and TC400. It can be seen that the Ti 2p XPS spectra of the TC composite nanorods demonstrated positive shifts in binding energy positions in comparison with the binding energy position of pristine TiO_2. The modification of TiO_2 nanorods with copper oxides described herein might have changed the electronic state of Ti in Ti-O because of the formation of heterojunctions between the n-type TiO_2 and p-type copper oxides. This has been demonstrated with $CuO@TiO_2$ powders and core–shell $N-TiO_2@CuOx$ heterojunction composites formed using ball milling [38,39]. Notably, the Cu/Ti atomic ratio of TC350 was evaluated to be 3.6. The investigation depth of XPS is usually below 10 nm. This Cu/Ti atomic ratio substantially differs from the Cu/Ti atomic ratios calculated from the EDS spectra of electron microscopes because of the different measurement depths of the various analysis methods.

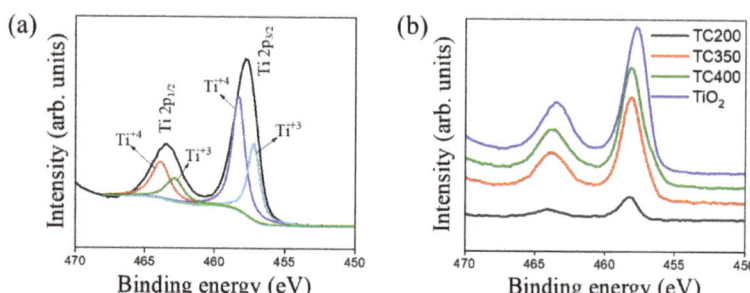

Figure 7. High-resolution XPS Ti 2p spectra: (**a**) TiO$_2$ and (**b**) comparison of Ti 2p spectra of various samples.

Figure 8a shows the optical absorption characteristics of pristine TiO$_2$ nanorods and TC composite nanorods. A sharp absorption drop appeared at approximately 410 nm for TiO$_2$ nanorod template, and this absorption edge was consistent with the inherent band-gap absorption of rutile TiO$_2$ [1]. Notably, the TC composite nanorods demonstrated a significant red-shift extension of the absorption edge in comparison with that of the pristine TiO$_2$. This can be attributed to the decoration of the TiO$_2$ nanorods with Cu$_2$O and CuO visible-light sensitizers. These visible-light sensitizers helped to absorb the longer wavelength spectra, making up for the inability of TiO$_2$ to absorb visible light, and enhanced the absorption in the visible light region. The higher annealing temperature resulted in a larger size for the red-shift of the absorption edge of the TC composite samples; this was associated with the fact that the CuO formed at the higher annealing temperature had a narrower band-gap energy than that of Cu$_2$O [40,41]. Figure 8b shows the Kubelka–Munk function (F(R)) vs. energy plots for various nanorod samples [42]. Notably, the TiO$_2$ and copper oxides used herein were expected to exhibit a direct transition in the band-gap measurements. Therefore, the band-gap energy of the TiO$_2$ nanorods and TC composite nanorods could be deduced from the (F(R)$h\nu$)^2 vs. $h\nu$ plots by extrapolating the straight portion of the curves to the energy axis. The TiO$_2$ nanorod template was estimated to have an energy gap of approximately 3.03 eV. The energy gap values of TC 200, TC300, TC350, and TC400 were estimated to be approximately 2.59 eV, 2.43 eV, 2.34 eV, and 2.27 eV, respectively. The phase evolution of the decorated layer from Cu$_2$O to CuO with increased annealing temperature visibly demonstrated a decreased energy gap in the TC composite nanorods. The band-gap energy variation in the copper oxides due to the phase evolution was consistent with a report on electrodeposited Cu$_2$O/CuO powder oxides [43]. The UV-vis analysis demonstrated that the energy gap size of the TC composite nanorods could be effectively tuned by varying the postannealing temperature. In addition, the energy gaps of single CuO and Cu$_2$O films were also estimated from the Tauc plot (Figure 8c). The energy gap values for CuO and Cu$_2$O were estimated to be 1.76 eV and 2.04 eV, respectively, by extrapolating the curve tangent to the energy axis in Figure 8c. These values are similar to those from previous work on Cu$_2$O formed with copper foil annealing and sputtering CuO [44,45].

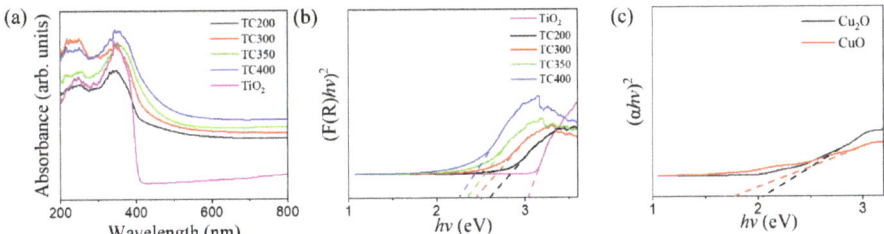

Figure 8. (a) Absorption spectra of various samples and Band-gap evaluations of (b) various nanorod samples, (c) pristine Cu_2O and CuO films.

Figure 9a shows the transient photoresponses of various samples. Irradiation was applied with the full-band spectrum, and a bias potential of 1.2 V was used to measure the photocurrent. Photocurrent generation occurred entirely as a result of the on and off responses to the irradiation. Seven cycles of on/off irradiation were repeated, as shown in Figure 9a, and all samples could obtain a stable photogenerated current when the irradiation was turned on, indicating that the samples were stable under cycling chopping irradiation. A higher photocurrent indicates better efficiency for the separation of photogenerated charges and better photocatalytic activity for the photoelectrode [46]. Comparatively, all the TC composite nanorods exhibited improved photoresponses compared to that of the pristine TiO_2. This was attributed to the fact that decoration with Cu_2O and CuO visible-light sensitizers enhanced the light-harvesting ability of the TiO_2 nanorod template, and the formation of heterojunctions in the composite system resulted in improved photogenerated carrier separation efficiency. Furthermore, compared to TC200, which had a single-phase Cu_2O decoration, TC400 (with single-phase CuO decoration) had a higher photocurrent, which can be attributed to the narrower energy gap in CuO compared to Cu_2O. This led to TC400 absorbing across a longer wavelength range than TC200, as revealed in the previous absorption analysis, thereby increasing light absorption and promoting the photoexcited carrier density. This has also been demonstrated in previous work on the photoactive performance of a Cu_2O/CuO system [32,47]. Notably, TC300 and TC350 displayed the best photoresponse abilities among the various nanorod samples, revealing that the composite nanorod system decorated with dual Cu_2O and CuO phases was a more efficient material combination for enhancing the photoactivity of the copper oxide–TiO_2 composite nanorods. Figure 9b presents the Nyquist plots of various samples measured at the frequency range from 100 kHz to 0.1 Hz and a potential amplitude of 10 mV. The radius of the semicircles in Nyquist plots is associated with the interfacial charge transfer resistance [48]. Notably, TC350 had the smallest semicircular radius, and the pure TiO_2 nanorod template exhibited the largest semicircular radius, indicating that TC350 had the smallest charge transfer resistance and TiO_2 the largest. The sizes of the semicircle radii from the Nyquist plots for various samples were ordered in the following trend: TiO_2 > TC200 > TC400 > TC300 > TC350. This result was also found with the transient photoresponse measurements. The multi-interface heterostructures consisting of TiO_2, CuO, and Cu_2O in TC350 and TC300 effectively helped to enhance the separation and transfer abilities of electron–hole pairs, as revealed in the previous I-t curves (Figure 9a). Similar coexistence of ternary phases leading to substantial improvements in PEC properties has also been demonstrated in $BiVO_4$/CdS/CoO_x core-shell composites [49]. These improvements can provide an opportunity to induce electron redistribution and synergistic effects at the interfaces for heterogeneous catalysis consisting of two or more components connected by well-defined interfaces [50,51]. The existence of multiple heterointerfaces in ternary phase composites improves their PEC properties. The charge transfer resistance can be estimated by fitting the arc radius of the Nyquist curves according to the proposed equivalent circuits in Figure 9c. Rs, CPE, and Rct represent the series resistance, constant phase element, and charge transfer resistance, respectively. Similar equivalent circuits have also been used in a ternary Fe_2O_3–MoS_2–Cu_2O nanofilm system to determine the Rct [52]. The representative fitting parameters for TC350 were Rct = 582 Ohm and Rs = 43.89 Ohm. After fitting the Nyquist plots using the proposed equivalent circuits, the Rct values for the other samples, TiO_2, TC200, TC300, and TC400, were 3653, 1832, 702, and 1284 Ohm, respectively. Notably, although the TC300 and TC350 were both ternary-phase composite nanorods, lower interfacial charge transfer resistance in TC350 was observed in comparison to that of TC300. This might have been associated with the fact that, as the annealing temperature increases, the crystallite size of the decorated copper oxides increased, and this could have reduced the grain boundaries in the decoration layer. Therefore, TC350 had a better charge transport ability than TC300, and this was evidenced in the Rct.

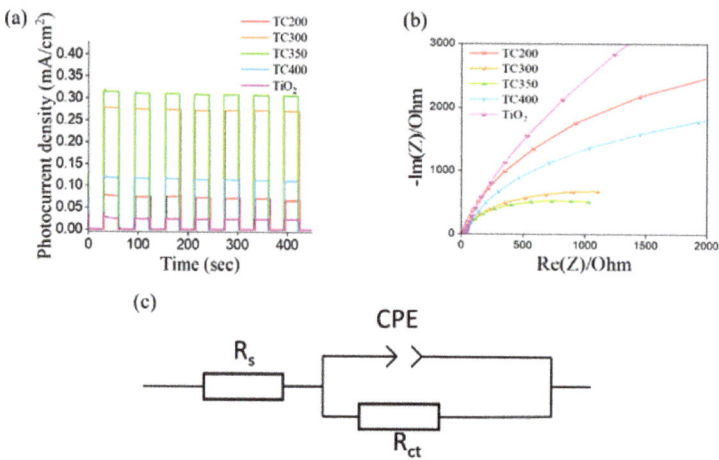

Figure 9. (a) Photocurrent density versus time curves for various samples at 1.2 V (vs. Ag/AgCl) under chopping illumination. (b) Nyquist plots for various samples under irradiation. (c) Possible equivalent circuits for Rct evaluation.

In order to further analyze and construct the energy-band structure of the composite nanorods, measurements of the flat-band potential of the TiO_2 nanorod template, Cu_2O film, and CuO film were carried out and presented in Figure 10a–c. The M-S curves exhibited a positive slope for the TiO_2 and negative slopes for the Cu_2O and CuO, revealing the n-type nature of the TiO_2 and the p-type nature of the Cu_2O and CuO. According to the M-S equation [53], when $1/C^2$ is extrapolated to a value equal to 0, the X-axis intercept is equal to the flat-band potential of the material [54]. The flat-band potential of pure TiO_2 was estimated to be about −0.11 eV (vs. NHE). The flat-band potential in n-type semiconductors is closer to the conduction band (CB) and the CB position of an n-type semiconductor is generally more negative (0.1 eV) than the flat-band potential [55]. After calculation, it was deduced that the CB of TiO_2 was −0.21 eV. In contrast, the flat-band potential of the p-type semiconductor is closer to its valence band (VB) [56,57]. The flat-band potentials of Cu_2O and CuO were estimated to be approximately 0.46 eV and 0.71 eV, respectively, as shown in Figure 10b,c. The VB positions of Cu_2O and CuO were further calculated to be 0.56 and 0.81 eV (vs. NHE), respectively. The VB positions assessed herein are close to previously reported results for Cu_2O and CuO [58,59]. Figure 10d shows the M-S curves for various TC composite nanorods. Inverted V-shaped M-S curves were observed for the all composite nanorods, demonstrating that the composites had both n-type and p-type electronic properties and confirming the formation of p-n junctions in the TC composite nanorods [60]. Construction of p-n junctions in composite systems has been posited to be a sensible strategy to enhance photocatalytic activity. The formation of a p-n junction with space charge regions at the heterointerface could induce the electric field-driven diffusion of electrons and holes and further inhibit the recombination of photogenerated charges [54,61].

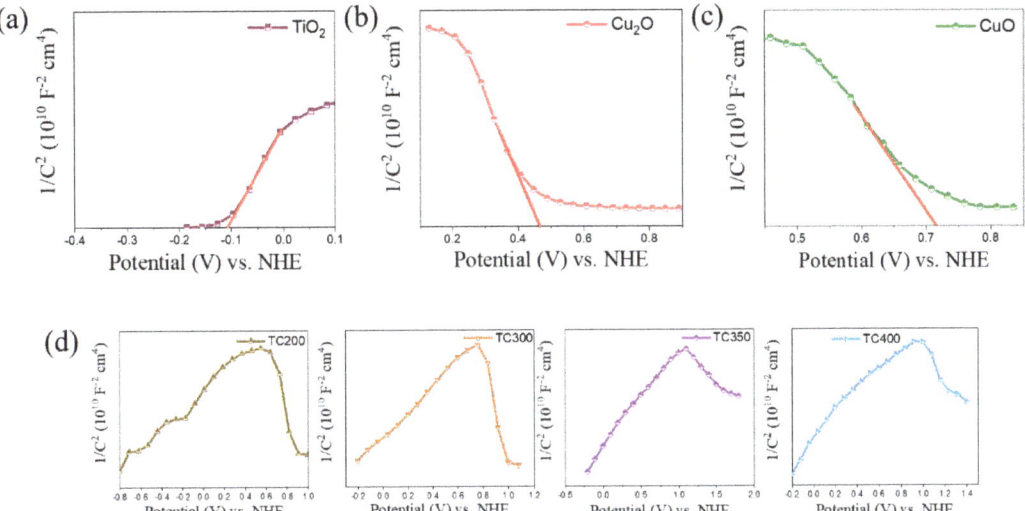

Figure 10. Mott–Schottky plots for various samples: (**a**) TiO_2, (**b**) Cu_2O, and (**c**) CuO. (**d**) A series of M−S plots for various composite nanorods.

The photocatalytic ability of the samples was further estimated by using the formula whereby the percentage degradation = C/Co, where Co is the initial concentration of RhB solution and C is the time-dependent concentration of RhB solution upon irradiation. Figure 11a presents C/Co vs. irradiation duration plots for RhB solution with different samples. Adsorption–desorption equilibrium was reached by placing the photocatalysts in the RhB solution for 45 min in the dark before starting the photodegradation experiments. Under dark equilibration conditions for 45 min, the C/Co values for TiO_2, TC200, TC300, TC350, and TC400 were approximately 3.1%, 4.2%, 6.8%, 7.9%, and 5%, respectively. This indicated that the TiO_2 nanorod template decorated with copper oxide had an improved surface dye absorption capacity. After offsetting with a dark adsorption contribution, the degradation rates of TC200, TC300, TC350, and TC400 were approximately 59%, 83%, 90%, and 70%, respectively, with 60 min irradiation. The TC composite nanorods exhibited improved photodegradation abilities towards RhB solution in comparison to the pristine TiO_2 nanorod template. Furthermore, among the various TC composites, TC350 had the highest photodegradation ability towards RhB solution under the given test conditions. In addition, the photodegradation kinetics of the RhB solution with all samples were also investigated and presented in Figure 11b. The pseudo-first-order kinetic equation is expressed as: kt = ln Co/C, where k represents the pseudo-first-order rate constant (min^{-1}) for the initial degradation [61]. All the TC composite samples displayed larger k values than that of the pristine TiO_2. Furthermore, TC350 had the highest k value of 0.04578 min^{-1}. The photodegradation abilities of the photocatalysts towards organic pollutants were significantly related to the separation efficiency for electrons and holes. The magnitude trends for the k values for the various samples investigated herein were consistent with the previously measured PEC and EIS experimental results. In addition, the photocatalytic reaction was closely related to the active species produced in the process. The role of these species in the degradation reaction was investigated by measuring the variation in the degradation performance of the RhB solution with TC350 through the addition of various radical scavengers after 60 min irradiation. The radical capture experiments were performed using tert-butanol (TBA) as a hydroxyl radical (·OH) scavenger, ammonium oxalate (AO) as a hole quencher, and benzoquinone (BQ) as a superoxide radical (·O_2^-) scavenger. As shown in Figure 11c, when 1 mM AO was added, the RhB degradation efficiency slightly decreased to 69%, indicating that holes played a minor role in the degradation process. In contrast, adding TBA or BQ scavengers resulted in a more intense decrease in the photodegradation level of the RhB solution. This shows that ·O_2^- and ·OH were the main radicals involved in the photodegradation process of the RhB solution with TC350. Comparatively, the removal of the ·O_2^- active species resulted in the most significant decrease in the degradation efficiency.

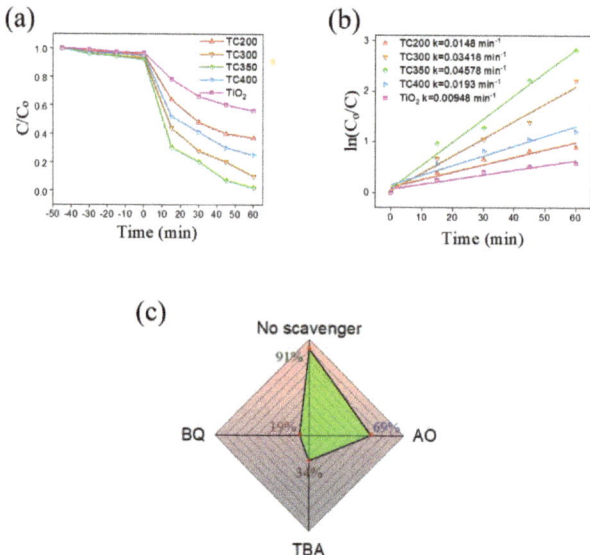

Figure 11. (a) C/Co vs. irradiation duration plots; (b) ln (Co/C) vs. irradiation duration plots; (c) degradation percentages of RhB solution with TC350 in the presence of various scavengers.

The band structures of pristine TiO_2, Cu_2O, and CuO were constructed according to the M-S measurements and the UV-vis analysis results, as shown in Figure 12. As shown in the previous scavenger experiments, the main active species involved in the TC350 photodegradation process with RhB solution were superoxide and hydroxyl radicals. Moreover, superoxide radicals demonstrated a greater contribution than hydroxyl radicals, as seen from the constructed band alignment in the ternary TC350 composite nanorods. If the electron–hole transfer route in the TC350 composite nanorods had followed the type II transfer mode, superoxide and hydroxyl radicals would not have been formed according to the relative band positions of the CB, VB, and redox potentials [62]. Therefore, none of the electrons/holes at the CB/VB positions would reach the required redox potential, so superoxide and hydroxyl active species would not have been produced with this mechanism. This contradicts the previous scavenger experiments. The Z-scheme mechanism shown in Figure 12 is more appropriate to explain the movement of photogenerated electrons/holes and the generation of active species for photodegradation. Under irradiation, photoinduced carriers form in the composite system (reaction 3). Through the movement of photogenerated carriers in the Z-scheme mechanism, the holes finally accumulated in the VB of TiO_2 (2.82 eV vs. NHE), which was significantly higher than the oxidation potential of water or (–OH) molecules, which is 2.4 eV. Therefore, the holes were able to react with water (or –OH) molecules and generate hydroxyl radicals (reaction 4) [63]. In contrast, electrons accumulated in the CB of Cu_2O (−1.48 eV vs. NHE). The electrons were located significantly lower than the reduction potential of oxygen (−0.33 eV), and electrons could react with oxygen to form superoxide radicals (reaction 5) [64]. These main reactive species could further react with RhB dye molecules and decompose into carbon dioxide and water (reaction 6) [65]:

$$TiO_2/CuO/Cu_2O + hv_{(UV-visible)} \rightarrow TiO_2\left(e_{CB}^- h_{VB}^+\right)/CuO\left(e_{CB}^- h_{VB}^+\right)/Cu_2O\left(e_{CB}^- h_{VB}^+\right) \quad (3)$$

$$TiO_2\left(h_{VB}^+\right) + H_2O \rightarrow \cdot OH + H^+ + TiO_2 \quad (4)$$

$$Cu_2O\left(e_{CB}^-\right) + O_2 \rightarrow \cdot O_2^- + Cu_2O \quad (5)$$

$$\cdot OH + \cdot O_2^- + RhB \rightarrow product\ (ex:\ H_2O + CO_2) \quad (6)$$

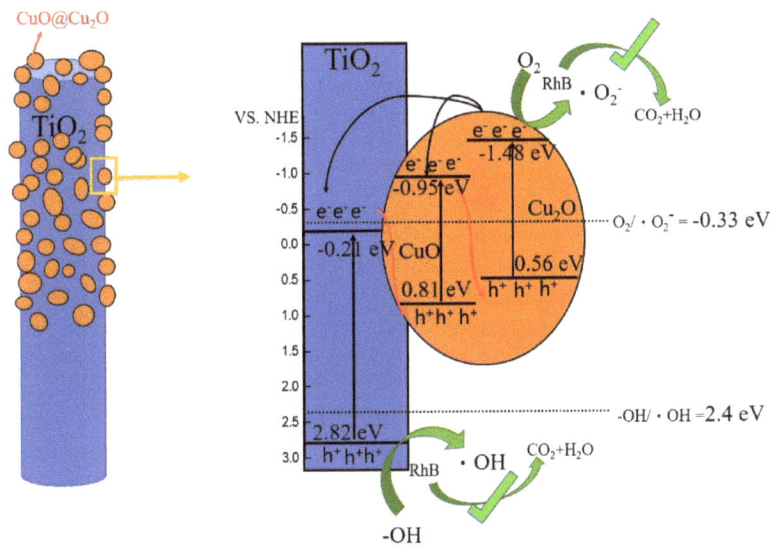

Figure 12. Possible mechanisms of charge transfer in TC350 under irradiation.

Notably, from among the band alignments proposed for the TiO$_2$, Cu$_2$O, and CuO, multiple photoinduced charger transfer routes could occur in the ternary TiO$_2$–Cu$_2$O–CuO composite system. The p-n junctions formed between the n-type TiO$_2$ and p-type copper oxides induced an internal electric field at the heterointerfaces, promoting charge separation under irradiation. The stepped-band edge arrangement in the composite system caused multiple Z-scheme transfer routes for the photoinduced charges. This finally resulted in the accumulation of holes at the VB of TiO$_2$ and of electrons at the CB of Cu$_2$O. A similar Z-scheme carrier movement was also exhibited in a ternary ZnO–Cu$_2$O–CuO photocatalyst system [66]. In the report by Wei et al., the composite material TiO$_2$–Cu$_2$O showed carrier movement with a Z scheme under irradiation [67]. These examples echo the carrier movement mechanism proposed in this work. In addition, the Z-scheme charge transfer in the composite nanorods had an important contribution in preventing the photocorrosion of Cu$_2$O. Photocorrosion has been demonstrated in previous work on single-phase Cu$_2$O photocatalysts [68]. The Cu$_2$O phase coupling with TiO$_2$ (TC200) or TiO$_2$–CuO (TC300 and TC350) in the composite system effectively guided the photoexcited electrons and holes accumulated in the Cu$_2$O and inhibited photocorrosion. Therefore, a stable photocurrent curve could be observed in the previous photoresponse plots. This is supported by work on introducing a protective layer of TiO$_2$ in Cu$_2$O–CuO heterojunction films to prevent the photocorrosion effect [69]. The multiple charge transfer routes shown in Figure 12 explain the superior photoactive performance of TiO$_2$–Cu$_2$O–CuO composite systems (TC350 and TC300) among the various TC composite nanorods. Finally, it should be mentioned that TC350 exhibited better photoactivity than that of TC300. This can be attributed to optical absorption and the microstructural differences between the TC300 and TC350. The TC350 exhibited a better light absorption ability than TC300, as revealed by the previous UV-vis absorption analysis, which enhanced the generation efficiency of photoexcited charges in TC350. Moreover, TC350 also had a larger surface particle size in the decorated copper oxide layer in comparison to that of TC300. A larger grain size reduces grain boundaries in the decorated copper oxide layer, resulting in enhanced charge transport [47]. The ternary phase and suitable microstructural and optical properties of TC350 mean that it has excellent photoactivity compared to the other TC composite samples.

4. Conclusions

The morphology of copper oxide decorated on a TiO_2 nanorod template changed from a continuous layer morphology to granular aggregates when the postannealing temperature was varied from 200 to 400 °C. The composite nanorods formed at 350 °C (TC350) exhibited superior photoactive performance compared to the other composite nanorods. The larger particle size resulting from the copper oxide modification in TC350 reduced the grain boundaries in the decorated layer, thereby increasing the charge transport ability. Moreover, the surface-modified Cu_2O–CuO mixed crystallites on the TiO_2 template could absorb sunlight more efficiently. These factors enhanced the photoactive performance of the TC350 composite nanorods studied herein. The scavenger tests demonstrated that the Z scheme was the possible carrier movement mechanism in TC350 under irradiation, and that result explains the high photocatalytic degradation ability of TC350 towards organic pollutants. The experimental results obtained herein demonstrate that regulation of the composition phase and microstructure of the modified copper oxide layer through control of the thermal annealing budget for the thin copper layer on TiO_2 nanorod templates is a promising approach to design copper oxide–TiO_2 composite nanorods with satisfactory photoactive performance.

Author Contributions: Methodology, Y.-C.L. and T.-H.L.; formal analysis, T.-H.L.; investigation, T.-H.L.; writing—original draft preparation, Y.-C.L. and T.-H.L.; supervision, Y.-C.L. All authors have read and agreed to the published version of the manuscript.

Funding: This research was funded by Ministry of Science and Technology of Taiwan, grant no. MOST 108-2221-E-019-034-MY3.

Data Availability Statement: Not applicable.

Conflicts of Interest: The authors declare no conflict of interest.

References

1. Liang, Y.-C.; Zhao, W.-C. Morphology-dependent photocatalytic and gas-sensing functions of three-dimensional TiO_2–ZnO nanoarchitectures. *CrystEngComm* **2020**, *22*, 7575–7589. [CrossRef]
2. Wei, N.; Cui, H.; Wang, C.; Zhang, G.; Song, Q.; Sun, W.; Song, X.; Sun, M.; Tian, J. Bi_2O_3 nanoparticles incorporated porous TiO_2 films as an effective p-n junction with enhanced photocatalytic activity. *J. Am. Ceram. Soc.* **2017**, *100*, 1339–1349. [CrossRef]
3. Liang, Y.-C.; Chiang, K.-J. Coverage Layer Phase Composition-Dependent Photoactivity of One-Dimensional TiO_2–Bi_2O_3 Composites. *Nanomaterials* **2020**, *10*, 1005. [CrossRef] [PubMed]
4. Ravishankar, T.N.; Vaz, M.D.O.; Teixeira, S.R. The effects of surfactant in the sol–gel synthesis of CuO/TiO_2 nanocomposites on its photocatalytic activities under UV-visible and visible light illuminations. *New J. Chem.* **2020**, *44*, 1888–1904. [CrossRef]
5. Wang, M.; Sun, L.; Lin, Z.; Cai, J.; Xie, K.; Lin, C. p–n Heterojunction photoelectrodes composed of Cu_2O-loaded TiO_2 nanotube arrays with enhanced photoelectrochemical and photoelectrocatalytic activities. *Energy Environ. Sci.* **2013**, *6*, 1211–1220. [CrossRef]
6. Liu, H.; Gao, L. Preparation and Properties of Nanocrystalline α-Fe_2O_3-Sensitized TiO_2 Nanosheets as a Visible Light Photocatalyst. *J. Am. Ceram. Soc.* **2006**, *89*, 370–373. [CrossRef]
7. Ye, M.; Pan, J.; Guo, Z.; Liu, X.; Chen, Y. Effect of ball milling process on the photocatalytic performance of CdS/TiO_2 composite. *Nanotechnol. Rev.* **2020**, *9*, 558–567. [CrossRef]
8. Lutz, T.; MacLachlan, A.; Sudlow, A.; Nelson, J.; Hill, M.S.; Molloy, K.C.; Haque, S.A. Thermal decomposition of solution processable metal xanthates on mesoporous titanium dioxide films: A new route to quantum-dot sensitised heterojunctions. *Phys. Chem. Chem. Phys.* **2012**, *14*, 16192–16196. [CrossRef]
9. Ge, M.; Cao, C.; Li, S.; Zhang, S.; Deng, S.; Huang, J.; Li, Q.; Zhang, K.; Al-Deyab, S.S.; Lai, Y. Enhanced photocatalytic performances of n-TiO_2 nanotubes by uniform creation of p–n heterojunctions with p-Bi_2O_3 quantum dots. *Nanoscale* **2015**, *7*, 11552–11560. [CrossRef]
10. Shi, W.; Wang, J.C.; Chen, A.; Xu, X.; Wang, S.; Li, R.; Zhang, W.; Hou, Y. Cu Nanoparticles Modified Step-Scheme Cu_2O/WO_3 Heterojunction Nanoflakes for Visible-Light-Driven Conversion of CO_2 to CH_4. *Nanomaterials* **2022**, *12*, 2284. [CrossRef]
11. Mizuno, K.; Izaki, M.; Murase, K.; Shinagawa, T.; Chigane, M.; Inaba, M.; Tasaka, Y.; Awakura, Y. Structural and Electrical Characterizations of Electrodeposited p-Type Semiconductor Cu_2O Films. *J. Electrochem. Soc.* **2005**, *152*, C179. [CrossRef]
12. Wu, F.; Myung, Y.; Banerjee, P. Unravelling transient phases during thermal oxidation of copper for dense CuO nanowire growth. *CrystEngComm* **2014**, *16*, 3264–3267. [CrossRef]
13. Wang, J.; Ji, G.; Liu, Y.; Gondal, M.; Chang, X. Cu_2O/TiO_2 heterostructure nanotube arrays prepared by an electrodeposition method exhibiting enhanced photocatalytic activity for CO_2 reduction to methanol. *Catal. Commun.* **2014**, *46*, 17–21. [CrossRef]

14. Shi, Q.; Ping, G.; Wang, X.; Xu, H.; Li, J.; Cui, J.; Abroshan, H.; Ding, H.; Li, G. CuO/TiO₂ heterojunction composites: An efficient photocatalyst for selective oxidation of methanol to methyl formate. *J. Mater. Chem. A* **2019**, *7*, 2253–2260. [CrossRef]
15. Park, S.M.; Razzaq, A.; Park, Y.H.; Sorcar, S.; Park, Y.; Grimes, C.A.; In, S.I. Hybrid CuxO–TiO₂ Heterostructured Composites for Photocatalytic CO₂ Reduction into Methane Using Solar Irradiation: Sunlight into Fuel. *ACS Omega* **2016**, *1*, 868–875. [CrossRef]
16. Diachenko, O.; Kováč, J., Jr.; Dobrozhan, O.; Novák, P.; Kováč, J.; Skriniarova, J.; Opanasyuk, A. Structural and Optical Properties of CuO Thin Films Synthesized Using Spray Pyrolysis Method. *Coatings* **2021**, *11*, 1392. [CrossRef]
17. Dai, M.-J.; Lin, S.-S.; Shi, Q.; Liu, F.; Wang, W.-X.; Chen, S.-C.; Kuo, T.-Y.; Sun, H. Transparent Conductive p-Type Cuprous Oxide Films in Vis-NIR Region Prepared by Ion-Beam Assisted DC Reactive Sputtering. *Coatings* **2020**, *10*, 473. [CrossRef]
18. Nair, M.; Guerrero, L.; Arenas, O.L.; Nair, P. Chemically deposited copper oxide thin films: Structural, optical and electrical characteristics. *Appl. Surf. Sci.* **1999**, *150*, 143–151. [CrossRef]
19. Mahendra, G.; Malathi, R.; Kedhareswara, S.P.; LakshmiNarayana, A.; Dhananjaya, M.; Guruprakash, N.; Hussain, O.M.; Mauger, A.; Julien, C.M. RF Sputter-Deposited Nanostructured CuO Films for Micro-Supercapacitors. *Appl. Nano* **2021**, *2*, 46–66. [CrossRef]
20. Valladares, L.D.L.S.; Salinas, D.H.; Dominguez, A.B.; Najarro, D.A.; Khondaker, S.I.; Mitrelias, T.; Barnes, C.H.W.; Aguiar, J.A.; Majima, Y. Crystallization and electrical resistivity of Cu₂O and CuO obtained by thermal oxidation of Cu thin films on SiO₂/Si substrates. *Thin Solid Film.* **2012**, *520*, 6368–6374. [CrossRef]
21. Serin, N.; Serin, T.; Horzum, Ş.; Celik, Y. Annealing effects on the properties of copper oxide thin films prepared by chemical deposition. *Semicond. Sci. Technol.* **2005**, *20*, 398–401. [CrossRef]
22. Khojier, K.; Behju, A. Annealing Temperature Effect On Nanostructure And Phase Transition Of Copper Oxide Thin Films. *Int. J. Nano Dimens.* **2012**, *2*, 185–190.
23. Liang, Y.-C.; Xu, N.-C.; Chiang, K.-J. Surface Morphology-Dependent Functionality of Titanium Dioxide–Nickel Oxide Nanocomposite Semiconductors. *Nanomaterials* **2019**, *9*, 1651. [CrossRef] [PubMed]
24. Vidyasagar, C.C.; Naik, Y.A.; Venkatesha, T.G.; Viswanatha, R. Solid-State Synthesis and Effect of Temperature on Optical Properties of CuO Nanoparticles. *Nano-Micro Lett.* **2012**, *4*, 73–77. [CrossRef]
25. Siqingaowa, Z.; Yao, H. Preparation and characterization of nanocrystalline ZnO by direct precipitation method. *Front. Chem. China* **2006**, *1*, 277–280. [CrossRef]
26. Neumann, J.P.; Zhong, T.; Chang, Y.A. The Cu—O (Copper-Oxygen) system. *Bull. Alloy. Phase Diagr.* **1984**, *5*, 136–140. [CrossRef]
27. Adilov, S.; Afanaciev, V.P.; Kashkul, I.N.; Kumekov, S.; Mukhin, N.V.; Terukov, E.I. Studying the Composition and Structure of Films Obtained by Thermal Oxidation of Copper. *Glass Phys. Chem.* **2017**, *43*, 272–275. [CrossRef]
28. Akkari, F.C.; Kanzari, M.; Rezig, B. Preparation and characterization of obliquely deposited copper oxide thin films. *Eur. Phys. J. Appl. Phys.* **2007**, *40*, 49–54. [CrossRef]
29. Johan, M.R.; Suan, M.S.M.; Hawari, N.L.; Ching, H.A. Annealing Effects on the Properties of Copper Oxide Thin Films Prepared by Chemical Deposition. *Int. J. Electrochem. Sci.* **2011**, *6*, 6094–6104.
30. Dong, K.; He, J.; Liu, J.; Li, F.; Yu, L.; Zhang, Y.; Zhou, X.; Ma, H. Photocatalytic performance of Cu₂O-loaded TiO₂/rGO nanoheterojunctions obtained by UV reduction. *J. Mater. Sci.* **2017**, *52*, 6754–6766. [CrossRef]
31. Datye, A.K.; Xu, Q.; Kharas, K.C.; McCarty, J.M. Particle size distributions in heterogeneous catalysts: What do they tell us about the sintering mechanism? *Catal. Today* **2006**, *111*, 59–67. [CrossRef]
32. Lim, Y.F.; Chua, C.S.; Lee, C.J.J.; Chi, D. Sol-gel deposited Cu₂O and CuO thin films for photocatalytic water splitting. *Phys. Chem. Chem. Phys.* **2014**, *16*, 25928–25934. [CrossRef]
33. Wang, Y.; Lü, Y.; Zhan, W.; Xie, Z.; Kuang, Q.; Zheng, L. Synthesis of Porous Cu₂O/CuO Cages using Cu-based Metal-Organic-Framework as Templates and their Gas-sensing Properties. *J. Mater. Chem. A* **2015**, *3*, 12796–12803. [CrossRef]
34. Khan, M.A.; Nayan, N.; Ahmad, M.K.; Soon, C.F. Surface Study of CuO Nanopetals by Advanced Nanocharacterization Techniques with Enhanced Optical and Catalytic Properties. *Nanomaterials* **2020**, *10*, 1298. [CrossRef]
35. Dubale, A.A.; Pan, C.-J.; Tamirat, A.G.; Chen, H.-M.; Su, W.-N.; Chen, C.-H.; Rick, J.; Ayele, D.W.; Aragaw, B.A.; Lee, J.-F.; et al. Heterostructured Cu₂O/CuO decorated with nickel as a highly efficient photocathode for photoelectrochemical water reduction. *J. Mater. Chem. A* **2015**, *3*, 12482–12499. [CrossRef]
36. Liang, Y.-C.; Liu, Y.-C. Design of Nanoscaled Surface Morphology of TiO₂–Ag₂O Composite Nanorods through Sputtering Decoration Process and Their Low-Concentration NO₂ Gas-Sensing Behaviors. *Nanomaterials* **2019**, *9*, 1150. [CrossRef]
37. Liang, Y.-C.; Wang, C.-C. Hydrothermally derived zinc sulfide sphere-decorated titanium dioxide flower-like composites and their enhanced ethanol gas-sensing performance. *J. Alloy. Compd.* **2018**, *730*, 333–341. [CrossRef]
38. Hamad, H.; Elsenety, M.M.; Sadik, W.; El-Demerdash, A.G.; Nashed, A.; Mostafa, A.; Elyamny, S. The superior photocatalytic performance and DFT insights of S-scheme CuO@TiO₂ heterojunction composites for simultaneous degradation of organics. *Sci. Rep.* **2022**, *12*, 2217. [CrossRef]
39. Wang, S.; Huo, R.; Zhang, R.; Zheng, Y.; Li, C.; Pan, L. Synthesis of core–shell N-TiO₂@CuOx with enhanced visible light photocatalytic performance. *RSC Adv.* **2018**, *8*, 24866–24872. [CrossRef]
40. Musa, A.; Akomolafe, T.; Carter, M. Production of cuprous oxide, a solar cell material, by thermal oxidation and a study of its physical and electrical properties. *Sol. Energy Mater. Sol. Cells* **1998**, *51*, 305–316. [CrossRef]
41. Deng, X.; Wang, C.; Shao, M.; Xu, X.; Huang, J. Low-temperature solution synthesis of CuO/Cu₂O nanostructures for enhanced photocatalytic activity with added H₂O₂: Synergistic effect and mechanism insight. *RSC Adv.* **2017**, *7*, 4329–4338. [CrossRef]

42. Nakatani, K.; Himoto, K.; Kono, Y.; Nakahashi, Y.; Anma, H.; Okubo, T.; Maekawa, M.; Kuroda-Sowa, T. Synthesis, Crystal Structure, and Electroconducting Properties of a 1D Mixed-Valence Cu(I)–Cu(II) Coordination Polymer with a Dicyclohexyl Dithiocarbamate Ligand. *Crystals* **2015**, *5*, 215–225. [CrossRef]
43. Balık, M.; Bulut, V.; Erdogan, I.Y. Optical, structural and phase transition properties of Cu_2O, CuO and Cu_2O/CuO: Their photoelectrochemical sensor applications. *Int. J. Hydrog. Energy* **2019**, *44*, 18744–18755. [CrossRef]
44. Hsu, Y.-K.; Yu, C.-H.; Chen, Y.-C.; Lin, Y.-G. Synthesis of novel Cu_2O micro/nanostructural photocathode for solar water splitting. *Electrochim. Acta* **2013**, *105*, 62–68. [CrossRef]
45. Masudy-Panah, S.; Radhakrishnan, K.; Tan, H.R.; Yi, R.; Wong, T.I.; Dalapati, G.K. Ten It Wong, Goutam Kumar Dalapati, Titanium doped cupric oxide for photovoltaic application. *Sol. Energy Mater. Sol. Cells* **2015**, *140*, 266–274. [CrossRef]
46. Dong, Y.; Tao, F.; Wang, L.; Lan, M.; Zhang, J.; Hong, T. One-pot preparation of hierarchical Cu_2O hollow spheres for improved visible-light photocatalytic properties. *RSC Adv.* **2020**, *10*, 22387–22396. [CrossRef]
47. Jeong, D.; Jo, W.; Jeong, J.; Kim, T.; Han, S.; Son, M.-K.; Jung, H. Characterization of Cu_2O/CuO heterostructure photocathode by tailoring CuO thickness for photoelectrochemical water splitting. *RSC Adv.* **2022**, *12*, 2632–2640. [CrossRef]
48. Liang, Y.C.; Chou, Y.H. Matrix phase induced boosting photoactive performance of ZnO nanowire turf-coated Bi_2O_3 plate composites. *J. Am. Ceram. Soc.* **2021**, *104*, 5432–5444. [CrossRef]
49. Kmentova, H.; Henrotte, O.; Yalavarthi, R.; Haensch, M.; Heinemann, C.; Zbořil, R.; Schmuki, P.; Kment, Š.; Naldoni, A. Nanoscale Assembly of $BiVO_4$/CdS/CoOx Core–Shell Heterojunction for Enhanced Photoelectrochemical Water Splitting. *Catalysts* **2021**, *11*, 682. [CrossRef]
50. Huang, W.; Li, W.-X. Surface and interface design for heterogeneous catalysis. *Phys. Chem. Chem. Phys.* **2019**, *21*, 523–536. [CrossRef]
51. Zheng, D.; Yu, L.; Liu, W.; Dai, X.; Niu, X.; Fu, W.; Shi, W.; Wu, F.; Cao, X. Structural advantages and enhancement strategies of heterostructure water-splitting electrocatalysts. *Cell Rep. Phys. Sci.* **2021**, *2*, 100443. [CrossRef]
52. Cong, Y.; Ge, Y.; Zhang, T.; Wang, Q.; Shao, M.; Zhang, Y. Fabrication of Z-Scheme Fe_2O_3–MoS_2–Cu_2O Ternary Nanofilm with Significantly Enhanced Photoelectrocatalytic Performance. *Nd. Eng. Chem. Res.* **2018**, *57*, 881–890. [CrossRef]
53. Liang, Y.-C.; Wang, Y.-P. Optimizing crystal characterization of WO_3–ZnO composites for boosting photoactive performance via manipulating crystal formation conditions. *CrystEngComm* **2021**, *23*, 3498–3509. [CrossRef]
54. Bengas, R.; Lahmar, H.; Redha, K.M.; Mentar, L.; Azizi, A.; Schmerber, G.; Dinia, A. Electrochemical synthesis of n-type ZnS layers on p-Cu_2O/n-ZnO heterojunctions with different deposition temperatures. *RSC Adv.* **2019**, *9*, 29056–29069. [CrossRef]
55. Zhang, Y.; Yi, Z.; Wu, G.; Shen, Q. Novel Y doped $BiVO_4$ thin film electrodes for enhanced photoelectric and photocatalytic performance. *J. Photochem. Photobiol. A Chem.* **2016**, *327*, 25–32. [CrossRef]
56. Jang, J.S.; Kim, H.G.; Lee, J.S. Heterojunction semiconductors: A strategy to develop efficient photocatalytic materials for visible light water splitting. *Catal. Today* **2012**, *185*, 270–277. [CrossRef]
57. Chen, H.; Leng, W.; Xu, Y. Enhanced Visible-Light Photoactivity of $CuWO_4$ through a Surface-Deposited CuO. *J. Phys. Chem. C* **2014**, *118*, 9982–9989. [CrossRef]
58. Peng, B.; Zhang, S.; Yang, S.; Chen, H.; Wang, H.; Yu, H.; Zhang, S.; Peng, F. Synthesis and characterization of g-C_3N_4/Cu_2O composite catalyst with enhanced photocatalytic activity under visible light irradiation. *Mater. Res. Bull.* **2014**, *56*, 19–24. [CrossRef]
59. Peng, B.; Zhang, S.; Yang, S.; Wang, H.; Yu, H.; Zhang, S.; Peng, F. The facile hydrothermal synthesis of CuO@ZnO heterojunction nanostructures for enhanced photocatalytic hydrogen evolution. *New J. Chem.* **2019**, *43*, 6794–6805.
60. Aguilera-Ruiz, E.; De La Garza-Galván, M.; Zambrano-Robledo, P.; Ballesteros-Pacheco, J.C.; Vazquez-Arenas, J.; Peral, J.; García-Pérez, U.M. Facile synthesis of visible-light-driven Cu_2O/$BiVO_4$ composites for the photomineralization of recalcitrant pesticides. *RSC Adv.* **2017**, *7*, 45885–45895. [CrossRef]
61. Liang, Y.C.; Chiang, K.J. Design and tuning functionality of rod-like titanium dioxide–nickel oxide composites via a combinational methodology. *Nanotechnology* **2020**, *31*, 195709. [CrossRef] [PubMed]
62. Jiang, X.; Lai, S.; Xu, W.; Fang, J.; Chen, X.; Beiyuan, J.; Zhou, X.; Lin, K.; Liu, J.; Guan, G. Novel ternary BiOI/g-C_3N_4/CeO_2 catalysts for enhanced photocatalytic degradation of tetracycline under visible-light radiation via double charge transfer process. *J. Alloy. Compd.* **2019**, *809*, 151804. [CrossRef]
63. Aguirre, M.E.; Zhou, R.; Eugene, A.J.; Guzman, M.I.; Grela, M.A. Cu_2O/TiO_2 heterostructures for CO_2 reduction through a direct Z-scheme: Protecting Cu_2O from photocorrosion. *Appl. Catal. B Environ. Vol.* **2017**, *217*, 485–493. [CrossRef]
64. Dasineh Khiavi, N.; Katal, R.; Kholghi Eshkalak, S.; Masudy-Panah, S.; Ramakrishna, S.; Jiangyong, H. Visible Light Driven Heterojunction Photocatalyst of CuO–Cu_2O Thin Films for Photocatalytic Degradation of Organic Pollutants. *Nanomaterials* **2019**, *9*, 1011. [CrossRef]
65. Xu, X.; Sun, Y.; Fan, Z.; Zhao, D.; Xiong, S.; Zhang, B.; Zhou, S.; Liu, G. Mechanisms for $\cdot O^{-2}$ and OH Production on Flowerlike $BiVO_4$ Photocatalysis Based on Electron Spin Resonance. *Front. Chem.* **2018**, *6*, 64. [CrossRef]
66. Yoo, H.; Kahng, S.; Kim, J.H. Z-scheme assisted ZnO/Cu_2O-CuO photocatalysts to increase photoactive electrons in hydrogen evolution by water splitting. *Sol. Energy Mater. Sol. Cells* **2020**, *204*, 110211. [CrossRef]
67. Wei, T.; Zhu, Y.N.; An, X.; Liu, L.M.; Cao, X.; Liu, H.; Qu, J. Defect Modulation of Z-Scheme TiO_2/Cu_2O Photocatalysts for Durable Water Splitting. *ACS Catal.* **2019**, *9*, 8346–8354. [CrossRef]

68. Huang, L.; Peng, F.; Yu, H.; Wang, H. Preparation of cuprous oxides with different sizes and their behaviors of adsorption, visible-light driven photocatalysis and photocorrosion. *Solid State Sci.* **2009**, *11*, 129–138. [CrossRef]
69. Wang, P.; Wen, X.; Amal, R.; Ng, Y.H. Introducing a protective interlayer of TiO_2 in Cu_2O–CuO heterojunction thin film as a highly stable visible light photocathode. *RSC Adv.* **2015**, *5*, 5231–5236. [CrossRef]

Article

Experimental Study on the Stability of a Novel Nanocomposite-Enhanced Viscoelastic Surfactant Solution as a Fracturing Fluid under Unconventional Reservoir Stimulation

Xiaodong Si [1,2], Mingliang Luo [1,2,*], Mingzhong Li [1,2], Yuben Ma [3], Yige Huang [1,2] and Jingyang Pu [1,2]

1. School of Petroleum Engineering, China University of Petroleum (East China), Qingdao 266580, China; sixiaodong0021@163.com (X.S.); limingzhong_upc@hotmail.com (M.L.); h17667746750@163.com (Y.H.); 20200099@upc.edu.cn (J.P.)
2. Key Laboratory of Unconventional Oil and Gas Development, China University of Petroleum (East China), Ministry of Education, Qingdao 266580, China
3. Oilfield Production Department, China Oilfield Services Limited, Tianjin 300451, China; mayb6@cnooc.com.cn
* Correspondence: mlluo@upc.edu.cn

Abstract: Fe_3O_4@ZnO nanocomposites (NCs) were synthesized to improve the stability of the worm-like micelle (WLM) network structure of viscoelastic surfactant (VES) fracturing fluid and were characterized by Fourier transform infrared spectrometry (FT-IR), scanning electron microscopy (SEM), energy dispersive spectrometry (EDS), X-ray diffraction (XRD) and vibrating sample magnetometry (VSM). Then, an NC-enhanced viscoelastic surfactant solution as a fracturing fluid (NC-VES) was prepared, and its properties, including settlement stability, interactions between NCs and WLMs, proppant-transporting performance and gel-breaking properties, were systematically studied. More importantly, the influences of the NC concentration, shear rate, temperature and pH level on the stability of NC-VES were systematically investigated. The experimental results show that the NC-VES with a suitable content of NCs (0.1 wt.%) shows superior stability at 95 °C or at a high shear rate. Meanwhile, the NC-VES has an acceptable wide pH stability range of 6–9. In addition, the NC-VES possesses good sand-carrying performance and gel-breaking properties, while the NCs can be easily separated and recycled by applying a magnetic field. The temperature-resistant, stable and environmentally friendly fracturing fluid opens an opportunity for the future hydraulic fracturing of unconventional reservoirs.

Keywords: nanocomposite; worm-like micelle; viscoelastic surfactant; fracturing fluid; stability

1. Introduction

In recent years, unconventional oil and gas resources, especially shale gas and tight oil, have become an important part of the world energy landscape and a substitute resource for conventional oil and gas [1–5]. Hydraulic fracturing technology is one of the key technologies for the efficient development of unconventional reservoirs [6–8]. During the fracturing operation, fracturing fluid (FF) is injected into the formation through the wellbore using a high-pressure pump, forming high well pressure, and resulting in the fracture of the reservoir rock, which opens high-conductivity fracture channels for hydrocarbons migrating from the reservoir to the wellbore [9,10]. Another function of the FF is to carry proppant (such as quartz sand) into these channels to maintain their opening [11,12]. Therefore, FF is a key component of the fracturing operation, and its performance directly affects the success of the fracturing operation. Traditional FFs are mainly water-based fracturing fluids based on polymer thickening compounds such as polyacrylamide (PAM), guar gum and their derivatives [13–15]. However, these polymer thickeners cannot be completely broken and degraded, and insoluble residues are preserved in the formation [16]. The residues block rock pores, reducing reservoir porosity and permeability, resulting in extremely unfavourable conditions for subsequent exploitation of oil and gas resources [17,18]. To effectively reduce damage to reservoirs, a viscoelastic surfactant (VES) fracturing fluid called clean fracturing fluid was developed [19].

The structure of VES is a worm-like micelle (WLM) network structure with hydrophilic groups facing outward and hydrophobic groups facing inward, which is self-assembled by small molecular surfactants under the action of some organic or inorganic salts (such as sodium salicylate, potassium chloride and sodium chloride) [20]. The viscoelastic surfactant solution as a fracturing fluid (VESFF) shows excellent viscoelasticity and has the advantages of good proppant-transporting performance, a strong drag reduction effect and low impact on the environment [21]. Under the action of oil, gas or formation fluids, the WLM network structure of VESFF can completely break without damaging the formation, which is ideal compared to traditional FFs [22]. However, in harsh in situ conditions of unconventional reservoirs such as high temperature and shear, the WLM network structure of VES is easily destroyed, which greatly weakens the performance of the clean FF [23,24]. Encouragingly, the rapid development and application of nanoparticles (NPs) provide new ideas to meet these challenges. Some NPs have been used to improve the structural stability of VESFF to withstand the complex environment during fracturing. Nettesheim et al. [25] found that SiO_2 NPs incorporated into a $NaNO_3$/cetyltrimethylammonium bromide (CTAB) WLM system could promote the entanglement of micelles to effectively improve the viscosity of the fracturing fluid. Similarly, Zhang et al. [26] proposed a SiO_2 NP-enhanced CTAB/NaSal (sodium salicylate) WLM structure, while SiO_2 NPs act as junctions of micelles to produce a cross-linked micellar system. Philippova et al. [27] further found that NP-WLM with a pseudo-crosslinking structure could increase the viscosity of the fracturing fluid by three orders of magnitude. García et al. [28] studied the influence of Al_2O_3 NPs on the rheological behaviour of nanoparticle-enhanced VES and found that the Al_2O_3 NPs increased the viscosity but not the elastic properties of the VES. In addition, some pyroelectric NPs (such as $BaTiO_3$ and ZnO NPs) were investigated to improve the rheological properties and temperature resistance of VESFF, and they show good proppant-transporting performance at high temperature [29,30]. However, the abovementioned NPs do not have external response characteristics and are easily detained in the formation pores, which will damage the formation. Moreover, these NPs mixed with oil and water are difficult to separate and reuse, which not only pollutes the environment but also increases the application cost and wastes nanomaterials [31–33]. Fortunately, magnetic NPs can respond to an external magnetic field, which is easily separated from oil and water for reuse [34]. Magnetic NPs (such as Fe_3O_4 NPs) have been reported in fracture and oil/water monitoring [35,36], fracturing drag reduction [37] and enhanced oil recovery [38]. What is more, the rheological properties of magnetic NPs /WLM systems can be improved with a magnetic field [39,40]. Therefore, a nanomaterial with both magnetic and pyroelectric effects may be more suitable to improve the stability, proppant-transporting capacity, environmental protection and reusability of VESFF with NPs in complex conditions for hydraulically fracturing unconventional reservoirs.

Accordingly, this study aims to prepare a novel Fe_3O_4@ZnO nanocomposite-enhanced viscoelastic surfactant solution as a fracturing fluid (NC-VES) using the interactions between nanoparticles and micelles, and gain further insights into the influence and mechanism of NC concentration, shear rate, temperature and pH on the stability of the NC-VES. First, Fe_3O_4@ZnO nanocomposites were synthesized and characterized by Fourier transform infrared spectrometry (FT-IR), scanning electron microscopy (SEM), energy dispersive spectrometry (EDS) and vibrating sample magnetometry (VSM). Then, the NC-VES was prepared, and its properties, including settlement stability, interactions between NCs and WLMs, proppant-transporting performance and gel-breaking property, were systematically studied.

2. Materials and Methods

2.1. Materials

Ferric chloride ($FeCl_3$) and ferrous sulfate heptahydrate ($FeSO_4 \cdot 7H_2O$) with purities of ≥99.5 wt.% were purchased from Shanghai Aladdin Biochemical Technology Co., Ltd. (Shanghai, China). Acetoxyzinc dihydrate ($Zn(Ac)_2 \cdot 2H_2O$), triethanolamine (TEOA) and sodium hydroxide (NaOH) with purities of ≥99.5 wt.% were manufactured by Shanghai Macklin Biochemical Co., Ltd. (Shanghai, China). Octadecyl trimethyl ammonium chloride (OTAC), ethyl alcohol and sodium salicylate (NaSal) with purities of ≥99 wt.% were provided by Sinopharm Chemical Reagent Co., Ltd. (Shanghai, China). The Krafft points of OTAC surfactant before and after addition of NaSal are 40 °C and 25 °C, respectively. Polyacrylamide (PAM) and guar gum were provided by PetroChina Changqing Downhole Technical Operation Company (Xi'an, China).

2.2. Synthesis of $Fe_3O_4@ZnO$ NCs

First, Fe_3O_4 NPs were prepared by a chemical coprecipitation method, where 1.62 g $FeCl_3$ and 2.08 g $FeSO_4 \cdot 7H_2O$ were dissolved in 150 mL deionized water. Then, 0.1 mol/L NaOH solution was slowly added to the mixed solution until the pH reached 10 and stirred at constant speed mixer at 300 r/min for 3 h. The Fe_3O_4 precipitate was then washed with ethanol and deionized water and dried at 60 °C for 10 h.

Second, ZnO NPs were synthesized by a hydrothermal method where 1.09 g $Zn(Ac)_2 \cdot 2H_2O$ was dissolved in 100 mL deionized water. A total of 50 mL of 0.2 mol/L NaOH solution was mixed in zinc acetate solution. Then, the mixed solution was placed in a high-pressure reactor at 160 °C with a mixing rate of 300 r/min for 8 h. The ZnO precipitate was subsequently washed with ethanol and deionized water and dried at 60 °C for 10 h.

Finally, 0.25 g Fe_3O_4 NPs were dispersed in 50 mL deionized water with ultrasonic vibration for 15 min. Then, 20 mL TEOA solution (1.6 mol/L) and 30 mL zinc acetate solution (0.02 mol/L) were added into the previous solution at 90 °C with stirring at 300 r/min for 10 h. Thus, the composite nanoparticles were obtained after repeated purification more than 5 times through centrifugation (Centrifuge ST16, Thermo Fisher Scientific Inc., Osterode, Lower Saxony, Germany), washing, drying and magnetic adsorption separation.

2.3. Preparation of the NC-VES

First, 2.87 mmol OTAC was dissolved in 80 mL deionized water at 40 °C. $Fe_3O_4@ZnO$ NCs were dispersed in the OTAC solution then ultrasonically vibrated for 15 min. Then, 20 mL NaSal solution (0.156 mol/L) was added into the previous suspension solution at a mixing rate of 180 r/min for 10 min. NC-VES was obtained after standing for 12 h at room temperature. Meanwhile, VES solution with Fe_3O_4 nanoparticles and without nanoparticles were prepared with the same concentration (OTAC: 0.287 mmol/L; NaSal: 0.312 mmol/L) using the same methods.

2.4. Characterization of Nanoparticles

The main functional groups of the samples were examined by FT-IR with a Nicolet 6700 FT-IR instrument (Thermo Fisher Scientific Inc., Waltham, MA, USA). The surface morphology of the nanoparticles was observed by scanning electron microscopy (JSM-5800, JEOL Ltd., Toyoshima, Tokyo, Japan). The elemental content of the nanoparticles was examined with an energy dispersive spectrometer (JEOL Ltd., Toyoshima, Tokyo, Japan). The size of the synthesized nanoparticles was measured by a Malvern laser particle size analyzer (ZS90, Malvern Panalytical, Malvern, UK). The crystal structure of the synthesized samples was measured by X-ray diffractometer (XRD, PANalytical B.V., Almelo, The Netherlands). The magnetizing ability of the samples was evaluated through a vibrating sample magnetometer (Lake Shore 7400, Lake Shore Cryotronics Inc., Westerville, OH, USA).

2.5. Property Tests of the NC-VES

If the network structure of the NC-VES system is unstable or seriously damaged, some or all of the NCs in the system will settle out due to the density difference between the NCs and liquid phase. In this test, the settlement rate of NCs was used to characterize the stability of the NC-VES system. The slower the settlement rate is, the better the stability of the NC-VES. The settlement rates (Equation (1)) were measured by the weighing method with a precision balance (the accuracy was 0.0001 g; Shanghai Fangrui Instrument CO., LTD., Shanghai, China). In this test, the settled NCs were separated from the liquid phase by a magnet as seen in Figure 1a-(2), while the magnet did not suck out the unsettled particles in the system as seen in Figure 1a-(1). Additionally, the effect of various factors on the settlement stability of NC-VES were studied by changing the nanoparticle concentration, shear rate, temperature and pH.

$$settlement\ rate = \frac{W_2 - W_1}{W_0} \times 100\% \qquad (1)$$

where W_0 is the total weight of the NCs added to the NC-VES, g; W_1 is the weight of the empty sample bottle, g; and is the total weight of the sample bottle and settled NCs after being dried, g.

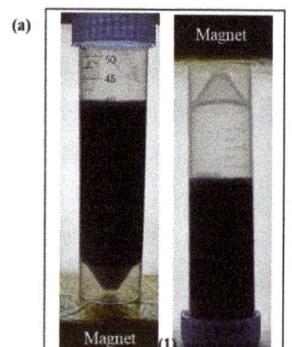
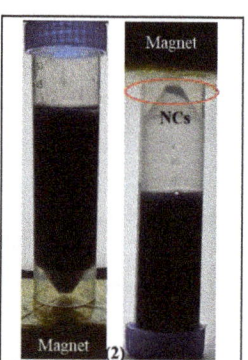
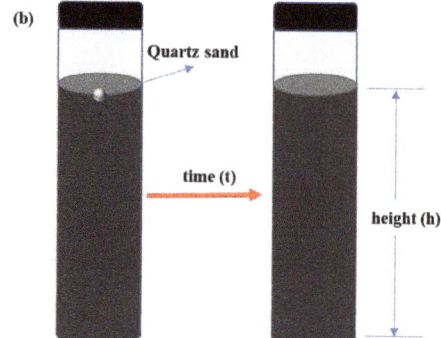

Figure 1. Test of the settlement stability (**a**) and proppant-transporting performance (**b**).

Proppant-transporting performance is one of the key properties of fracturing fluid. In the test, the settling velocity of proppant (quartz sand) in the NC-VES is used to characterize the proppant-transporting performance. As shown in Figure 1b, the settling velocity was calculated by the Formula $v = h/t$, where h and t were recorded as the vertical distance (cm) and time (s) taken for quartz sand as proppant to settle from the surface to the bottom of 40 mL NC-VES.

3. Results and Discussion

3.1. Characterization of Fe_3O_4@ZnO Nanocomposites

Figure 2 shows the comparison of the chemical structures of Fe_3O_4, ZnO and Fe_3O_4@ZnO based on the FT-IR spectra obtained over the wavenumber range of 400–4000 cm^{-1}. The stretching vibration peak at approximately 3410 cm^{-1} in all the IR spectra is attributed to hydroxyl groups (O–H) due to the adsorbed water molecules on the surface of the samples. The peak at approximately 570 cm^{-1} represents Fe-O bonds in spectra Figure 2a,b. The stretching peak at approximately 450 cm^{-1} is related to Zn-O bonds in spectra Figure 2a,c. Both the Fe-O bond and Zn-O bond existing in the synthesized Fe_3O_4@ZnO composite indicate that ZnO was adsorbed on the surface of the Fe_3O_4 particles.

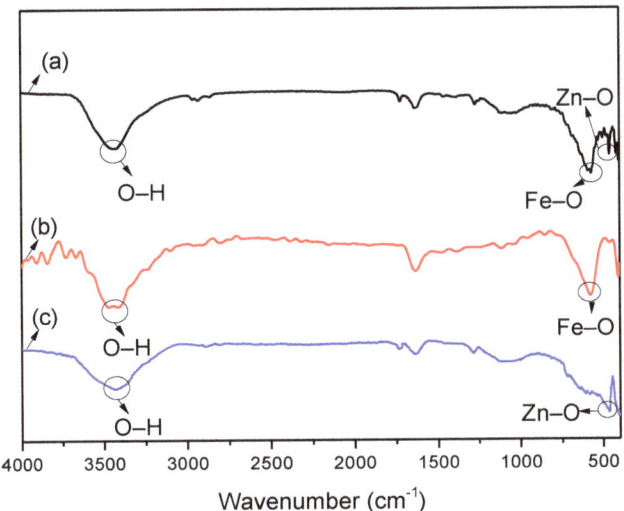

Figure 2. FT-IR spectra of the samples, (**a**) Fe_3O_4@ZnO, (**b**) Fe_3O_4, (**c**) ZnO.

The SEM images with EDS analysis of the Fe_3O_4 (a), ZnO (b) and Fe_3O_4@ZnO samples (c) are depicted in Figure 3. As seen from the SEM images, the average sizes of the synthesized Fe_3O_4 NPs, ZnO NPs and Fe_3O_4@ZnO NCs are approximately 25 nm, 10 nm and 60 nm, respectively. As shown in Figure 4, the medium diameters of the three nanoparticles with a narrow size distribution, were consistent with the observation results of the SEM images. The pure Fe_3O_4 NPs appear to exhibit an octahedral structure (Figure 3a), while the ZnO NPs display a spherical-like structure with smaller dimensions (Figure 3b). Figure 3c clearly shows that ZnO NPs are successfully adsorbed on the surface of Fe_3O_4 NPs. Additionally, the comparison results of the EDS analysis further showed that the composite nanoparticles were successfully synthesized. Meanwhile, the mass ratio of Fe_3O_4 and ZnO (5.14:1) in the Fe_3O_4@ZnO NCs was obtained from the EDS analysis. Besides, in all EDS diagrams, the oxygen content higher than the theoretical value was mainly attributed to the adsorption of water molecules.

XRD analysis of synthesized Fe_3O_4, ZnO and Fe_3O_4@ZnO nanoparticles and Rietveld refinement of the XRD data of the Fe_3O_4@ZnO composite were shown in Figure 5. As shown in Figure 5a, the characteristic peaks and Bragg lattice planes reported at 30.08° (220), 35.42° (311), 43.08° (400), 53.56° (422), 56.98° (511) and 62.63° (440) are related to Fe_3O_4 structure, while the characteristic peaks and Bragg lattice planes reported at 31.72° (100), 34.44° (002), 36.21° (101), 47.49° (102), 56.51° (110), 62.81° (103) and 67.86° (112) are related to ZnO structure, which are in accordance with the patterns of standard pure Fe_3O_4 and ZnO, respectively [41,42]. Almost no obvious impurity peak was observed, which confirmed the high purity for the synthesized samples. As shown in Figure 5b, Rietveld refinement with good fit (goodness of fit $x^2 = 1.19$) was carried out to obtain more crystal phase information of the composite. Parameters for Fe_3O_4 (crystal structure: cubic; lattice constant: a = 8.372 Å; space group: Fd$\bar{3}$m) and parameters for ZnO (lattice constant: a = 3.249 Å, c = 5.208 Å; crystal structure: hexagonal; space group: P63mc) were obtained. Meanwhile, the mass ratio of Fe_3O_4 and ZnO (4.91:1) in the Fe_3O_4@ZnO NCs from the Rietveld refinement is consistent with the EDS test within the range of allowable error. The results revealed that Fe_3O_4@ZnO NCs with a high purity were successfully synthesized.

Figure 3. SEM (**left**) and EDS analysis (**right**) of Fe_3O_4 (**a**), ZnO (**b**) and Fe_3O_4@ZnO (**c**).

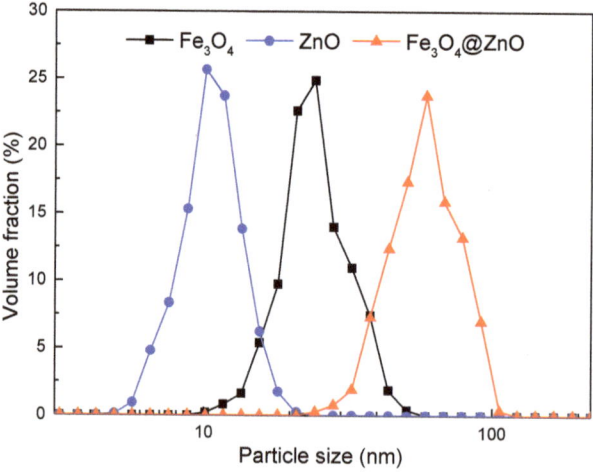

Figure 4. Particle size distribution of Fe_3O_4, ZnO and Fe_3O_4@ZnO nanoparticles.

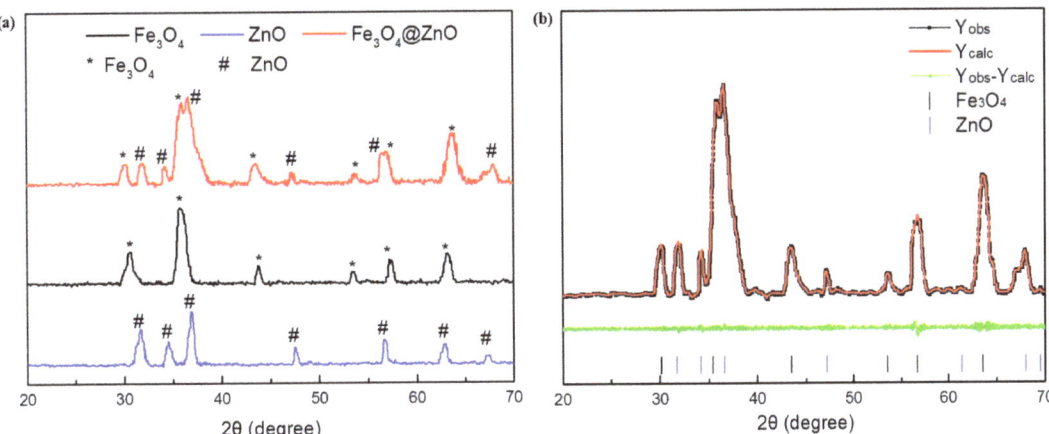

Figure 5. XRD patterns for synthesized nanoparticles (**a**) and Rietveld refinement of the XRD data of the Fe$_3$O$_4$@ZnO composite (**b**).

Figure 6 shows the magnetization of the Fe$_3$O$_4$ and Fe$_3$O$_4$@ZnO samples. It is clearly observed that the magnetization curves for the measured samples are S-shaped. When the magnetic field is zero, the remanence and coercive force are close to zero, indicating that the synthesized Fe$_3$O$_4$ NPs and Fe$_3$O$_4$@ZnO NCs possess good soft magnetic and practically superparamagnetic properties. The saturation magnetization measured for pure Fe$_3$O$_4$ NPs is 76.03 emu/g, while the saturation magnetization of Fe$_3$O$_4$@ZnO NCs decreases slightly but still maintains a high value of 62.26 emu/g. The high magnetization ensures that the Fe$_3$O$_4$@ZnO NCs have strong magnetic response ability and are easily separated from the liquid phase by magnetic field during quantitatively measuring the settlement rate of NCs (seen in Section 2.5). On the other hand, Fe$_3$O$_4$@ZnO NCs is simply recycled from the dispersed solution by using a magnet (seen in Section 3.3). What is more, the mass ratio of Fe$_3$O$_4$ and ZnO calculated from the comparative saturation magnetization is 5.02:1, consistent with the above tests (EDS analysis and XRD analysis), which further proves the purity of the synthetic Fe$_3$O$_4$@ZnO NCs.

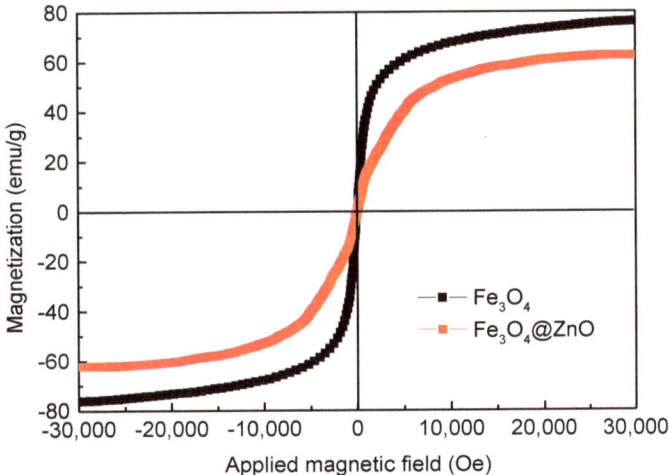

Figure 6. Magnetization of the Fe$_3$O$_4$ NPs and Fe$_3$O$_4$@ZnO NCs.

3.2. Stability of the NC-VES

3.2.1. Effects of Shear Rate and NC Concentration

The shear rate-dependent viscosities of the VESFF for different NC concentrations was studied with shear rates ranging from 0.1 to 1000 s^{-1} at 25 °C (tested with an Anton Paar rheometer, Physica MCR 302, Anton Paar GmbH, Graz, Austria). Figure 7 shows that the viscosity remains unchanged at low shear rates, while a notably reduced slope is observed under high shear rates, and the shear-thinning phenomenon occurs in all samples. The dependence of viscosity on shear rate is usually explained by shear banding behaviour, which has been reported in previous research results [43–46]. It is found that the NCs can improve the viscosity of the VES system. When the concentration of the NCs is low (0.01 wt.%), the viscosity increases but is not clear compared with the VES without NCs. When the NC concentration reaches 0.1 wt.%, the viscosity at 170 s^{-1} of the NC-VES is higher than VES without NCs. However, when the NC concentration is higher (0.3 wt.%), the viscosity of NC-VES decreases sharply at high shear rates.

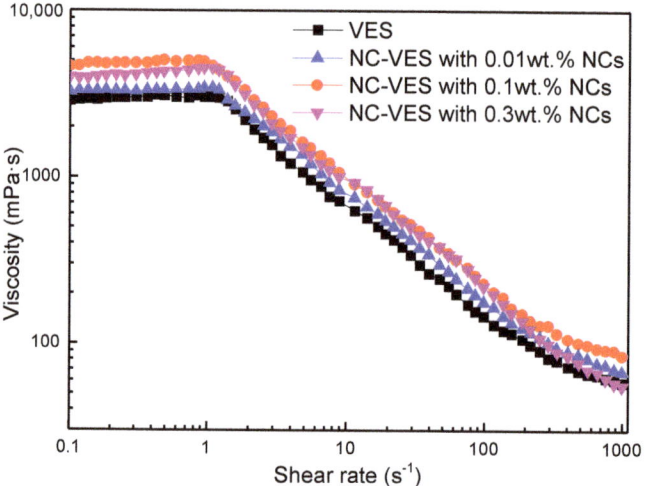

Figure 7. Viscosity as a function of shear rate at different NC concentrations.

Dynamic modulus (storage modulus G′ and loss modulus G″) as a function of frequency is shown in Figure 8. Within the measured test frequency range, the storage modulus remains almost unchanged, while the loss modulus increases gradually. For NC-VES systems, the storage modulus and loss modulus are greater than those of the VES system, suggesting that the WLM network in VES is strengthened by NCs. Here, the nanoparticles are incorporated into worm-like micellar systems to form nanoparticle-micelle junctions as the connection point of WLMs, which improves the structural stability and viscoelasticity of the WLM systems [47–49]. Combined with Figures 7 and 8, this indicates that the NC concentration (0.1 wt.%) enables the NC-VES system to possess sufficient stability and high viscosity, maintaining good proppant-transporting performance during fracturing operations. Therefore, subsequent research on the influencing factors of fracturing fluid focuses on the NC-VES system with a 0.1 wt.% NC concentration.

3.2.2. Effect of Temperature

All samples were heated to the corresponding experimental temperature by a water bath and held for 10 h, then the settlement rate of NCs in each sample was obtained by weighing and calculation, as shown in Figure 9. The results show that when the temperature increased from 25 to 65 °C, the settlement rate of NCs increased from 0 to 0.1%, while almost no sedimentation phenomenon occurred, indicating that the NC-WLM network structure of the system was very stable in the low to medium temperature range. As the temperature continued to rise, the sedimentation rate increased slightly, but even at a high temperature of 95 °C, the settlement rate was only about 2%, indicating that the temperature resistance of the system was very good. Conventional VESFF is a WLM network structure formed by the spontaneous aggregation and self-assembly of high-concentration surfactant molecules, which has a specific viscoelasticity and can meet the needs of fracturing in medium- and low–temperature reservoirs. However, the WLM structure in traditional VESFF is easily broken at high temperatures, because the interaction between molecules such as van der Waals forces and hydrogen bonding forces weakens and the stability of aggregates decreases [23], leading to a sharp decrease in the viscosity of the system, which severely weakens the performance of the fracturing fluid.

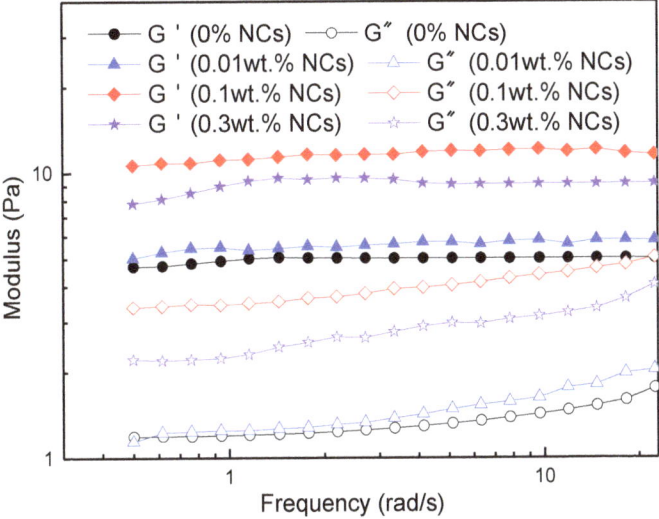

Figure 8. Variations of storage modulus (G′) and loss modulus (G″) with frequency at 25 °C.

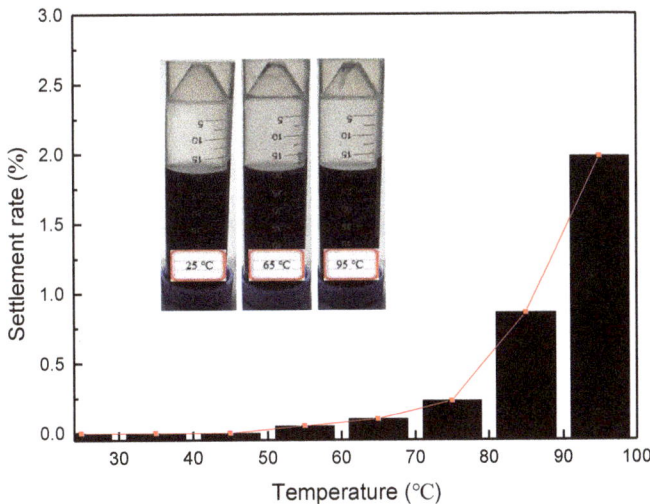

Figure 9. Settlement rate as a function of temperature, in which the insertions from left to right refer to the stable state of the NC-WLM system at 25 °C, 65 °C and 95 °C respectively.(OTAC: 0.287 mmol/L; NaSal: 0.312 mmol/L).

The viscosities as a function of time at 95 °C with a constant heating rate of 3 °C/min and a shear rate of 170 s^{-1} are displayed in Figure 10. When the temperature is lower than 60 °C, a slight increase in viscosity was observed in all three curves. However, as the temperature continues to increase to 95 °C, the viscosity of the samples decreases sharply. The viscosity of VES with Fe$_3$O$_4$ NPs is lower than that without NPs at 95 °C. The reason is that a certain amount of surfactant molecules are adsorbed on the surface of Fe$_3$O$_4$ NPs, resulting in a decrease in the surfactant concentration involved in WLM structure in the liquid phase [50]. Compared with the other two systems, the high-temperature resistance of the NC-VES system is mainly attributed to the ZnO NPs on the surface of the NCs. As shown in Figure 11, the ZnO nanoparticles adsorbed on the surface of NCs possess a pyroelectric effect, releasing charges with increasing temperature. The charged NCs easily adsorb micelles and play a role in the junction of the WLM network structure. The electrostatic screening of charged WLMs promotes further growth of wormlike micelles, and maintains good stability of the WLM network structure [29]. Therefore, the NCs incorporated into the WLM network act as a skeleton-like structure, greatly improving the stability of the NC-WLM system at high temperature.

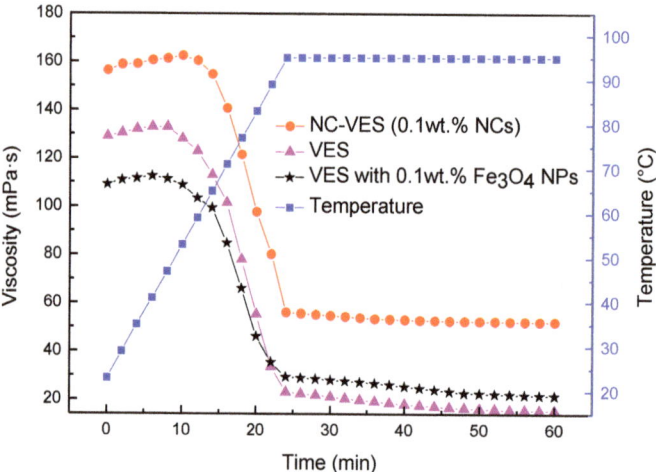

Figure 10. Viscosity as a function of time at high temperature (heated to 95 °C).

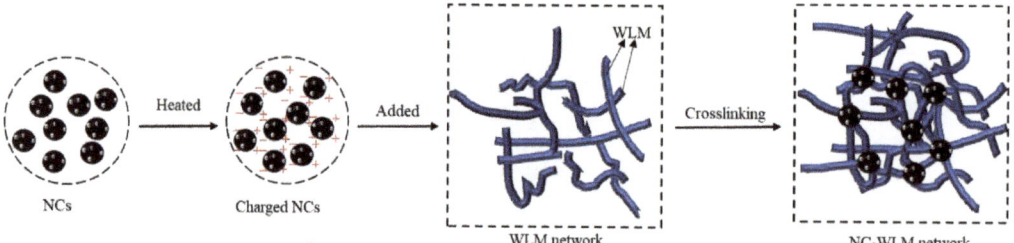

Figure 11. Schematic diagram of the stability of the NC-WLM network structure at high temperatures.

3.2.3. Effect of pH

As shown in Figure 12a, in a highly acidic environment (pH = 1–5), the settlement phenomenon of the system is very clear, while the settlement rate is as high as 94.9%. This shows that the system cannot exist stably in strong acids, which can be attributed to two aspects. On the one hand, in the presence of a strong acid, some NCs react with H$^+$ (Fe$_3$O$_4$ + 8H$^+$ = Fe^{2+} + 2Fe^{3+} + 4H$_2$O, ZnO + 2H$^+$ = Zn^{2+} + H$_2$O) and are dissolved in the solution. The solution will then exhibit the crimson colour of the Fe^{2+} and Fe^{3+} aqueous solutions. On the other hand, excessive H^+ leads to an increase in the repulsive force between the head groups of the OTAC molecule and destroys the WLM network structure, which results in large settlement of NCs [51,52]. From the results, when pH = 1, the settlement rate decreases, which does not mean that the settlement stability of the NC-VES becomes better, but that more NCs have been dissolved in the solution. Therefore, it can be considered that the stronger the acidity, the worse the settlement stability of the system. When the pH is weakly acidic–neutral–weakly alkaline (pH = 6–9), the settlement rate of the system is very small. Meanwhile, Figure 12b shows that it is difficult for NCs to form stable suspension in water without WLMs even under neutral conditions, because the particle density (about 5.26 g/cm^3) is much greater than that of the liquid. In the strongly alkaline environment (pH = 10–14), the settlement rate of NCs increases rapidly with increasing pH. When pH $\geq$ 13, a large number of NCs settle out. On the one hand, ZnO is dissolved by alkaline solution (ZnO + 2OH$^-$ + H$_2$O = [Zn(OH)$_4$]$^{2-}$). On the other hand, an excessive OH$^-$ destroys the self-assembly mechanism of surfactant molecules, leading to micelles unsuccessful for connecting with each other and assembling into a WLM network structure [53]. Therefore, the pH environment has a strong impact on the stability of the NC-VES, with the system being able to maintain a good stability in a wide range of pH = 6–9.

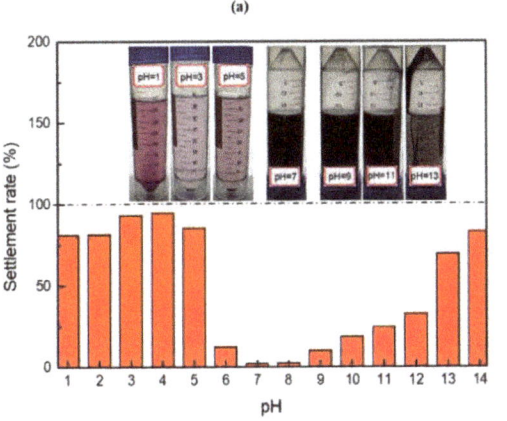 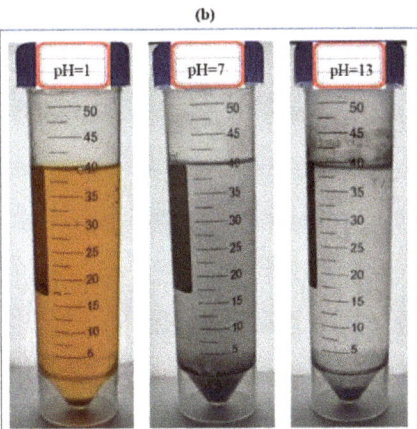

Figure 12. Settlement rate as a function of pH (**a**) and behaviour of NCs in deionized water without of WLMs (**b**).

3.3. Proppant-Transporting Performance and Gel-Breaking Property

In the process of a hydraulic fracturing operation, fracturing fluid is pumped into the formation, cracking the rock formation and forming a fracture channel to improve the oil and gas production efficiency. To prevent fracture closure, proppants (such as quartz sand) are transported by fracturing fluid into the formation to prop fractures, keeping the fracture open for a long time under formation pressure. Therefore, it is very important that the fracturing fluid possess good proppant-transporting performance, which is one of the key factors for the success of fracturing operations. In this work, the proppant-transporting performance of four fracturing fluids was compared, and the results are shown in Table 1. Compared with conventional fracturing fluids (Guar FF and PAM FF), the VES systems (NC-VES and VESFF) show better proppant-transporting performance. The settlement velocity of quartz sand in the NC-VES is only 0.52×10^{-3} cm/s, far less than 0.08 cm/s, which displays superior proppant-transporting performance [54]. In addition, by adding a small amount of kerosene to NC-VES, gel breaking occurs rapidly with no residue, and the NCs can be easily recycled using a magnet to apply a magnetic field, as shown in Figure 13.

The results imply that the NC-VES system is a low-damage, environmentally friendly and cost-saving fracturing fluid.

Table 1. Settlement velocity of different FF samples at room temperature.

Samples	Constituents (per 100 mL Water)	Viscosity (mPa·s)	Settlement Velocity (cm/s)
NC-VES	1 wt.% OTAC + 0.5 wt.% NaSal + 0.1 wt.% NCs	158	0.52×10^{-3}
VESFF	1 wt.% OTAC + 0.5 wt.% NaSal	130	7.29×10^{-3}
Guar FF	0.35 wt.% Guar	51	0.18
PAM FF	0.35 wt.% PAM	59	0.11

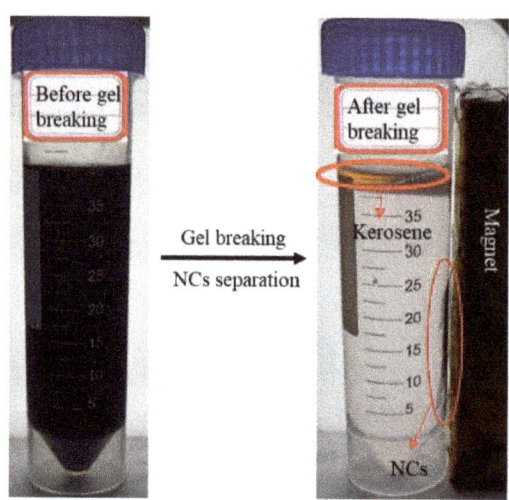

Figure 13. Gel-breaking test and the separation of NCs by a magnet.

4. Conclusions

In this study, the influence and mechanism of NC concentration, shear rate, temperature and pH on the settlement stability of the NC-VES fracturing fluid was systematically investigated. First, NCs with good magnetic response ability were synthesized and acted as junctions of micelles to improve the stability of the WLM network structure. The results showed that the NC-VES system with the optimal concentration of 0.1 wt.% possesses good shear and temperature resistance. At high temperatures (such as 95 °C), the ZnO NPs with a pyroelectric effect on the surface of NCs can effectively reduce the decomposition of the WLM network structure to prevent the NCs from settling out of the fracturing fluid. Strong acid and strong alkaline solutions seriously damage the NC-VES network structure by dissolving NCs and hindering the self-assembly behaviour of surfactant molecules, resulting in an acceptable pH range of 6~9. In addition, the settling velocity of quartz sand in the system at room temperature was only 0.52×10^{-3} cm/s, indicating that the NC-VES has good proppant-transporting performance. Finally, no residue was found after gel breaking of the NC-VES and the recovery of NCs by magnetic adsorption, implying no damage to the formation and the environment. A new type of temperature-resistant, stable and environmentally friendly fracturing fluid was thus provided for hydraulic fracturing of unconventional reservoirs.

Author Contributions: Conceptualization, M.L. (Mingliang Luo) and M.L. (Mingzhong Li); methodology, X.S., M.L. (Mingzhong Li) and M.L. (Mingliang Luo); experimental, X.S., Y.H. and Y.M.; writing, X.S. and M.L. (Mingliang Luo); analytical studies, X.S. and M.L. (Mingliang Luo); visualization, X.S.; funding acquisition, M.L. (Mingliang Luo) and J.P. All authors have read and agreed to the published version of the manuscript.

Funding: This research was funded by the National Natural Science Foundation of China (grant numbers 51874334 and 52104057) and the China Postdoctoral Science Foundation (grant number 2021M693506).

Institutional Review Board Statement: Not applicable.

Informed Consent Statement: Not applicable.

Data Availability Statement: Data are contained within the article.

Conflicts of Interest: The authors declare no conflict of interest.

References

1. Luo, M.L.; Si, X.D.; Zhang, Y.; Yuan, Z.H.; Yang, D.Y.; Gong, J. Performance evaluation of water control with nanoemulsion as pre-pad fluid in hydraulically fracturing tight gas formations. *Energy Fuels* **2017**, *31*, 3698–3707. [CrossRef]
2. Andrés, J.C.; Omar, J.G.; Lazaros, G.P.; Gintaras, V.R. Disclosing water-energy-economics nexus in shale gas development. *Appl. Energy* **2018**, *225*, 710–731. [CrossRef]
3. Tiffany, L.; Doug, D.; Shinji, M. Comparison of the degree of fouling at various flux rates and modes of operation using forward osmosis for remediation of produced water from unconventional oil and gas development. *Sci. Total Environ.* **2019**, *675*, 73–80. [CrossRef]
4. Luo, M.L.; Jia, X.H.; Si, X.D.; Luo, S.; Zhan, Y.P. A novel polymer encapsulated silica nanoparticles for water control in development of fossil hydrogen energy-tight carbonate oil reservoir by acid fracturing. *Int. J. Hydrogen Energy* **2021**, *46*, 31191–31201. [CrossRef]
5. Apergis, N.; Mustafa, G.; Dastidar, S.G. An analysis of the impact of unconventional oil and gas activities on public health: New evidence across Oklahoma counties. *Energy Econ.* **2021**, *97*, 105223. [CrossRef]
6. Vishkai, M.; Gates, I. On multistage hydraulic fracturing in tight gas reservoirs: Montney Formation, Alberta, Canada. *J. Pet. Sci. Eng.* **2019**, *174*, 1127–1141. [CrossRef]
7. Li, Y.Y.; Hu, W.; Zhang, Z.H.; Zhang, Z.B.; Shang, Y.J.; Han, L.L.; Wei, S.Y. Numerical simulation of hydraulic fracturing process in a naturally fractured reservoir based on a discrete fracture network model. *J. Struct. Geol.* **2021**, *147*, 104331. [CrossRef]
8. Liu, J.R.; Sheng, J.J.; Emadibaladehi, H.; Tu, J.W. Experimental study of the stimulating mechanism of shut-in after hydraulic fracturing in unconventional oil reservoirs. *Fuel* **2021**, *300*, 120982. [CrossRef]
9. Javadpour, F.; McClure, M.; Naraghi, M.E. Slip-corrected liquid permeability and its effect on hydraulic fracturing and fluid loss in shale. *Fuel* **2015**, *160*, 549–559. [CrossRef]
10. Etoughe, P.; Siddhamshetty, P.; Cao, K.Y. Incorporation of sustainability in process control of hydraulic fracturing in unconventional reservoirs. *Chem. Eng. Res. Des.* **2018**, *139*, 62–76. [CrossRef]
11. Zhao, X.; Guo, J.C.; Peng, H.; Pan, R.; Aliu, A.O.; Lu, Q.L.; Yang, J. Synthesis and evaluation of a novel clean hydraulic fracturing fluid based on star-dendritic polymer. *J. Nat. Gas Sci. Eng.* **2017**, *43*, 179–189. [CrossRef]
12. Xu, T.; Mao, J.C.; Zhang, Y.Z.; Yang, X.J.; Lin, C.; Du, A.Q.; Zhang, H. Application of gemini viscoelastic surfactant with high salt in brine-based fracturing fluid. *Colloids Surf. A* **2021**, *611*, 125838. [CrossRef]
13. Baruah, A.; Pathak, A.K.; Ojha, K. Phase behaviour and thermodynamic properties of lamellar liquid crystal developed for viscoelastic surfactant based fracturing fluid. *Chem. Eng. Sci.* **2015**, *131*, 146–154. [CrossRef]
14. Chauhan, G.; Verma, A.; Hazarika, A.; Ojha, K. Rheological, structural and morphological studies of Gum Tragacanth and its inorganic SiO_2 nanocomposite for fracturing fluid application. *J. Taiwan Inst. Chem. Eng.* **2017**, *80*, 978–988. [CrossRef]
15. Yang, B.; Zhao, J.Z.; Mao, J.C.; Tan, H.Z.; Zhang, Y.; Song, Z.F. Review of friction reducers used in slickwater fracturing fluids for shale gas reservoirs. *J. Nat. Gas Sci. Eng.* **2019**, *62*, 302–313. [CrossRef]
16. Wang, J.; Holditch, S.A.; McVay, D.A. Effect of gel damage on fracture fluid cleanup and long-term recovery in tight gas reservoirs. *J. Nat. Gas Sci. Eng.* **2012**, *9*, 108–118. [CrossRef]
17. Liang, X.Y.; Zhou, F.J.; Liang, T.B.; Wang, C.Z.; Wang, J.; Yuan, S. Impacts of low harm fracturing fluid on fossil hydrogen energy production in tight reservoirs. *Int. J. Hydrogen Energy* **2020**, *45*, 21195–21204. [CrossRef]
18. Yan, Z.H.; Dai, C.L.; Zhao, M.W.; Sun, Y.P.; Zhao, G. Development, formation mechanism and performance evaluation of a reusable viscoelastic surfactant solution as fracturing fluid. *J. Ind. Eng. Chem.* **2016**, *37*, 115–122. [CrossRef]
19. Samuel, M.M.; Card, R.J.; Nelson, E.B.; Brown, J.E.; Vinod, P.S.; Temple, H.L.; Qu, Q.; Fu, D.K. Polymer-free fluid for fracturing applications. *SPE Drill. Complet.* **1999**, *14*, 240–246. [CrossRef]
20. García, B.F.; Saraji, S. Mixed in-situ rheology of viscoelastic surfactant solutions using a hyperbolic geometry. *J. Non-Newton. Fluid Mech.* **2019**, *270*, 56–65. [CrossRef]
21. Shibaev, A.V.; Aleshina, A.L.; Arkharova, N.A.; Orekhov, A.S.; Kuklin, A.I.; Philippova, O.E. Disruption of cationic/anionic viscoelastic surfactant micellar networks by hydrocarbon as a basis of enhanced fracturing fluids clean-up. *Nanomaterials* **2020**, *10*, 2353. [CrossRef] [PubMed]
22. Mushi, S.J.; Kang, W.L.; Yang, H.B.; Wang, P.X.; Hou, X.Y. Viscoelasticity and microstructural properties of zwitterionic surfactant induced by hydroxybenzoate salt for fracturing. *J. Mol. Liq.* **2020**, *301*, 112485. [CrossRef]
23. Wu, X.P.; Song, Z.H.; Zhen, J.W.; Wang, H.B.; Yao, L.S.; Zhao, M.W.; Dai, C.L. A smart recyclable VES fluid for high temperature and high pressure fracturing. *J. Pet. Sci. Eng.* **2020**, *190*, 107097. [CrossRef]

24. Wu, X.P.; Zhang, Y.; Sun, X.; Huang, Y.P.; Dai, C.L.; Zhao, M.W. A novel CO_2 and pressure responsive viscoelastic surfactant fluid for fracturing. *Fuel* **2018**, *229*, 79–87. [CrossRef]
25. Nettesheim, F.; Liberatore, M.W.; Hodgdon, T.K. Influence of nanoparticle addition on the properties of wormlike micellar solutions. *Langmuir* **2008**, *24*, 7718–7726. [CrossRef]
26. Zhang, Y.; Dai, C.L.; Qian, Y.; Fan, X.Q.; Wu, Y.N.; Wu, X.P. Rheological properties and formation dynamic filtration damage evaluation of a novel nanoparticle-enhanced VES fracturing system constructed with wormlike micelles. *Colloids Surf. A* **2018**, *533*, 244–252. [CrossRef]
27. Philippova, O.E.; Molchanov, V.S. Enhanced rheological properties and performance of viscoelastic surfactant fluids with embedded nanoparticles. *Curr. Opin. Colloid Interface Sci.* **2019**, *43*, 52–62. [CrossRef]
28. García, B.F.; Saraji, S. Linear rheology of nanoparticle-enhanced viscoelastic surfactants. *J. Mol. Liq.* **2020**, *300*, 112215. [CrossRef]
29. Luo, M.L.; Jia, Z.L.; Sun, H. Rheological behavior and microstructure of an anionic surfactant micelle solution with pyroelectric nanoparticle. *Colloids Surf. A* **2012**, *395*, 267–275. [CrossRef]
30. Baruah, A.; Khilendra, K.; Pathak, A.K.; Ojha, K. Study on rheology and thermal stability of mixed (nonionic–anionic) surfactant based fracturing fluids. *AICHE J.* **2016**, *62*, 2177–2187. [CrossRef]
31. Díez, R.; Medina, O.E.; Giraldo, L.J.; Cortés, F.B.; Franco, C.A. Development of nanofluids for the inhibition of formation damage caused by fines migration: Effect of the interaction of quaternary amine (CTAB) and MgO nanoparticle. *Nanomaterials* **2020**, *10*, 928. [CrossRef] [PubMed]
32. Franco, C.A.; RichardZabala, R.; Cortés, F.B. Nanotechnology applied to the enhancement of oil and gas productivity and recovery of Colombian fields. *J. Pet. Sci. Eng.* **2017**, *157*, 39–55. [CrossRef]
33. Giraldo, L.J.; Diez, R.; Acevedo, S.; Cortés, F.B.; Franco, C.A. The effects of chemical composition of fines and nanoparticles on inhibition of formation damage caused by fines migration: Insights through a simplex-centroid mixture design of experiments. *J. Pet. Sci. Eng.* **2021**, *203*, 108494. [CrossRef]
34. Dong, P.; Chen, X.; Guo, M.T.; Wu, Z.Y. Heterogeneous electro-Fenton catalysis with self-supporting CFP@MnO_2-Fe_3O_4/C cathode for shale gas fracturing flowback wastewater. *J. Hazard. Mater.* **2021**, *412*, 125208. [CrossRef]
35. Sengupta, S. An innovative approach to image fracture dimensions by injecting ferrofluids. In Proceedings of the Abu Dhabi International Petroleum Conference and Exhibition, Abu Dhabi, United Arab Emirates, 11 November 2012. [CrossRef]
36. Rahmani, A.R.; Bryant, S.L.; Huh, C. Crosswell magnetic sensing of superparamagnetic nanoparticles for subsurface applica-tions. *SPE J.* **2015**, *20*, 1067–1082. [CrossRef]
37. Luo, M.L.; Si, X.D.; Li, M.Z.; Jia, X.H.; Yang, Y.L.; Zhan, Y.P. Experimental study on the drag reduction performance of clear fracturing fluid using wormlike surfactant micelles and magnetic nanoparticles under a magnetic field. *Nanomaterials* **2021**, *11*, 885. [CrossRef]
38. Ali, A.M.; Yahya, N.; Qureshi, S. Interactions of ferro-nanoparticles (hematite and magnetite) with reservoir sandstone: Implications for surface adsorption and interfacial tension reduction. *Pet. Sci.* **2020**, *17*, 1037–1055. [CrossRef]
39. Molchanov, V.S.; Pletneva, V.A.; Klepikov, I.A.; Razumovskaya, I.V.; Philippova, O.E. Soft magnetic nanocomposites based on adaptive matrix of wormlike surfactant micelles. *RSC Adv.* **2018**, *8*, 11589–11597. [CrossRef]
40. Claracq, J.; Sarrazin, J.; Montfort, J.P. Viscoelastic properties of magnetorheological fluids. *Rheol. Acta* **2004**, *43*, 38–49. [CrossRef]
41. Christus, A.; Ravikumar, A.; Panneerselvam, P.; Radhakrishnan, K. A novel Hg(II) sensor based on Fe_3O_4@ZnO nanocomposite as peroxidase mimics. *Appl. Surf. Sci.* **2018**, *449*, 669–676. [CrossRef]
42. Nadafan, M.; Sabbaghan, M.; Sofalgar, P.; Anvari, J.Z. Comparative study of the third-order nonlinear optical properties of ZnO/Fe_3O_4 nanocomposites synthesized with or without Ionic Liquid. *Opt. Laser Technol.* **2020**, *131*, 106435. [CrossRef]
43. Croce, V.; Cosgrove, T.; Dreiss, C.A.; King, S.; Maitland, G.; Hughes, T. Giant micellar worms under shear: A rheological study using SANS. *Langmuir* **2005**, *21*, 6762–6768. [CrossRef]
44. Chu, Z.L.; Feng, Y.J.; Su, X.; Han, Y.X. Wormlike micelles and solution properties of a C22-tailed amidosulfobetaine surfactant. *Langmuir* **2010**, *26*, 7783–7791. [CrossRef]
45. Gao, Z.B.; Dai, C.L.; Sun, X.; Huang, Y.P.; Gao, M.W.; Zhao, M.W. Investigation of cellulose nanofiber enhanced viscoelastic fracturing fluid system: Increasing viscoelasticity and reducing filtration. *Colloids Surf. A* **2019**, *582*, 123938. [CrossRef]
46. Yin, H.Y.; Feng, Y.J.; Li, P.X.; Doutch, J.; Han, Y.X.; Mei, Y.J. Cryogenic viscoelastic surfactant fluids: Fabrication and application in a subzero environment. *J. Colloid Interface Sci.* **2019**, *551*, 89–100. [CrossRef]
47. Fanzatovich, I.I.; Aleksandrovich, K.D.; Rinatovich, I.A. Supramolecular system based on cylindrical micelles of anionic surfactant and silica nanoparticles. *Colloids Surf. A* **2016**, *507*, 255–260. [CrossRef]
48. Zhu, J.Y.; Yang, Z.Z.; Li, X.G.; Hou, L.L.; Xie, S.Y. Experimental study on the microscopic characteristics of foams stabilized by viscoelastic surfactant and nanoparticles. *Colloids Surf. A* **2019**, *572*, 88–96. [CrossRef]
49. Zhang, Y.; Dai, C.L.; Wu, X.P.; Wu, Y.N.; Li, Y.Y.; Huang, Y.P. The construction of anhydride-modified silica nanoparticles (AMSNPs) strengthened wormlike micelles based on strong electrostatic and hydrogen bonding interactions. *J. Mol. Liq.* **2019**, *277*, 372–379. [CrossRef]
50. Helgeson, M.E.; Hodgdon, T.K.; Kaler, E.W.; Wagner, N.J.; Vethamuthu, M.; Ananthapadmanabhan, K.P. Formation and rheology of viscoelastic "double networks" in wormlike micelle–nanoparticle mixtures. *Langmuir* **2010**, *26*, 8049–8060. [CrossRef]
51. Jiao, W.X.; Wang, Z.; Liu, T.Q.; Li, X.F.; Dong, J.F. pH and light dual stimuli-responsive wormlike micelles with a novel Gemini surfactant. *Colloids Surf. A* **2021**, *618*, 126505. [CrossRef]

52. Fu, H.R.; Duan, W.M.; Zhang, T.L.; Xu, K.; Zhao, H.F.; Yang, L.; Zheng, C.C. Preparation and mechanism of pH and temperature stimulus-responsive wormlike micelles. *Colloids Surf. A* **2021**, *624*, 126788. [CrossRef]
53. Liu, F.; Liu, D.J.; Zhou, W.J.; Wang, S.; Chen, F.; Wei, J.J. Weakening or losing of surfactant drag reduction ability: A coarse-grained molecular dynamics study. *Chem. Eng. Sci.* **2020**, *219*, 115610. [CrossRef]
54. Wu, H.R.; Zhou, Q.; Xu, D.R.; Sun, R.X.; Wang, P.Y. SiO_2 nanoparticle-assisted low-concentration viscoelastic cationic surfactant fracturing fluid. *J. Mol. Liq.* **2018**, *266*, 864–869. [CrossRef]

Wettability Improvement in Oil–Water Separation by Nano-Pillar ZnO Texturing

Xiaoyan Liu [1], Shaotong Feng [1], Caihua Wang [1], Dayun Yan [2,*], Lei Chen [3] and Bao Wang [1,3,*]

[1] School of Mechanical Science and Engineering, Northeast Petroleum University, Daqing 163318, China; liu_xydq@163.com (X.L.); fengshaotong0513@163.com (S.F.); wch.dqsy@163.com (C.W.)
[2] Department of Mechanical and Aerospace Engineering, The George Washington University, Washington, DC 20052, USA
[3] State Key Laboratory of Tribology, Tsinghua University, Beijing 100084, China; leichen16@mail.tsinghua.edu.cn
* Correspondence: ydy2012@gwmail.gwu.edu (D.Y.); wb09@tsinghua.org.cn (B.W.)

Abstract: The nanostructure-based surface texturing can be used to improve the materials wettability. Regarding oil–water separation, designing a surface with special wettability is as an important approach to improve the separation efficiency. Herein, a ZnO nanostructure was prepared by a two-step process for sol–gel process and crystal growth from the liquid phase to achieve both a superhydrophobicity in oil and a superoleophobic property in water. It is found that the filter material with nanostructures presented an excellent wettability. ZnO-coated stainless-steel metal fiber felt had a static underwater oil contact angle of 151.4° ± 0.8° and an underoil water contact angle of 152.7° ± 0.6°. Furthermore, to achieve water/oil separation, the emulsified impurities in both water-in-oil and oil-in-water emulsion were effectively intercepted. Our filter materials with a small pore (~5 μm diameter) could separate diverse water-in-oil and oil-in-water emulsions with a high efficiency (>98%). Finally, the efficacy of filtering quantity on separation performance was also investigated. Our preliminary results showed that the filtration flux decreased with the collection of emulsified impurities. However, the filtration flux could restore after cleaning and drying, suggesting the recyclable nature of our method. Our nanostructured filter material is a promising candidate for both water-in-oil and oil-in-water separation in industry.

Keywords: nanostructured surface; wettability; metal fiber felt; oil–water separation; emulsion

1. Introduction

Surface wettability is an important material property, and it is determined by surface structure and surface chemical composition [1,2]. Based on existing theoretical methods, a special material wettability can be achieved by modification of surface texture and surface chemical composition [3,4]. During the last decades, there has been increasing interest in the materials with unique functions due to their promising application in various fields such as self-cleaning, antimicrobial, anti-icing, biocompatible materials, or oil transportation [4–9]. Particularly, depending on the different interfacial energy between organic compounds and water in immiscible oil–water mixtures, the materials with special wettability have gained widespread attention in oil–water separation [10,11]. As a simple preparation method, the meshed membrane owned many advantages, such as high permeability of pore space, good oil–water separation efficacy, and promising corrosion resistance [12,13]. Membrane emulsification method attracted plenty of attention. The factors to affect the demulsification performance have been systematically investigated, which included membrane pore size, membrane thickness, transmembrane pressure, as well as emulsion composition [14–17]. In contrast, traditional methods such as gravity-driven separation, biological treatment, adsorption, and electrochemical techniques have many flaws, such as low separation efficiency, low separation capacity, long time cost, and noticeable secondary pollution [18–20].

Furthermore, the structural parameters of filter membranes may play pivotal roles in the performance of oil–water separation, such as pore size, porosity, and membrane thickness [21–23]. To date, some simulation and experimental studies have provided clues for filter membrane design. Some representative examples were introduced here. You et al. utilized electrospinning techniques to prepare two fibrous membranes with different nanopore diameters. They found that the physical cutting effect dominated the coalescence process when oil droplets crossed a smaller pore on the membrane [24]. Zhu et al. found that the oil droplets continuously separated emulsion via using squeezing coalescence demulsification (SCD) within a narrow pore size. Their hydrogel nanofiber membranes had a sustained separation capacity. More importantly, these results demonstrated that the pore size, the membrane wettability, and the interfacial tension coefficient of oil–water could control the SCD's efficiency [25]. Recently, Wei et al. fabricated highly hydrophilic and underwater oleophobic polytetrafluoroethylene membranes, which showed a high oil–water separation efficiency and achieved a high oil–water separation flux. They found that the emulsion separation may be an interfacial problem and may be related to the pore structure of a filtration membrane [26]. Furthermore, Xi et al. manufactured underwater superoleophobic paper-based materials with excellent wet intensity through green papermaking techniques. They also demonstrated that the water flux could be increased by controlling the average membrane pores' size [27]. These preliminary studies could provide an important basis for the design of filter membranes.

Among these structural parameters, the diameter of the liquid distributor and the hole shape dominated the oil–water separation. To achieve a high demulsification efficiency, the filtering precision should be improved accordingly [28,29]. For the highly precise filtering materials at the submicron scale, a nanostructured membrane is an ideal candidate to improve separation efficacy [30]. In addition, surface energy, a key material characteristic, could only be modified by the coating on that material. Due to the limitation of coating technology, these treatment approaches have been used to promote surface wettability without achieving a satisfied performance in terms of coating combinations [31].

In addition, membrane filtration modified by organic polymer could achieve effective separation of oil–water mixtures. Due to its poor stability in the aqueous medium, however, it is hard to use the common polymer membranes with a special wettability in practice [32]. Therefore, the stable materials such as the inorganic membranes are promising alternatives. To date, all previous studies just focused on the fabrication of hierarchical surface structures without giving adequate attention to the substrate, which determines the material strength and stability.

ZnO nanostructured pillars could be used in complex industrial oil–water mixtures [33–36]. In this study, the stable ZnO nano-pillars were prepared on a high-intensity metal fiber filter by crystal growth, which had a good combination with the substrate. Therefore, the modified filter material possessed stability of mineral coating and strength. Our method promoted the surface wettability and achieved both a superhydrophobicity in oil and a superoleophobic property in water. The filter membrane modified by nano-pillars could be stably used in both water-in-oil and oil-in-water conditions. The filtration flux variation with the collection of emulsified impurities was also investigated. This study provided a novel strategy to achieve stability and strength of a mineral coating simultaneously and might have wide application in separating water-in-oil and oil-in-water emulsions.

2. Materials and Methods

2.1. Preparation of ZnO Seed Layers

The stainless-steel fiber felts with the filtration precision of 5 μm were used as a substrate in this study. To eliminate surface contamination, all substrate samples were ultrasonically cleaned with absolute ethanol and ultra-pure water for 10 min. $(CH3COO)_2Zn$ (0.75 M $(CH3COO)_2Zn$) (Sinopharm Chemical Reagent Co., Ltd., Shanghai, China) and Monoethanolamine ($(CH3COO)_2Zn$ and Monoethanolamine in the proportion 1:1) (Aladdin Biochemical Technology Co., Ltd., Shanghai, China) were dissolved in a 2-methoxyethanol solution (Aladdin Biochemical Technology Co., Ltd., Shanghai, China). This solution was magnetically stirred and heated in a temperature-controlled water bath (70 °C, 60 min). The seed solutions were obtained by standing at room temperature for 24 h. Substrates are submerged in the seed solution and lifted repeatedly. Finally, the substrates with the ZnO nanostructured coating were sintered (350 °C, 20 min) to stabilize the coating.

2.2. Preparation of ZnO Nano-Pillar Coated Substrates

As shown in Figure 1, the seed layer coated substrates were first immersed in $Zn(NO_3)_2$ (Sinopharm Chemical Reagent Co., Ltd., Shanghai, China)/hexamethylenetetramine (HMTA) (Sinopharm Chemical Reagent Co., Ltd., Shanghai, China)/DAP (Macklin Inc., Shanghai, China) (0.05 M $Zn(NO_3)_2$, $Zn(NO_3)_2$, and DAP in the proportion 1:6) aqueous solution (10 min) and then was heated in a constant temperature water bath (90 °C, 3 h). After the reaction, the substrates were separated from the solution and washed with ultra-pure water to obtain the substrates with ZnO nanostructures.

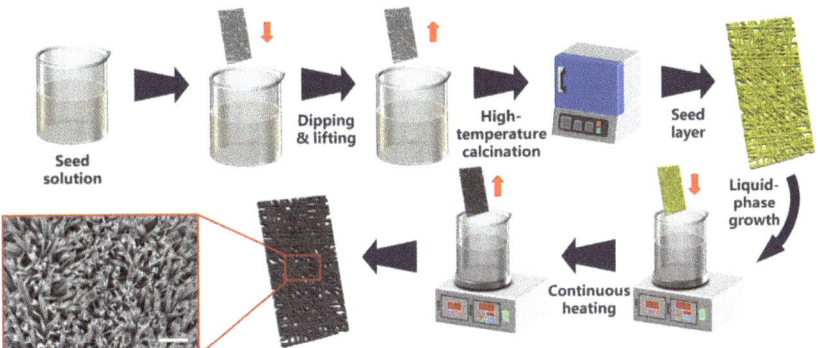

Figure 1. Schematic illustration of manufacturing ZnO modified metal fiber felt.

2.3. Characteristic Analysis

A field emission scanning electron microscope was used to examine the surface morphology of the samples as they were developed (SEM, HITACHI-SU8220, Hitachi, Ltd., Tokyo, Japan). Chemical component was characterized by energy dispersive spectrometry (EDS, Bruker, Billerica, MA, USA). Data-physics was used to characterize surface wettability, including water contact angle (WCA), oil contact angle (OCA), underoil–water contact angle (UWCA), and underwater–oil contact angle (UOCA) (Data-Physics, DataPhysics Instrumente GmbH., Filderstadt, Germany). The liquid droplets utilized for WCA, UWCA, CA and UOCA analyses had a volume of 4 µL. The amount of water remained in the collected filtrates was determined by Karl Fischer Titrator (TP653, TimePower Measure and Control Equipment Co. Ltd., Beijing, China). Total organic carbon (TOC) equipment was used to measure the oil content of the filtrates (TOC-L, Shimadzu (Shanghai) Global Laboratory Consumables Co. Ltd., Shanghai, China). The Zetasizer Nano ZS (Malvern Instruments Ltd., Malvern, UK) was used to investigate the dispersion of water droplets in oil and oil droplets in water. The microscopy pictures of W/O and O/W emulsions were acquired by using an inverted microscope (Keyence Corporation, Osaka, Japan).

2.4. Oil–Water Separation

Here, 99.6 g of infused oils (Diesel, Decane, Dodecane, and N-tetradecane) were combined with 0.4 g of water and ultrasonically agitated for 1 min to create water-in-oil(W/O) emulsions. To prepare oil-in-water (O/W) emulsions, 0.4 g of oil (Diesel, Decane, Dodecane, and N-tetradecane) was combined with 99.6 g of ultra-pure water and ultrasonically agitated for 10 min. The experiment of separation was performed by using a home-made filter device that sandwiched the created membrane between two glass tubes. Before assembling the separation device, the prepared filter membranes were placed in ultrapure water or oil solutions for 30 s, dependent on the types of emulsions. The pre-wetted filter membranes were used in the separation experiments. A glass cylinder tube containing metal fiber felt as a filter was placed at the position shown in Figure 2 to realize filtration of oil–water emulsions under gravity. The filtrate was collected at the bottom of the vessel after the emulsion was poured through the open end of the glass cylinder. The oil–water emulsion was quickly poured into the home-made separation device. The liquid column was sustained at a depth of 10 cm during the separation. The separation efficiency was estimated by the following equation:

$$\eta = 1 - \frac{C_1}{C_0} \tag{1}$$

where C_0 represented the dispersed phase content in the initial W/O and O/W emulsions (ppm); C_1 represented the water/oil content in the collected filtrate (ppm); and η represented the separation efficiency. Meanwhile, the filtration flux (L) was calculated by:

$$L = \frac{m}{\rho \pi r^2 t} \tag{2}$$

where m was the mass of collected filtrate measured with a balance scale; ρ was the density of water/oil ($\rho = 10^3$ kg/m^3); r was the radius of a glass tube (the radius of the glass tube in this study was 8 mm); t was the filtration duration, with the filtration flux computed by measuring the quantity of filtrate 1 min per time.

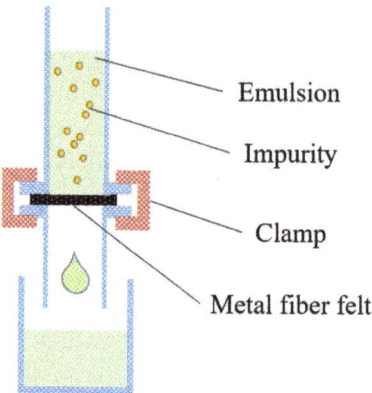

Figure 2. Schematic diagram of experimental setup.

3. Results and Discussion

3.1. Surface Morphology and Chemical Composition of ZnO Coated Metal Fiber Felt

The seed layer was coated on a metal fiber felt substrate using the sol-gel method. We fabricated ZnO-coated metal fiber felts by immersing the metal fiber felts with the seed layer into the precursor solution for the liquid phase growth process (Figure 1). At the initial growth stage, HMTA tardily decomposed to liberate OH$^-$ ions. The generated OH$^-$ ions form insoluble ZnO precipitated at supersaturated Zn(OH)$_4{}^{2-}$ ions or Zn^{2+} ions. As a polar crystal, ZnO (0001) exhibited a positive charge on the crystalline surface and attracted OH$^-$ ions and Zn(OH)$_4{}^{2-}$ ions in the solution. As a result, ZnO had the highest growth rate along the C-axis compared to other crystal planes and eventually forms ZnO nanostructures [37–40]. The overall chemical reactions were presented as follows:

$$C_6H_{12}N_4 + 6H_2O \rightarrow 6HCHO + 4NH_3$$

$$NH_3 + H_2O \rightarrow NH_4^+ + OH^-$$

$$Zn^{2+} + 2OH^- \rightarrow ZnO(s) + H_2O$$

SEM images indicated that the raw stainless-steel fiber felt was composed of metal fibers with an intricate arrangement and the surface of the metal fibers was smooth (Figure 3a,b). The average metal fiber diameter of the fabricated substrates increased, and the average pore size decreased after being coated with ZnO nanostructures (Figure 3c,d). The zoomed-in imaging showed that ZnO had nano pillar-like structures. Furthermore, the ZnO nanopillars in our growth experiments had hexagonal cross section with an average top diameter of 185 ± 81 nm. Finally, ZnO nanostructures wrapped metal fibers' surface without blocking pores.

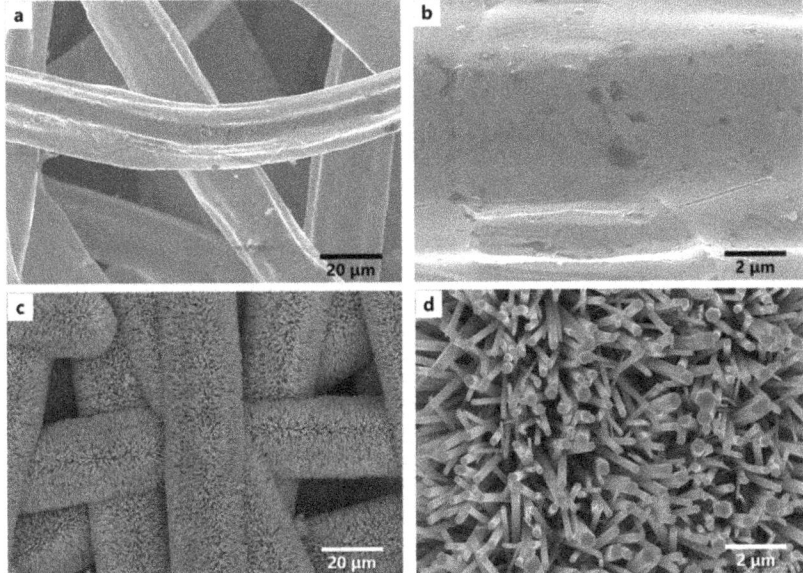

Figure 3. SEM images of original filter membrane (**a,b**). SEM images of ZnO-coated filter membrane (**c,d**).

The liquid-phase ZnO growth had two stages: dissolution and crystallization. The growth solution concentration was one of the critical parameters for ZnO's nucleation and growth. The effect of growth solution concentration on the formation of ZnO coating on filter membrane was shown in Figure 4. We found that the surface was accumulated by dense ZnO structures at low growth solution concentrations (Figure 4a). The nanostructures had a slight pitch and low ratio of length to neck diameter. During crystal growth, the concentration of growth solution determined the solutions' concentration at the center of crystal growth face. As the concentration of growth solution increased, the supersaturation of Zn-ions in the solution was enhanced. It led to apparent orientation growth behavior and eventually formed nanoarray structures (Figure 4b). When the concentration increased to 0.05 mol/L, the filter membrane surface presented arranged pillar-like structures. A large gap between nanostructures facilitated the droplets' immersion into nanostructures. The average diameter of nano-pillars decreased from 314 ± 52 nm to 91 ± 24 nm, and the average length of the nano-pillars just increased slightly. ZnO nanostructure transformed from nano-pillars to nano-needles (Figure 4c). The decrease in the diameter of the nanostructures increased the size of structural interstices, resulting in a coating deficiency on metal fibers and weakened stability of ZnO coatings. Energy-dispersive X-ray spectroscopy (EDS) data were shown in Figure 4d–f. Zn and O element distribution and measurements in coated metal fiber confirmed that the final coating was constituted of ZnO with an excellent coverage density.

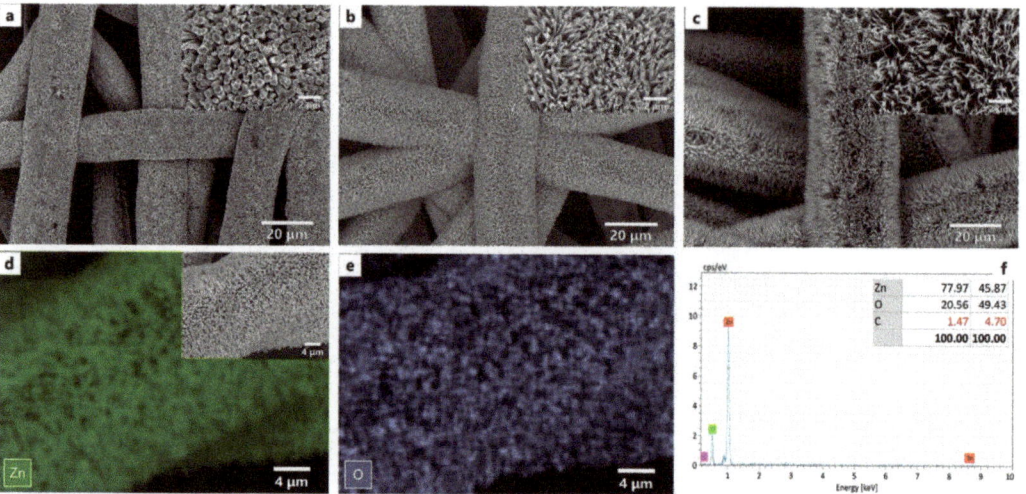

Figure 4. SEM imaging of ZnO-coated filter membranes manufactured with various growth solution concentrations: (**a**) 0.025 mol/L (MF-0.025); (**b**) 0.05 mol/L (MF-0.05); (**c**) 0.1 mol/L (MF-0.1) (The length of the white scalebar in (**a**–**c**) is 20 μm, and the size of scalebar in the zoomed picture is 2 μm); (**d**,**e**) the original EDS mapping of Zn and O elements in a ZnO coated filter membrane; (**f**) EDS elemental analysis.

3.2. Surface Wettability

The contact angle measurements were accustomed to analyzing the surface wettability of as-prepared filter membranes, as illustrated in Figure 5. Additionally, the wettability of the studied surface layers was evaluated using MF-0.05 as a test sample. A droplet of water was dropped onto a ZnO-coated filter membrane in the air. It diffused rapidly within 80 ms, which suggested that the coated filter membrane had satisfactory hydrophilicity (Figure 5a–c). Furthermore, the oil droplet swiftly spread on the as-prepared filter membrane in the air, demonstrating a superoleophilicity (Figure 5e–g). However, the oil droplet could not spread on the as-prepared filter membrane in the water, illustrating that the coated filter membrane had an excellent oil repellency once submerged (Figure 5d). A drop of water was also unable to spread on the filter membrane previously prepared, which demonstrated the filter membrane's outstanding repellence towards water underoil (Figure 5h). ZnO coated filter membrane had a static underwater oil CA of 151.4° ± 0.8° and underoil water CA of 152.7° ± 0.6°, respectively, exhibiting both superhydrophobicity in oil and superoleophobic properties in water.

3.3. Separation of Oil-in-Water Emulsions

In contrast to a stratified oil–water mixture, an oil-in-water emulsion containing emulsified oil droplets with various sizes was difficult to be separated. To prevent emulsified droplet penetration and achieve separation of O/W emulsions, the micropore size was one of the important characteristics of separation membranes. Here, the capacity of ZnO coated filter membrane to separate in the O/W emulsion was demonstrated. To analyze the separation efficacy of an as-prepared filter membrane, emulsions of diesel/water, N-decane/water, dodecane/water, and N-tetradecane/water were invented. Figure 2 illustrated that the apparatus utilized separate O/W emulsions. The cleaned water was gravity-fed through the membrane and collected in the beaker underneath, while the as-prepared emulsions were poured into the top tube. The electronic balance recorded the mass of the collected filtrate in real-time. The separation findings of dodecane/water emulsion were illustrated in Figure 6. As shown in the microscopic imaging, the prepared water/dodecane emulsion contained a large amount of micron-sized oil droplets with an average size distribution of 8 µm. The collected water was transparent after filtration across the as-prepared filter membrane and there were no oil droplets revealed when observed under an inverted microscope.

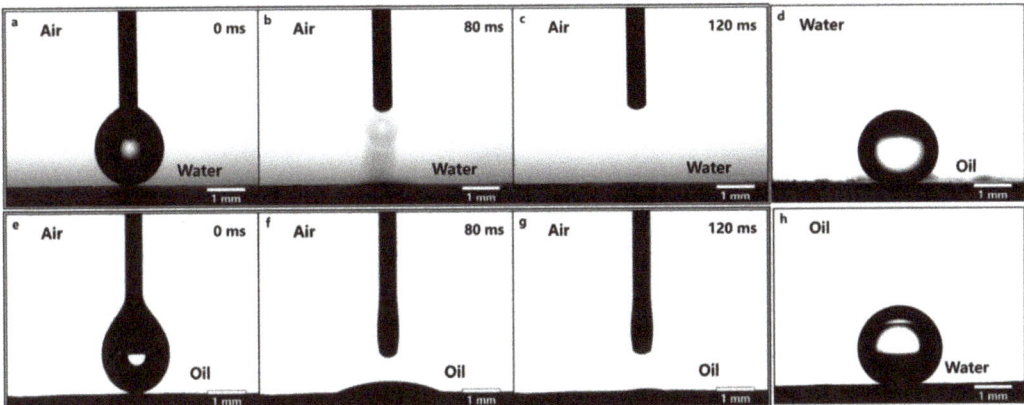

Figure 5. (a–c) Imaging of a water droplet spreading through coated filter membrane; (d) imaging of an underwater oil droplet on a filter membrane with a 151.4° contact angle; (e–g) imaging of oil droplets spreading through coated filter membrane; (h) imaging of an underoil water droplet on a filter membrane with a 152.7° contact angle.

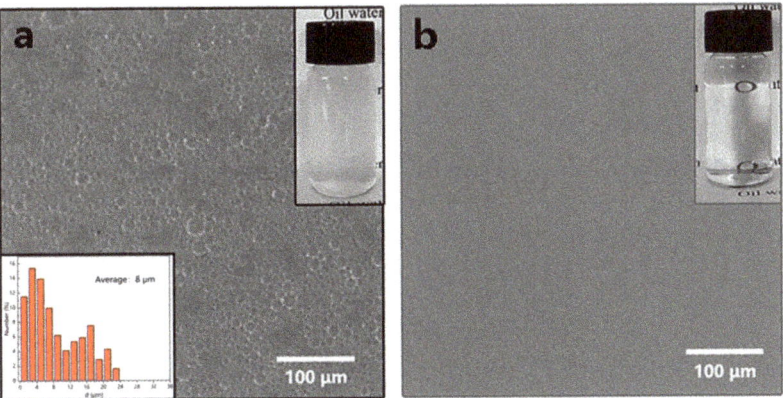

Figure 6. The microscopic and optical images of O/W emulsions (a) before separation and average size of oil droplets; (b) filtered with filter membrane coated with ZnO.

Furthermore, other emulsions such as diesel/water, N-decane/water, and N-tetradecane/water were successfully separated (Figure 7a,b). As expected, the as-prepared filter membrane successfully separated the four emulsions, but only left minimal residual oil in the filtrate. All the oil-in-water emulsions exhibited the promising permeation fluxes. The highest flux of 3139 $L·m^{-2}·h^{-1}$ was obtained for diesel-in-water emulsion. These findings suggested that the ZnO-coated filter membrane could separate a variety of O/W emulsions.

Though the separation efficiency obtained in this study was similar with previous studies using inorganic coating materials (Nano-ZnO and Nano-TiO$_2$) to separate oil-in-water emulsions, we obtained a higher separation flux performance than these studies [41–43]. In addition, the as-prepared filter membrane was evaluated by a ten-times recycling separation of the dodecane/water emulsion to indicate the characteristic of reusability. The membrane was ultrasonic-assisted cleaned with absolute ethyl alcohol for 5 min after each cycle of the experiment (collected filtrate volume was 80 mL), and it was immediately used for the following filtration test. After 10 recycles, the TOC concentration was less than 72 mg/L and the separation efficiency remained higher than 98%, confirming its capability for standing oil–water separation application (Figure 7c). Moreover, our results demonstrated an excellent cyclic stability of the filter membrane coated with ZnO, although the variations of flux in each cycle decreased with the increase in filtration duration (Figure 7d).

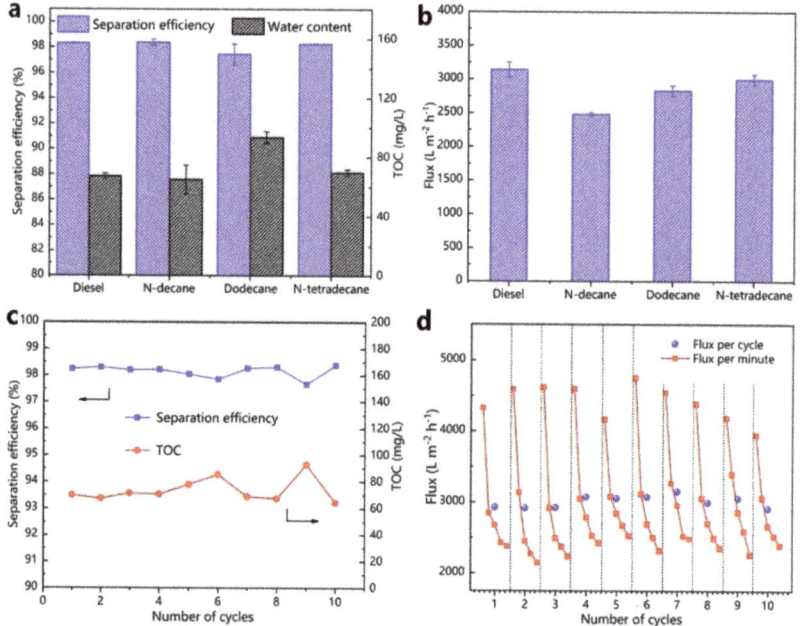

Figure 7. (a) The separation efficiency and TOC in the filtrates of oil-in-water emulsions; (b) various oil-in-water emulsion separation fluxes; (c) the recycling separation of the as-prepared filter membrane for O/W emulsions; (d) separation flux stability in 10-cycle experiments.

3.4. Separation of Water-in-Oil Emulsions

The ZnO-coated filter membrane could separate O/W emulsion and O/W emulsion. The water content in the filtrate, separation flux, and corresponding separation efficiency were used to analyze the separation performance of the as-prepared filter membrane. As demonstrated in Figure 8a, the appearance of the W/O emulsion was cloudy and contained sufficient micron-sized water droplets with an approximate average size of 10 μm. There were no water droplets in the liquid filtered by the ZnO-coated filter membrane (Figure 8b).

Figure 9a demonstrated that the as-prepared filter membrane was capable of accurately separating various W/O emulsions, such as water/diesel, water/N-decane, water/dodecane, and water/N-tetradecane. The efficiency of separation of the as-prepared filter membrane for various water-in-oil emulsions were more than 98%. Furthermore, the water concentration for the filtrates of different emulsions was less than 80 ppm. The following Equation (2) was used to estimate the filtration flux of the as-prepared filter membrane, and fluxes (L) for different emulsions were evaluated for each minute. The separation fluxes for water/diesel, water/N-decane, water/dodecane, and water/N-tetradecane emulsions were about 392, 663, 960, and 744 L·m^{-2}·h^{-1}, respectively (Figure 9b). The separation efficiency obtained in this study was similar to existing studies about inorganic coating materials, which were used to separate water-in-oil emulsions. However, the separation flux obtained here was lower than in all these studies [44,45].

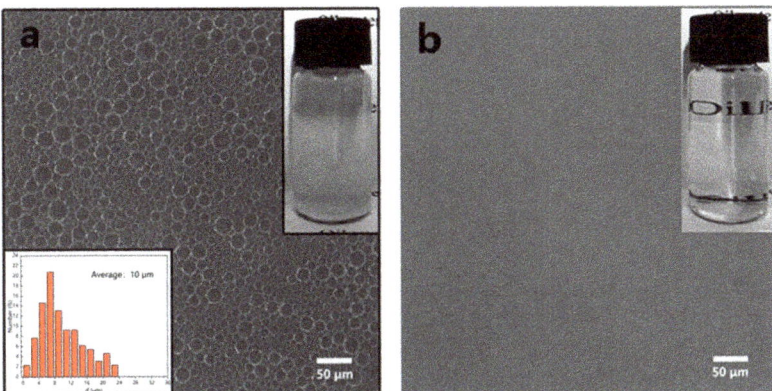

Figure 8. The microscopic and optical images of surfactant-stabilized water-in-oil emulsions (**a**) before separation average size of water droplets; (**b**) filtered with filter membrane coated with ZnO.

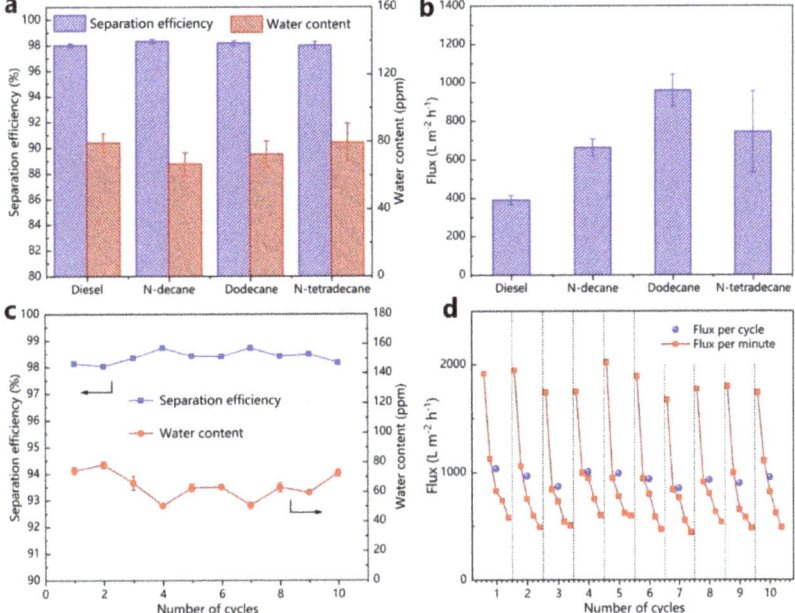

Figure 9. The separation efficiency and water content in (**a**) water-in-oil emulsions filtrates; (**b**) filtration fluxes of various water-in-oil emulsions; (**c,d**) the recycling separation performance of ZnO coated filter membrane for water-in-oil emulsions.

As an example, the reusable separation capability of the as-prepared filter membrane was accessed by the separation experiment of water/N-decane. The separation of each cycle lasted for 8 min, followed by a 5 min ultrasonically cleaned with absolute ethyl alcohol. Figure 9c showed that even after 10 cycles of use, the efficiency of separation of the as-prepared filter membrane was always greater than 98%, and the water content in the filtrates was lower than 80 ppm, suggesting that the as-prepared filter membrane was more recyclable. Furthermore, during the reusability experiment, the filtration flux did not change appreciably, suggesting the filter membrane had excellent separation efficiency and splendid stability (Figure 9d).

3.5. Demulsification Mechanism in Oil-in-Water and Water-in-Oil Emulsion

ZnO nanostructures had achieved both a superhydrophobicity in oil and a superoleophobic property in water. As shown in Figure 10a–c, the separation mechanism could be concluded as follows: it was due to the excellently amphiphilic property in air; the as-prepared filter membrane was extensively wetted by the water phase when it interacted with an oil-in-water emulsion, allowing the water phase to easily infiltrate through. The emulsified oil droplets were then intercepted and adsorbed on the filter membrane in the second step. Finally, oil droplets attracted and coalesced with other emulsified droplets, allowing oil-in-water emulsions to be separated efficiently. Moreover, these water droplets captured by the membrane during the separation process were rapidly coalesced and demulsified to create bigger droplets (Figure 10e–g).

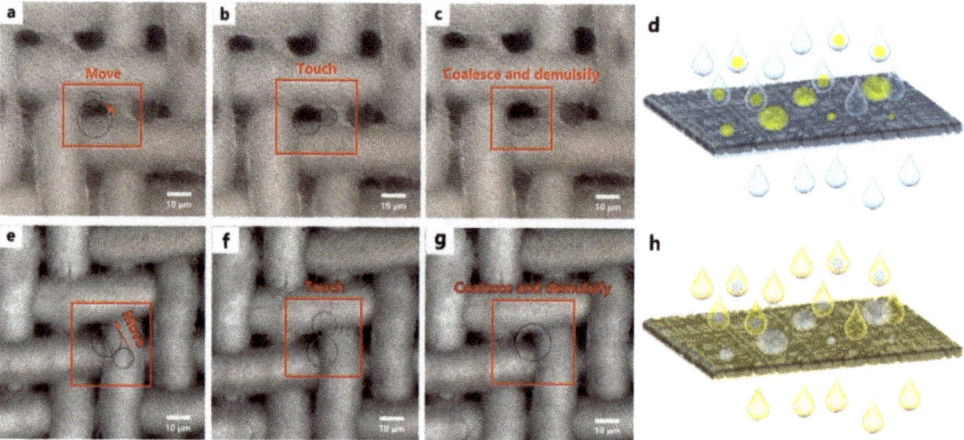

Figure 10. Mechanism of demulsification process of oil-in-water and water-in-oil emulsions; (**a–c**) coalescence and demulsification of emulsified oil droplets; (**e–g**) coalescence and demulsification of emulsified water droplets; (**d,h**) suppositional separation process of oil-in-water and water-in-oil emulsions.

Despite the above results and analysis, the flux change in separation process deserves further study. The oil droplets constantly agglomerated and adsorbed to the fibers during oil-in-water emulsion separation. As illustrated in Figure 11a,b, the oil phase was intercepted to access the filter membrane's surface. The oil–water flux through the membrane was determined by membrane porosity, pore size, as well as pore distribution. Current study found that relatively large pore size leads to a high separation flux. Some parts of porous structures would be blocked by the absorbed oil droplets on the membrane, which resulted in a decreased porosity and a shortened effective separation area of the membrane. In other words, the oil–water separation was saturated with the duration, which significantly decreased the oil–water separation flux. As envisioned, our results suggest that the variations of permeation flux decreased with the increase in filtration duration. The flux decline was in the range of 45–55% (Figure 11c). The same trend was observed during the separation of a water-in-oil emulsion (Figure 11d–f). In addition, the fluxes in separation process varied with the time (Figure 11c,f). As the accumulation of droplets increased on the surface, the pore size decreased and eventually changed the influx during the emulsion separation process. Our observations were consistent with the previous studies with similar focuses. For example, the separation flux of membranes with different pore sizes was reported at a range between 250 and 4300 $L \cdot m^{-2} \cdot h^{-1}$ [15,41]. We need to point out that the data obtained from a single moment may not accurately describe the flux change in the separation process. It will be more practical if we focus on a general description of the entire separation process or the average data shown in Figures 7d and 9d. Finally, it was needed to point out that the filter membrane could continue to adsorb emulsified droplets even after a simple cleaning.

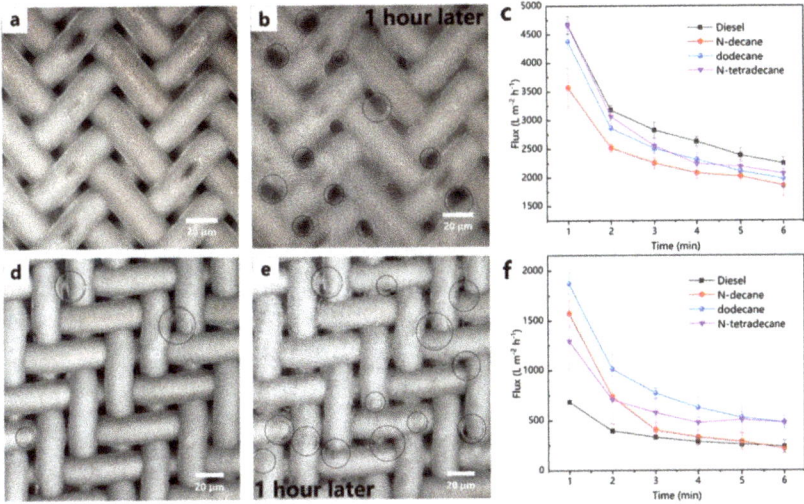

Figure 11. Image of metal fiber felt surface in the separation process of (**a**) oil-in-water emulsions and (**d**) water-in-oil emulsions; (**b,e**) the performance oil–water separation through one hour; (**c,f**) variation of separation fluxes with each minute.

4. Conclusions

Herein, the surfaces with special wettability were achieved by ZnO nanostructures, which could improve oil/water separation efficiency. In this study, ZnO nanostructures were fabricated by a two-step process for the sol-gel process and crystal growth from the liquid phase to achieve both a superhydrophobicity in oil and a superoleophobic property in water. Our study demonstrated an excellent wettability of nanostructured stainless-steel fiber felt. The ZnO coated filter membrane had a static underwater oil contact angle of 151.4° ± 0.8° and underoil water contact angle of 152.7° ± 0.6°, respectively.

In oil–water separation, the emulsified impurities in both water-in-oil and oil-in-water emulsion liquid were effectively intercepted to achieve an oil–water separation. The filter material with small pores (~5 μm diameter) could separate diverse water-in-oil and oil-in-water emulsions with a separation efficiency higher than 98%. Based on the oil–water separation mechanism, we could predict that the blocking of filter materials was caused by the collection of emulsified impurities. Thus, the effect of filtering quantity on the separation efficacy was investigated to analyze the influence of impurities. The investigation results showed that the filtration flux decreased with the collection of emulsified impurities. However, the filtration flux could restore after cleaning and drying, suggesting the recyclable nature of our strategy. The nanostructured filter material is a promising candidate for separating both water-in-oil and oil-in-water emulsion in industry. The proposed filter materials in this study may be repetitively used without any treatments, secondary pollution, and low energy consumption. Therefore, our methods may have wide application in the future.

Author Contributions: Conceptualization, X.L., D.Y. and B.W.; methodology, X.L.; validation, X.L., S.F., C.W. and L.C.; writing—original draft preparation, X.L. and S.F.; writing—review and editing, D.Y., L.C. and B.W.; supervision, D.Y. and B.W. All authors have read and agreed to the published version of the manuscript.

Funding: This research was funded by National Key R&D Program, grant number 2019YFC1907602; National Natural Science Foundation of China, grant number 52076036; Talent introduction and scientific research start-up funded project of Northeast Petroleum University, grant numbers 2019KQ and 1305021891; and Northeast Petroleum University Youth Science Foundation of China, grant number 2019YDQ-01.

Data Availability Statement: The data presented in this study that support the findings are available on reasonable request from the corresponding author.

Conflicts of Interest: The authors declare no conflict of interest.

References

1. Wang, Y.; Wang, Q.; Wang, B.; Tian, Y.; Di, J.; Wang, Z.; Jiang, L.; Yu, J. Modulation of solid surface with desirable under-liquid wettability based on molecular hydrophilic-lipophilic balance. *Chem. Sci.* **2021**, *12*, 6136–6142. [CrossRef] [PubMed]
2. Wong, W.S.Y. Surface chemistry enhancements for the tunable super-liquid repellency of low-surface-tension liquids. *Nano Lett.* **2019**, *19*, 1892–1901. [CrossRef] [PubMed]
3. Chanut, J.; Wang, Y.; Cin, I.D.; Ferret, E.; Gougeon, R.D.; Bellat, J.-P.; Karbowiak, T. Surface properties of cork: Is cork a hydrophobic material? *J. Colloid Interface Sci.* **2022**, *608*, 416–423. [CrossRef] [PubMed]
4. Cao, Q.; Wang, Z.; He, W.; Guan, Y. Fabrication of super hydrophilic surface on alumina ceramic by ultrafast laser microprocessing. *Appl. Surf. Sci.* **2021**, *557*, 149842. [CrossRef]
5. Atta, A.M.; Abdullah, M.M.S.; Al-Lohedan, H.A.; Mohamed, N.H. Novel superhydrophobic sand and polyurethane sponge coated with silica/modified asphaltene nanoparticles for rapid oil spill cleanup. *Nanomaterials* **2019**, *9*, 187. [CrossRef]
6. El-Khatib, E.M.; Ali, N.F.; Nassar, S.H.; El-Shemy, N.S. Functionalization of natural fibers properties by using TiO(2) nanoparticles to improve its antimicrobial activity. *Biointerface Res. Appl. Chem.* **2022**, *12*, 4177–4191.
7. Li, R.; Chen, T.; Pan, X. Metal–organic-framework-based materials for antimicrobial applications. *ACS Nano* **2021**, *15*, 3808–3848. [CrossRef]
8. Zhao, W.; Xiao, L.; He, X.; Cui, Z.; Fang, J.; Zhang, C.; Li, X.; Li, G.; Zhong, L.; Zhang, Y. Moth-eye-inspired texturing surfaces enabled self-cleaning aluminum to achieve photothermal anti-icing. *Opt. Laser Technol.* **2021**, *141*, 107115. [CrossRef]
9. Axinte, D.; Guo, Y.; Liao, Z.; Shih, A.J.; M'Saoubi, R.; Sugita, N. Machining of biocompatible materials—Recent advances. *CIRP Ann.* **2019**, *68*, 629–652. [CrossRef]
10. Chen, H.; Wang, R.; Meng, W.; Chen, F.; Li, T.; Wang, D.; Wei, C.; Lu, H.; Yang, W. Three-dimensional superhydrophobic hollow hemispherical MXene for efficient water-in-oil emulsions separation. *Nanomaterials* **2021**, *11*, 2866. [CrossRef]
11. Mahdi, N.; Kumar, P.; Goswami, A.; Perdicakis, B.; Shankar, K.; Sadrzadeh, M. Robust polymer nanocomposite membranes incorporating discrete TiO_2 nanotubes for water treatment. *Nanomaterials* **2019**, *9*, 1186. [CrossRef]
12. Cheng, J.; Huang, Q.; Huang, Y.; Yu, S.; Xiao, C.; Hu, Q. Pore structure design of NFES PTFE membrane for membrane emulsification. *J. Membr. Sci.* **2020**, *611*, 118365. [CrossRef]
13. Sai, H.; Jin, Z.; Wang, Y.; Fu, R.; Wang, Y.; Ma, L. Facile and green route to fabricate bacterial cellulose membrane with superwettability for oil–water separation. *Adv. Sustain. Syst.* **2020**, *4*, 2000042. [CrossRef]
14. Na, X.; Zhou, W.; Li, T.; Hong, D.; Li, J.; Ma, G. Preparation of double-emulsion-templated microspheres with controllable porous structures by premix membrane emulsification. *Particuology* **2019**, *44*, 22–27. [CrossRef]
15. Chen, C.; Chen, L.; Chen, S.; Yu, Y.; Weng, D.; Mahmood, A.; Wang, G.; Wang, J. Preparation of underwater superoleophobic membranes via TiO_2 electrostatic self-assembly for separation of stratified oil/water mixtures and emulsions. *J. Membr. Sci.* **2020**, *602*, 117976. [CrossRef]
16. Oliveira Neto, G.L.; Oliveira, N.G.; Delgado, J.M.; Nascimento, L.P.; Magalhães, H.L.; Oliveira, P.L.d.; Gomez, R.S.; Farias Neto, S.R.; Lima, A.G. Hydrodynamic and performance evaluation of a porous ceramic membrane module used on the water–oil separation process: An investigation by CFD. *Membranes* **2021**, *11*, 121. [CrossRef]
17. Qiu, L.; Zhang, J.; Guo, Z.; Liu, W. Asymmetric superwetting stainless steel meshes for on-demand and highly effective oil-water emulsion separation. *Sep. Purif. Technol.* **2021**, *273*, 118994. [CrossRef]
18. Li, F.; Bhushan, B.; Pan, Y.; Zhao, X. Bioinspired superoleophobic/superhydrophilic functionalized cotton for efficient separation of immiscible oil-water mixtures and oil-water emulsions. *J. Colloid Interface Sci.* **2019**, *548*, 123–130. [CrossRef]
19. Luo, S.; Dai, X.; Sui, Y.; Li, P.; Zhang, C. Preparation of biomimetic membrane with hierarchical structure and honeycombed through-hole for enhanced oil–water separation performance. *Polymer* **2021**, *218*, 123522. [CrossRef]
20. Lu, J.; Li, F.; Miao, G.; Miao, X.; Ren, G.; Wang, B.; Song, Y.; Li, X.; Zhu, X. Superhydrophilic/superoleophobic shell powder coating as a versatile platform for both oil/water and oil/oil separation. *J. Membr. Sci.* **2021**, *637*, 119624. [CrossRef]

21. Zheng, H.; Lehtinen, M.J.; Liu, G. Hydrophobic modification of sintered glass filters for the separation of organic solvents and gasoline from water as well as emulsified water. *J. Environ. Chem. Eng.* **2021**, *9*, 106449. [CrossRef]
22. Mino, Y.; Hasegawa, A.; Shinto, H.; Matsuyama, H. Lattice-Boltzmann flow simulation of an oil-in-water emulsion through a coalescing filter: Effects of filter structure. *Chem. Eng. Sci.* **2018**, *177*, 210–217. [CrossRef]
23. Woo, S.; Park, H.R.; Park, J.; Yi, J.; Hwang, W. Robust and continuous oil/water separation with superhydrophobic glass microfiber membrane by vertical polymerization under harsh conditions. *Sci. Rep.* **2020**, *10*, 21413. [CrossRef]
24. You, X.; Liao, Y.; Tian, M.; Chew, J.W.; Wang, R. Engineering highly effective nanofibrous membranes to demulsify surfactant-stabilized oil-in-water emulsions. *J. Membr. Sci.* **2020**, *611*, 118398. [CrossRef]
25. Zhu, X.; Zhu, L.; Li, H.; Zhang, C.; Xue, J.; Wang, R.; Qiao, X.; Xue, Q. Enhancing oil-in-water emulsion separation performance of polyvinyl alcohol hydrogel nanofibrous membrane by squeezing coalescence demulsification. *J. Membr. Sci.* **2021**, *630*, 119324. [CrossRef]
26. Wei, W.; Sun, M.; Zhang, L.; Zhao, S.; Wu, J.; Wang, J. Underwater oleophobic PTFE membrane for efficient and reusable emulsion separation and the influence of surface wettability and pore size. *Sep. Purif. Technol.* **2017**, *189*, 32–39. [CrossRef]
27. Xi, J.; Lou, Y.; Jiang, S.; Fang, G.; Wu, W. Robust paper-based materials for efficient oil–water emulsion separation. *Cellulose* **2021**, *28*, 10565–10578. [CrossRef]
28. Rangel-Muñoz, N.; González-Barrios, A.F.; Pradilla, D.; Osma, J.F.; Cruz, J.C. Novel bionanocompounds: Outer membrane protein a and laccase co-immobilized on magnetite nanoparticles for produced water treatment. *Nanomaterials* **2020**, *10*, 2278. [CrossRef]
29. Nguyen, D.C.; Bui, T.T.; Cho, Y.B.; Kim, Y.S. Highly hydrophobic polydimethylsiloxane-coated expanded vermiculite sorbents for selective oil removal from water. *Nanomaterials* **2021**, *11*, 367. [CrossRef] [PubMed]
30. Saththasivam, J.; Wubulikasimu, Y.; Ogunbiyi, O.; Liu, Z. Fast and efficient separation of oil/saltwater emulsions with anti-fouling ZnO microsphere/carbon nanotube membranes. *J. Water Process Eng.* **2019**, *32*, 100901. [CrossRef]
31. Huang, J.; Zhang, Z.; Weng, J.; Yu, D.; Liang, Y.; Xu, X.; Qiao, Z.; Zhang, G.; Yang, H.; Wu, X. Molecular understanding and design of porous polyurethane hydrogels with ultralow-oil-adhesion for oil-water separation. *ACS Appl. Mater. Interfaces* **2020**, *12*, 56530–56540. [CrossRef]
32. Wu, J.X.; Zhang, J.; Kang, Y.L.; Wu, G.; Chen, S.C.; Wang, Y.Z. Reusable and recyclable superhydrophilic electrospun nanofibrous membranes with in situ co-cross-linked polymer–chitin nanowhisker network for robust oil-in-water separation. *ACS Sustain. Chem. Eng.* **2018**, *6*, 1753–1762. [CrossRef]
33. Li, Y.; He, Y.; Zhuang, J.; Shi, H. An intelligent natural fibrous membrane anchored with ZnO for switchable oil/water separation and water purification. *Colloids Surf. A Physicochem. Eng. Asp.* **2022**, *634*, 128041. [CrossRef]
34. Zhou, F.; Wang, Y.; Dai, L.; Xu, F.; Qu, K.; Xu, Z. Anchoring metal organic frameworks on nanofibers via etching-assisted strategy: Toward water-in-oil emulsion separation membranes. *Sep. Purif. Technol.* **2022**, *281*, 119812. [CrossRef]
35. Wang, K.; He, H.; Wei, B.; Zhang, T.C.; Chang, H.; Li, Y.; Tian, X.; Fan, Y.; Liang, Y.; Yuan, S. Multifunctional switchable nanocoated membranes for efficient integrated purification of oil/water emulsions. *ACS Appl. Mater. Interfaces* **2021**, *13*, 54315–54323. [CrossRef]
36. Ji, S.M.; Tiwari, A.P.; Oh, H.J.; Kim, H.Y. ZnO/Ag nanoparticles incorporated multifunctional parallel side by side nanofibers for air filtration with enhanced removing organic contaminants and antibacterial properties. *Colloids Surf. A Physicochem. Eng. Asp.* **2021**, *621*, 126564. [CrossRef]
37. Chen, Z.; Gao, L. A facile route to ZnO nanorod arrays using wet chemical method. *J. Cryst. Growth* **2006**, *293*, 522–527. [CrossRef]
38. Li, P.; Wei, Y.; Liu, H.; Wang, X. A simple low-temperature growth of ZnO nanowhiskers directly from aqueous solution containing Zn(OH)$_4^{2-}$ ions. *Chem. Commun.* **2004**, *24*, 2856–2857. [CrossRef]
39. Liu, B.; Zeng, H.C. Room temperature solution synthesis of monodispersed single-crystalline ZnO nanorods and derived hierarchical nanostructures. *Langmuir* **2004**, *20*, 4196–4204. [CrossRef]
40. You, Q.; Ran, G.; Wang, C.; Zhao, Y.; Song, Q. A novel superhydrophilic–underwater superoleophobic Zn-ZnO electrodeposited copper mesh for efficient oil/water separation. *Sep. Purif. Technol.* **2018**, *193*, 21–28. [CrossRef]
41. Huang, A.; Chen, L.H.; Kan, C.C.; Hsu, T.Y.; Wu, S.E.; Jana, K.K.; Tung, K.L. Fabrication of zinc oxide nanostructure coated membranes for efficient oil/water separation. *J. Membr. Sci.* **2018**, *566*, 249–257. [CrossRef]
42. Yang, P.; Yang, J.; Wu, Z.; Zhang, X.; Liu, Y.; Lu, M. Facile fabrication of superhydrophilic and underwater superoleophobic surfaces on cotton fabrics for effective oil/water separation with excellent anti-contamination ability. *Colloids Surf. A Physicochem. Eng. Asp.* **2021**, *628*, 127290. [CrossRef]
43. Li, C.; Ren, L.; Zhang, C.; Xu, W.; Liu, X. TiO$_2$ Coated polypropylene membrane by atomic layer deposition for oil–water mixture separation. *Adv. Fiber Mater.* **2021**, *3*, 138–146. [CrossRef]
44. Wei, Y.; Xie, Z.; Qi, H. Superhydrophobic-superoleophilic SiC membranes with micro-nano hierarchical structures for high-efficient water-in-oil emulsion separation. *J. Membr. Sci.* **2020**, *601*, 117842. [CrossRef]
45. Fan, X.; Wang, W.; Su, J.; Wang, P. Mechanically robust superhydrophobic mesh for oil/water separation by a seed free hydrothermal method. *Mater. Res. Express* **2018**, *6*, 015026. [CrossRef]

Article

Effect of Cu₂O Substrate on Photoinduced Hydrophilicity of TiO₂ and ZnO Nanocoatings

Maria V. Maevskaya [1], Aida V. Rudakova [1], Alexei V. Emeline [1] and Detlef W. Bahnemann [1,2,*]

[1] Laboratory "Photoactive Nanocomposite Materials", Saint-Petersburg State University, Ulianovskaia str. 1, Peterhof, 198504 Saint-Petersburg, Russia; maevskaya.mv@gmail.com (M.V.M.); Aida.Rudakova@spbu.ru (A.V.R.); alexei.emeline@spbu.ru (A.V.E.)
[2] Institut fuer Technische Chemie, Gottfried Wilhelm Leibniz Universitaet Hannover, Callinstrasse 3, D-30167 Hannover, Germany
* Correspondence: bahnemann@iftc.uni-hannover.de

Abstract: The effect of a Cu_2O substrate on the photoinduced alteration of the hydrophilicity of TiO_2 and ZnO surfaces was studied. It was demonstrated that the formation of heterostructures Cu_2O/TiO_2 and Cu_2O/ZnO strongly changed the direction of the photoinduced alteration of surface hydrophilicity: while both TiO_2 and ZnO demonstrate surface transition to superhydrophilic state under UV irradiation and no significant alteration of the surface hydrophilicity under visible light irradiation, the formation of Cu_2O/TiO_2 and Cu_2O/ZnO heterostructures resulted in photoinduced decay of the surface hydrophilicity caused by both UV and visible light irradiation. All observed photoinduced changes of the surface hydrophilicity were compared and analyzed in terms of photoinduced alteration of the surface free energy and its polar and dispersive components. Alteration of the photoinduced hydrophilic behavior of TiO_2 and ZnO surfaces caused by formation of the corresponding heterostructures with Cu_2O are explained within the mechanism of electron transfer and increasing of the electron concentration on the TiO_2 and ZnO surfaces.

Keywords: photoinduced hydrophilicity; surface energy; heterostructures; charge transfer; work function; adsorbed water

1. Introduction

Surface wettability is an important property of modern functional materials [1,2]. The self-cleaning, anti-fogging and anti-corrosive action of photoactive coatings and photocatalytic materials is based on a light-controlled surface hydrophilicity effect [3]. The photoinduced alteration of the surface wettability possesses a number of advantages compared to other methods based on the application of electric potential, mechanical stress, thermal or chemical action on the surface, etc. Photostimulated alteration of the surface hydrophilicity is easily controllable, environmentally friendly, energetically beneficial, safe and non-destructive.

Since the discovery of the effect of photoinduced superhydrophilicity for the titanium dioxide surface [4], fundamental studies of the mechanism and key factors of the process of the photoinduced alteration of the surface hydrophilicity are still in progress. Nowadays, it is established that electronic photoexcitation of the coating material plays the key role in this surface process, which is confirmed by numerous experimental data obtained by different research groups [4–11]. Particularly, it is assumed that the trapping of photogenerated charge carriers by active surface sites leads to further rearrangement of the hydroxyl-hydrated layer. As a result of such reconstruction, the surface free energy (SFE) changes. Then, the hydrophilicity of the surface changes accordingly: the higher the SFE is, the more hydrophilic the surface becomes.

Recently, we proposed a mechanism of photoinduced hydrophilic conversion, which includes elementary steps associated with the photoactivation and photodeactivation of the active surface sites responsible for the transition of the surface from one hydrophilic state to another [5,6]. It is represented as follows:

$$S + h(e) \rightarrow S^* \quad k_1, \quad (1)$$

$$S^* + e(h) \rightarrow S \quad k_2, \quad (2)$$

$$S^* \rightarrow H \quad k_3, \quad (3)$$

$$H + e(h) \rightarrow S \quad k_4, \quad (4)$$

$$H \rightarrow S \quad k_5, \quad (5)$$

and a solution of the corresponding set of differential kinetic Equations (1)–(5) gives the expression for the rate of the photoinduced hydrophilic conversion (Equation (6)):

$$\Delta H(t) = H(t) - H_0 = \left(\frac{A}{B}S_0 - H_0\right)\left(1 - e^{-(\frac{C}{B})t}\right), \quad (6)$$

where $A = k_1 \cdot k_3 \cdot n_1$, $B = k_2 \cdot n_2 + k_3$ and $C = k_1 \cdot k_3 \cdot n_1 + k_2 \cdot k_4 \cdot n_2{}^2 - k_3 \cdot k_4 \cdot n_2$; k_i is the rate constant of the i-th stage; S_0 is the initial concentration of the surface sites (S) acting as either hole or electron trap; and n_1 and n_2 are the surface concentrations of the photocarriers (electrons and holes), participating in activation (1) and deactivation (2) of the active surface sites.

The direction of the photoinduced alteration of the surface hydrophilicity is determined by the sign of the $\Delta H(t)$ value: for a positive value, the surface becomes more hydrophilic, and, for a negative value, it becomes more hydrophobic. Analyzing the kinetic Equation (6), one can see that the rate and direction of the process on the surface of the same material is dictated by the ratio of the concentration of photocarriers of opposite sign (n_1/n_2) [5,6]. Upon photoexcitation of a solid, the ratio between concentrations of electrons and holes can be changed by two means.

The first method is achieved by varying the spectral composition of the acting light, namely by the photoexcitation of the material in the region of its intrinsic or extrinsic absorption. The change in surface wettability with a change in the spectral composition of the irradiating light has been demonstrated for various coatings [5,7,9,12–15].

The second way to change the n_1/n_2 ratio is to create a layered heterostructured coating, the components of which form a type II heterostructure. In the literature on self-cleaning materials, the formation of composite, or heterostructured, coatings is primarily mentioned as an effective method for improving their self-cleaning properties due to the enhanced photocatalytic oxidative ability [3,16–18]. It is well known that the formation of the type II heterojunctions promotes the charge separation, which reduces electron–hole recombination, thus improving the photocatalytic efficiency of such systems compared to single photocatalysts [3,15,19–21]. At the same time, it has been shown that this approach is also promising and productive both for fundamental studies of photoinduced superhydrophilicity of the surface and for application of self-cleaning coatings with controlled wettability [5,6,22–24].

In this study, the effect of the Cu_2O substrate on the photoinduced hydrophilic behavior of TiO_2 and ZnO thin films was studied and interpreted using the proposed approach (Equations (1)–(6)). For this purpose, the "layer-by-layer" thin films of TiO_2/Cu_2O and ZnO/Cu_2O, with TiO_2 and ZnO layers on the top of the sandwich-like heterostructures, were synthesized. These pairs of photoactive metal oxides were chosen due to suitable relative alignment of their conduction and valence band edges to form type II, staggered heterojunctions: the conduction-band edge and the valence-band edge of cuprous oxide are higher in energy than the corresponding band edges of either titanium dioxide or zinc oxide. As a result, for both TiO_2/Cu_2O and ZnO/Cu_2O heterostructures, electrons are confined in the titanium and zinc oxides, while holes can be accumulated in Cu_2O. Due to such an effective charge separation, the considered composite materials have already demonstrated their high efficiency in both photoelectrochemical and photocatalytic processes [25–37].

In this study, we demonstrated that the formation of TiO_2/Cu_2O and ZnO/Cu_2O heterojunctions drastically changes the surface hydrophilic properties of TiO_2 and ZnO surfaces under UV irradiation. In addition, it was shown that selective photoexcitation of Cu_2O with visible light also affects the surface hydrophilicity of the TiO_2 and ZnO components due to electron transfer in heterostructures.

2. Materials and Methods

The individual ZnO and Cu_2O films were formed by a sol-gel dip coating method (KSV Nima dip coater, Espoo, Finland) on SiO_2-coated glass substrates to prevent the diffusion of sodium ions from the glass during thermal treatments. The velocity of withdrawing from solution was 100 mm/min for all coatings.

For the copper oxide sol, 15 mL of diethanolamine (99%, Fluka, Seelze, Germany), used as a stabilizer, were intensively stirred in 150 mL of isopropanol (99.8%, Ecos-1, Moscow, Russia), and then 14.9 g of copper (II) acetate hydrate $Cu(OAc)_2 \cdot H_2O$ (99.0%, Vekton, Saint-Petersburg, Russia) were added at room temperature. A dark blue solution with a concentration of 0.5 M was kept for 24 h before further procedure [38]. To form Cu_2O films, the obtained Cu(II)-containing layers were annealed at 300 °C on a hot plate in a nitrogen atmosphere for 40 min.

For the zinc oxide sol, 2.5 mL of ethylene glycol (99.5%, LenReactiv, Saint-Petersburg, Russia) and 10 g of zinc acetate dihydrate $Zn(OAc)_2 \cdot 2H_2O$ (99.0%, Vekton, Saint-Petersburg, Russia) were mixed in a round-bottomed flask and heated at 100 °C for 15 min to obtain a uniform transparent mixture. After cooling down to room temperature, 150 mL of isopropanol (99.8%, Ecos-1, Moscow, Russia) and 6.4 mL of triethylamine (99.5%, PanReac AppliChem, Darmstadt, Germany) were added to the mixture to promote the hydrolysis of the zinc acetate. Then, 0.5 mL of glycerin (99.5%, LenReactiv, Saint-Petersburg, Russia) were added dropwise to improve the film quality. The obtained clear and homogeneous solution previously stirred at 60 °C for 1 h was aged for 24 h at room temperature [39]. The formed ZnO layers were annealed at 280 °C in ambient atmosphere for 60 min.

The TiO_2 thin film was formed by the atomic layer deposition (ALD) method ("Nanosurf" installation produced by "Nanoengineering Ltd.", Saint-Petersburg, Russia, RC "Centre for Innovative Technologies of Composite Nanomaterials", Research Park, Saint-Petersburg State University) on SiO_2-coated glass substrate. Titanium tetrachloride (CAS №7550-45-0, quality level MQ200, Merck, Darmstadt, Germany) and deionized water were used as titanium precursor and hydrolysis agent, respectively. The deposition was carried out on the substrate at 200 °C.

The "layer-by-layer" TiO_2/Cu_2O and ZnO/Cu_2O systems were formed by deposition of TiO_2 and ZnO layers, respectively, on a thin Cu_2O film formed as described above.

The surface morphology and film thickness of all synthesized coatings were explored by scanning electron microscopy (Zeiss Supra 40 VP system, Oberkochen, Germany). The smoothness of the film surface was assessed by the AFM method. X-ray diffraction measurements with Bruker "D8 DISCOVER" high-resolution diffractometer (CuKa X-ray radiation, within the angle range of $20° \leq 2\theta \leq 80°$ with a scanning speed 5.0°/min, Germany) were used for the crystal phase determination. Structural reference data were taken from the ICSD database. The transmittance spectra were recorded in the 250–800 nm spectral range at ambient conditions using Lambda 650S spectrophotometer (PerkinElmer, Inc., Shelton, CT, USA). The XPS spectra were recorded using a Thermo Fisher Scientific Escalab 250Xi spectrometer (Thermo Scientific™, Waltham, MA, USA).

Work function measurements were performed with a scanning Kelvin probe system SKP5050 (KP Technology, Wick, Scotland) versus a golden reference probe electrode (probe area 2 mm^2). The probe oscillation frequency was 74 Hz, and the back potential was 7000 mV. Work function values were obtained by averaging 50 data points for four different sites of each sample. Estimated experimental error does not exceed ±0.06 eV.

The contact angle values were measured using optical tensiometer (Bioline Theta Lite, Biolin Scientific, Gothenburg, Sweden). The surface energy was calculated by the Owens–Wendt–Rabel–Kaelble (OWRK)/Fowkes approach using the two-liquid method (water contact angle versus methylene iodide contact angle) [40]. Ultrapure water has initial pH of 5.5. An experimental error of contact angle measurements was determined using 5 data points measured at different spots of the coating and did not exceed 2°.

The work function and contact angle were measured after each step of the surface treatment procedure. After annealing at 200 °C for 30 min, the state of the film surface is designated as "as prepared". After wetting in the ultrapure water with pH of 5.5 and drying at 80 °C, the surface state is denoted as either "after wetting" or "initial state". The third step was irradiation of the coatings: the surface state of the samples irradiated by ultraviolet (UV) or visible light is mentioned as "after UV irradiation" or "after Vis irradiation", respectively.

The irradiation of the films by UV or visible light was carried out using 150 W Xenon lamp (LOMO) equipped with a water filter and UV band pass (250 nm < λ_{pass} < 400 nm) or Vis cutoff color filter (λ_{cut} = 420 nm). The irradiance was 1.19 and 12.3 mW for UV and Vis irradiation, respectively. For all sample surfaces, the kinetics of the photoinduced water contact angle alteration was presented as a dependence of the water contact angle on the irradiation time.

3. Results

3.1. Surface Characterization

The analysis of experimental data obtained by XRD, XPS, SEM and AFM methods (see the Supplementary Materials) confirmed the formation of coatings with desired structures. Indeed, XRD patterns of metal oxide coatings indicate the presence of the corresponding crystal phases of TiO_2 (anatase), ZnO (zincite) and Cu_2O (cuprite) (see Figures S1 and S2).

SEM images demonstrate that coatings were formed by closely packed nanoparticles with average sizes about 50 nm for TiO_2 and about 15 nm for ZnO (see Figures S3 and S4). AFM data show that the roughness of the coating surfaces does not exceed ±5 nm (see Figures S5 and S6), which indicates a nice smoothness of the surface, and, therefore, the surface hydrophilicity of the coatings is not significantly affected by the surface profile. All major characteristics of the prepared nanocoating surfaces are summarized in Table 1.

Table 1. Characterization of the morphology.

Sample	Crystalline Phase	Thickness, nm	Particle Diameter, nm	Smoothness, nm
TiO_2	Anatase	55	50	3
TiO_2/Cu_2O	Anatase/Cuprite	45/85	40	3
ZnO	Zincite	100	15	4
ZnO/Cu_2O	Zincite/Cuprite	120/80	15	5

3.2. Electronic Properties of the Nanocoating Components

XPS spectra recorded in low binding energy region (see Figure 1) were used to determine positions of the valence bands of the components of heterostructured coatings. Corresponding values in vacuum scale are given in Table 2.

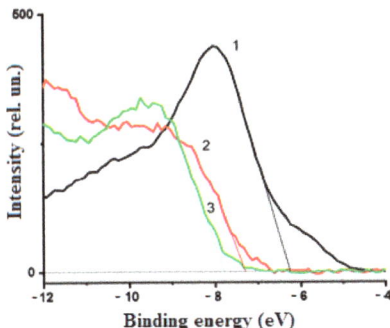

Figure 1. Low binding energy XPS spectra of the components of heterostructured coatings: 1, Cu_2O; 2, TiO_2; and 3, ZnO.

Table 2. Positions of the conduction and valence bands and band gap energies of the coating components with respect to vacuum energy level.

Sample	E_{bg} *, eV	E_{VB}, eV	E_{CB}, eV	E_F, eV
TiO_2	3.2	−7.4	−4.2	−5.24
ZnO	3.3	−7.8	−4.5	−5.14
Cu_2O	2.2	−6.2	−4.0	−5.04

* The E_{bg} values were determined using transmittance spectra of nanocoatings and corresponding Tauc plots (see Figures S7 and S8).

The Kelvin probe method was applied to measure work function characteristics of the nanocoatings to estimate the positions of the corresponding Fermi levels. The results of the measurements are given in Table 2. Based on the obtained parameters, one can sketch the energetic diagrams of the components forming heterostructured nanocoating, particularly, assuming that energy level of the bottom of conduction band (E_{CB}) is a sum of the energies corresponding to the top of the valence band (E_{VB}) and optical band gap energy (E_{bg}) of the corresponding components:

$$E_{CB} = E_{VB} + E_{bg} \tag{7}$$

The corresponding energy diagrams of the semiconductor components forming the heterostructured nanocoatings including Fermi level positions are presented in Figure 2.

According to the energy diagrams shown in Figure 2, the formation of Cu_2O/TiO_2 and Cu_2O/ZnO heterostructures should result in electron transfer from Cu_2O to either TiO_2 or ZnO, both in the dark and under photoexcitation.

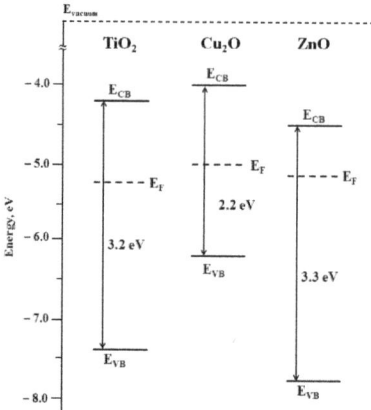

Figure 2. Energy diagrams of the positions of the top of the valence bands and the bottom of the conduction bands and Fermi level positions of the individual components of the heterostructured coatings.

3.3. Effect of Light Irradiation on the Photoinduced Hydrophilicity of Heterostructured Nanocoating Surfaces

Electron transfer from Cu_2O to either TiO_2 or ZnO in the dark caused by formation of the corresponding heterostructures should lead to the establishment of new equilibrium states of the heterostructures characterizing by new Fermi level positions located between the Fermi level positions of the individual components. Experimental data of new Fermi level positions of Cu_2O/TiO_2 and Cu_2O/ZnO heterostructures indicate a successful formation of heterostructures (see Figure 2 and the experimental data in Table 3). The effect of photoinduced alteration of the surface hydrophilicity of nanocoatings was studied for strongly hydrated surface. As noted in the Experimental Section, such state of the surface is characterized by good reproducibility and considered as the "initial" surface state for studies of photostimulated alteration of the surface hydrophilicity.

Table 3. Work function values (WF) after different treatments for all studied coatings.

Coatings	WF, eV as Prepared	WF, eV after Wetting	WF, eV after UV-Irradiation	WF, eV after Vis Irradiation
Cu_2O	5.04	4.94	4.99	5.14
TiO_2	5.24	6.79	5.40	6.92
TiO_2/Cu_2O	5.17	5.64	4.59	4.98
ZnO	5.14	5.49	4.99	5.64
ZnO/Cu_2O	5.09	5.35	5.06	5.36

Particularly, strong surface wetting leads to alteration of the work function values due to the formation of the multi-layered hydroxyl-hydrated structure of adsorbed water on the surfaces. As evident from experimental data, the effect of surface wetting decreases as follows: TiO_2 > TiO_2/Cu_2O > ZnO > ZnO/Cu_2O > Cu_2O. That indicates that the most significant interaction between adsorbed water and metal oxide surface resulting in formation of the surface dipole moment affecting work function takes place on the TiO_2 surface, while the Cu_2O surface is practically unaffected by wetting.

Irradiation of the "initial" state of the coating surfaces with either UV or visible light results in alteration of the surface hydrophilicity (see Figures 3 and 4). To observe the effect of Cu_2O substrate on photoinduced alteration of the surface hydrophilicity of TiO_2 and ZnO, the kinetics of water contact angle alteration measured for heterostructured coating are compared with the corresponding kinetics for pristine TiO_2 and ZnO coatings.

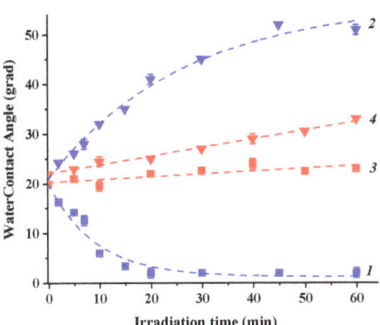

Figure 3. Kinetics of alteration of the water contact angle on the surfaces of TiO$_2$ (1 and 3) and TiO$_2$/Cu$_2$O (2 and 4) coatings under UV (1 and 2) and visible (3 and 4) light irradiation.

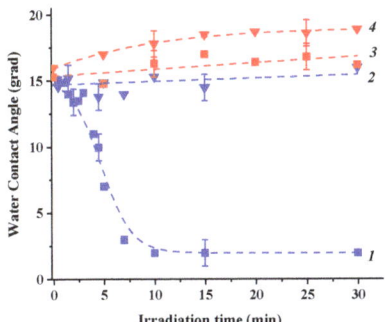

Figure 4. Kinetics of alteration of the water contact angle on the surfaces of ZnO (1 and 3) and ZnO/Cu$_2$O (2 and 4) coatings under UV (1 and 2) and visible (3 and 4) light irradiation.

As evident from the presented dependencies, both TiO$_2$ and ZnO single component coatings demonstrate a surface conversion to superhydrophilic states under UV irradiation and no significant alteration of the surface hydrophilicity under visible light irradiation. However, the presence of the Cu$_2$O substrate in the heterostructured coatings drastically changes the photoinduced hydrophilic behavior of both TiO$_2$ and ZnO surfaces. Indeed, under UV irradiation, both surfaces do not demonstrate any tendency to transform to superhydrophilic state.

Conversely, both surfaces show decrease of the surface hydrophilicity: stronger for TiO$_2$ and weaker for ZnO. At the same time, a pronounced decrease of the surface hydrophilicity is also observed under visible light irradiation for both surfaces when neither TiO$_2$ nor ZnO is photoexcited but Cu$_2$O only (see data on their band gaps in Table 2). In general, a main reason and a driving force for alteration of the surface hydrophilicity is the alteration of the surface energy. Therefore, we estimated the total surface energy and its polar and dispersion components using the "two-liquid" approach. The corresponding values of the total surface free energy and its polar and dispersion components for the "initial" surface states of the heterostructured coatings and ZnO and TiO$_2$ coatings and after UV and visible light irradiation are presented in Table 4. The graphical presentation of the tendencies of SFE alteration caused by irradiation is shown in Figures S9–S11.

Table 4. Total (t), polar (p), and dispersive (d) surface free energies (SFE) for all coatings studied after different treatments.

SFE, mJ/m²	After Wetting			After UV Irradiation			After Vis Irradiation		
	t	p	d	t	p	d	t	p	d
TiO_2	73.4	42.9	31.4	79.7	47.3	32.4	74.1	42.5	31.6
TiO_2/Cu_2O	74.4	41.6	32.8	56.3	30.3	26	67.8	42.9	24.9
ZnO	74.0	46.2	27.8	78.3	44.2	34.1	71.5	45.4	26.1
ZnO/Cu_2O	75.3	44.2	31.1	73.7	47.2	26.5	71.4	39.6	31.8

Thus, as evident from the presented data, the transformation of TiO_2 and ZnO surfaces to superhydrophilic state caused by UV irradiation is accompanied by an increase of the total surface free energy with a major impact from its polar component for TiO_2 and from its dispersive component for ZnO, while visible light irradiation induced only minor alteration of the total surface free energy and its components, which results in very weak changes in the surface hydrophilicity. At the same time, the surfaces of both Cu_2O/TiO_2 and Cu_2O/ZnO heterostructured coatings demonstrate a significant decay of the total surface energy caused by both UV and visible light irradiation, resulting in a decrease of the surface hydrophilicity. Note that the impact of the alteration of polar and dispersive components on changes of the total surface energy and surface hydrophilicity for heterostructured coatings is different for ZnO and TiO_2 surfaces and depends on the spectral region of photoexcitation. Indeed, UV irradiation of the Cu_2O/TiO_2 system results in decay of both polar and dispersive components of TiO_2 total surface energy, while, for Cu_2O/ZnO, the major impact on the surface energy decay originates from the decrease of the dispersive component, which completely compensates the slight increase of the polar component. At the same time, visible light irradiation (resulting in photoexcitation of Cu_2O only) causes mainly the decrease of the dispersive component for Cu_2O/TiO_2 heterostructure and the decay of the polar component for Cu_2O/ZnO system.

In turn, the alteration of the work function values induced by either UV or visible light irradiation (see Table 3 and Figure S12) demonstrates strong decrease of the work function for Cu_2O/TiO_2 coating and practically no significant alteration for Cu_2O/ZnO system.

4. Discussion

Both Cu_2O/TiO_2 and Cu_2O/ZnO systems represent type II heterostructures, which provides a condition for charge separation at the corresponding heterojunctions, particularly, electron transfer from Cu_2O to either TiO_2 or ZnO (see Figures 5 and 6).

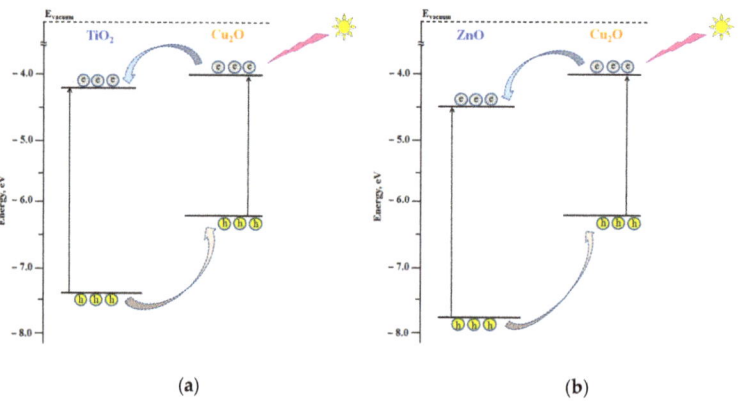

Figure 5. Schemes of the electron transfer in heterostructured coatings under UV light photoexcitation: (a) Cu_2O/TiO_2 and (b) Cu_2O/ZnO.

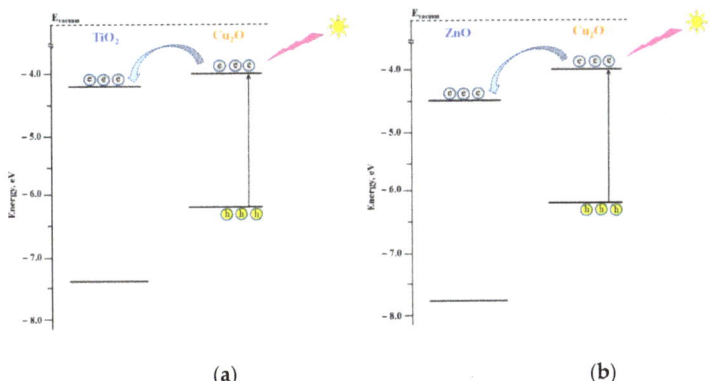

Figure 6. Schemes of the electron transfer in heterostructured coatings under visible light photoexcitation: (**a**) Cu_2O/TiO_2 and (**b**) Cu_2O/ZnO.

Consequently, both UV and visible light irradiation of heterostructures results in the increase of the electron concentration compared to the concentration of holes in both TiO_2 and ZnO. According to our proposed model described in the Introduction (see Equations (1)–(6)), the alteration of the ratio between electrons and holes at the surface, n_1/n_2, can result in a significant alteration of the surface hydrophilicity and turn it, particularly, from a more hydrophilic state to a less hydrophilic state. Therefore, we may conclude that a higher concentration of electrons on the TiO_2 and ZnO surfaces induced by electron transfer from Cu_2O results in the surface transformation to a less hydrophilic state.

Remarkably, the formation of Cu_2O/TiO_2 and Cu_2O/ZnO heterostructures also strongly affects the surface free energy behavior under their photoexcitation. Indeed, UV irradiation of the individual TiO_2 and ZnO surfaces leads to an increase of the SFE and, particularly, its polar component, which is a main reason for the surface transformation to the superhydrophilic state. However, photoexcitation of heterostructures and electron transfer from Cu_2O to either TiO_2 or ZnO results in the decay of the total SFE and its polar component. This observation may indicate that excess of electrons at the coating surfaces can destroy the order of the hydroxyl-hydrated multi-layer structure of adsorbed water and, thus, decrease the total surface energy of the system, resulting in a decreasing of the surface hydrophilicity. This conclusion was confirmed by the effect of visible light irradiation when only the Cu_2O component of the heterostructures was excited, and, therefore, only electron transfer from Cu_2O to TiO_2 and ZnO components could affect both SFE and surface hydrophilicity. Photoinduced destruction of the order in the hydroxyl-hydrated multi-layer structure of adsorbed water is also (indirectly) confirmed by the decrease of the work function of the heterostructures under irradiation, which might be caused by a decrease of the surface dipole moment due to the disorder of the adsorbed water structure in the hydroxyl-hydrated multilayer.

Summing up, we may conclude that the electron transfer realized in Cu_2O/TiO_2 and Cu_2O/ZnO heterostructures under photoexcitation leads to an increase of the concentration of electrons on the TiO_2 and ZnO surfaces that causes a higher disorder in the hydroxyl-hydrated multi-layer structure of the adsorbed water, thus resulting in the decay of the SFE and the surface hydrophilicity.

5. Conclusions

We demonstrated the effect of Cu_2O substrate on the effect of the photoinduced alteration of the surface hydrophilicity of TiO_2 and ZnO surfaces under both UV and visible light irradiation. Particularly, it was shown that under UV light both surfaces can be transformed to the superhydrophilic state, while both UV and visible light irradiation of the heterostructured coating results in a decrease of the surface hydrophilicity. This effect of the Cu_2O substrate on the photoinduced hydrophilic behavior of TiO_2 and ZnO surfaces is explained in terms of charge separation at corresponding heterojunctions and enrichments of TiO_2 and ZnO surfaces with electrons transferred from Cu_2O. Consequently, we may conclude that an increase of the free electron concentration on the surface of both TiO_2 and ZnO leads to the decay of their surface hydrophilicity. Thus, the purposeful formation of a certain type heterostructures may help to control the surface hydrophilicity at the desired level and the direction of its photostimulated alteration.

Supplementary Materials: The following are available online at https://www.mdpi.com/article/10.3390/nano11061526/s1, Figure S1: XRD patterns of Cu_2O, TiO_2 and TiO_2/Cu_2O heterostructured coatings, Figure S2: XRD patterns of Cu_2O, ZnO and ZnO/Cu_2O heterostructured coatings, Figure S3: SEM images of the TiO_2 surface and TiO_2/Cu_2O cross-section of the TiO_2/Cu_2O coating, Figure S4: SEM images of the TiO_2 surface and TiO_2/Cu_2O cross-section of the TiO_2/Cu_2O coating, Figure S5: AFM image of the TiO_2 surface and roughness profile of the TiO_2/Cu_2O coating, Figure S6: AFM image of the ZnO surface and roughness profile of the ZnO/Cu_2O coating, Figure S7: Transmittance spectra of: (a) TiO_2/Cu_2O heterostructured coating and its components; and (b) ZnO/Cu_2O heterostructured coating and its components, Figure S8: Tauc plots for 1 – Cu_2O, 2 – ZnO, and 3 – TiO_2., Figure S9: Alteration of the total SFE caused by irradiation with visible and UV light, Figure S10: Alteration of the SFE polar component caused by irradiation with visible and UV light, Figure S11: Alteration of the SFE dispersive component caused by irradiation with visible and UV light, Figure S12: Alteration of the work function of nanocoatings induced by wetting and UV and visible light irradiation.

Author Contributions: Conceptualization, A.V.R. and D.W.B.; methodology, A.V.E.; software, M.V.M.; validation, A.V.R. and A.V.E.; formal analysis, M.V.M.; investigation, M.V.M. and A.V.E.; resources, D.W.B.; writing—original draft preparation, M.V.M. and A.V.E.; writing—review and editing, A.V.E., A.V.R. and D.W.B.; visualization, M.V.M.; supervision, D.W.B.; project administration, A.V.E.; and funding acquisition, A.V.E. and D.W.B. All authors have read and agreed to the published version of the manuscript.

Funding: The reported study was funded by RFBR, project number 19-32-90111.

Acknowledgments: AVE and MVM are grateful to Russian Foundation for Basic Research (RFBR) for grant N 19-32-90111 to support this research. Experimental studies were performed in laboratory "Photoactive nanocomposite materials" supported by Saint-Petersburg State University (ID: 73032813). We are thankful to the Resource Center (RC) "Nanophotonics", RC "Centre for Innovative Technologies of Composite Nanomaterials", RC "X-ray Diffraction Studies", RC "Centre for Physical Methods of Surface Investigation" and RC "Nanotechnology" of the Research Park at the Saint-Petersburg State University for helpful assistance in the preparation and characterization of the samples.

Conflicts of Interest: The authors declare no conflict of interest.

References

1. Hozumi, A.; Jiang, L.; Lee, H.; Shimomura, M. (Eds.) Stimuli-Responsive Dewetting/Wetting Smart Surfaces and Interfaces. In *Bio-Logically-Inspired Systems (V. 11)*; Springer: Cham, Switzerland, 2018; p. 464. [CrossRef]
2. He, J. Self-Cleaning Coatings: Structure, Fabrication and Application. In *RSC Smart Materials (V. 21)*; RSC: London, UK, 2016; p. 327.
3. Banerjee, S.; Dionysiou, D.D.; Pillai, S.C. Self-cleaning applications of TiO_2 by photo-induced hydrophilicity and photocatalysis. *Appl. Catal. B Environ.* **2015**, *176-177*, 396–428. [CrossRef]
4. Wang, R.; Hashimoto, K.; Fujishima, A.; Chikuni, M.; Kojima, E.; Kitamura, A.; Shimohigoshi, M.; Watanabe, T. Light-induced amphiphilic surfaces. *Nat. Cell Biol.* **1997**, *388*, 431–432. [CrossRef]
5. Rudakova, A.V.; Emeline, A.V. Photoinduced Hydrophilicity of Surfaces of Thin Films. *Colloid J.* **2021**, *83*, 20–48. [CrossRef]
6. Rudakova, A.V.; Emeline, A.V.; Bahnemann, D.W. Effect of the TiO_2–ZnO Heterostructure on the Photoinduced Hydrophilic Conversion of TiO_2 and ZnO Surfaces. *J. Phys. Chem. C* **2019**, *123*, 8884–8891. [CrossRef]
7. Rudakova, A.V.; Oparicheva, U.G.; Grishina, A.E.; Maevskaya, M.; Emeline, A.V.; Bahnemann, D.W. Dependences of ZnO Photoinduced Hydrophilic Conversion on Light Intensity and Wavelengths. *J. Phys. Chem. C* **2015**, *119*, 9824–9828. [CrossRef]
8. Emeline, A.V.; Rudakova, A.; Sakai, M.; Murakami, T.; Fujishima, A. Factors Affecting UV-Induced Superhydrophilic Conversion of a TiO_2 Surface. *J. Phys. Chem. C* **2013**, *117*, 12086–12092. [CrossRef]

9. Yan, X.; Abe, R.; Ohno, T.; Toyofuku, M.; Ohtani, B. Action spectrum analyses of photoinduced superhydrophilicity of titania thin films on glass plates. *Thin Solid Films* **2008**, *516*, 5872–5876. [CrossRef]
10. Fujishima, A.; Zhang, X.; Tryk, D.A. TiO_2 photocatalysis and related surface phenomena. *Surf. Sci. Rep.* **2008**, *63*, 515–582. [CrossRef]
11. Fujishima, A.; Rao, T.N.; Tryk, D.A. Titanium dioxide photocatalysis. *J. Photochem. Photobiol. C Photochem. Rev.* **2000**, *1*, 1–21. [CrossRef]
12. Wang, S.; Feng, X.; Yao, J.; Jiang, L. Controlling Wettability and Photochromism in a Dual-Responsive Tungsten Oxide Film. *Angew. Chem. Int. Ed.* **2006**, *45*, 1264–1267. [CrossRef]
13. Lim, H.S.; Kwak, D.; Lee, D.Y.; Lee, S.G.; Cho, K. UV-Driven Reversible Switching of a Roselike Vanadium Oxide Film between Superhydrophobicity and Superhydrophilicity. *J. Am. Chem. Soc.* **2007**, *129*, 4128–4129. [CrossRef]
14. Rudakova, A.V.; Maevskaya, M.; Emeline, A.; Bahnemann, D.W. Light-Controlled ZrO_2 Surface Hydrophilicity. *Sci. Rep.* **2016**, *6*, srep34285. [CrossRef]
15. Rudakova, A.V.; Emeline, A.V.; Bulanin, K.; Chistyakova, L.; Maevskaya, M.; Bahnemann, D.W. Self-cleaning properties of zirconium dioxide thin films. *J. Photochem. Photobiol. A Chem.* **2018**, *367*, 397–405. [CrossRef]
16. Zhang, L.; Dillert, R.; Bahnemann, D.W.; Vormoor, M. Photo-induced hydrophilicity and self-cleaning: Models and reality. *Energy Environ. Sci.* **2012**, *5*, 7491–7507. [CrossRef]
17. Gaminian, H.; Montazer, M. Enhanced Self-Cleaning Properties on Polyester Fabric under Visible Light Through Single-Step Synthesis of Cuprous Oxide Doped Nano-TiO_2. *Photochem. Photobiol.* **2015**, *91*, 1078–1087. [CrossRef] [PubMed]
18. Upadhaya, D.; Kumar, P.; Purkayastha, D.D. Superhydrophilicity of photocatalytic ZnO/SnO_2 heterostructure for self-cleaning applications. *J. Sol-Gel Sci. Technol.* **2019**, *92*, 575–584. [CrossRef]
19. Serpone, N.; Emeline, A. Semiconductor Photocatalysis—Past, Present, and Future Outlook. *J. Phys. Chem. Lett.* **2012**, *3*, 673–677. [CrossRef]
20. Marschall, R. Semiconductor Composites: Strategies for Enhancing Charge Carrier Separation to Improve Photocatalytic Activity. *Adv. Funct. Mater.* **2014**, *24*, 2421–2440. [CrossRef]
21. Wang, Z.; Wu, W.; Xu, Q.; Li, G.; Liu, S.; Jia, X.; Qin, Y.; Wang, Z.L. Type-II hetero-junction dual shell hollow spheres loaded with spatially separated cocatalyst for enhancing visible light hydrogen evolution. *Nano Energy* **2017**, *38*, 518–525. [CrossRef]
22. Rudakova, A.V. How we can manipulate the photoinduced surface hydrophilicity. Book of Abstract. In Proceedings of the 7th International Conference on Semiconductor Photochemistry, SP7, Milan, Italy, 11–14 September 2019.
23. Jiang, X.; Chen, X. Crystallization behavior and hydrophilic performances of V_2O_5–TiO_2 films prepared by sol–gel dip-coating. *J. Cryst. Growth* **2004**, *270*, 547–552. [CrossRef]
24. Miyauchi, M.; Nakajima, A.; Watanabe, T.; Hashimoto, K. Photoinduced Hydrophilic Conversion of TiO_2/WO_3 Layered Thin Films. *Chem. Mater.* **2002**, *14*, 4714–4720. [CrossRef]
25. Janczarek, M.; Kowalska, E. On the Origin of Enhanced Photocatalytic Activity of Copper-Modified Titania in the Oxidative Reaction Systems. *Catalysts* **2017**, *7*, 317. [CrossRef]
26. Huang, M.H.; Madasu, M. Facet-dependent and interfacial plane-related photocatalytic behaviors of semiconductor nanocrystals and heterostructures. *Nano Today* **2019**, *28*, 100768. [CrossRef]
27. Xiong, L.; Yu, H.; Ba, X.; Zhang, W.; Yu, Y. Cu_2O-based nanocomposites for environmental protection: Relationship between structure and photocatalytic activity, application, and mechanism. In *Nanomaterials for Environmental Protection*; Chapter: 3 (Part I: Remediation with Use of Metals, Metal Oxides, Complexes and Composites); Kharisov, B.I., Kharissova, O.V., Rasika Dias, H.V., Eds.; John Wiley & Sons, Inc.: Hoboken, NJ, USA, 2014; p. 30. [CrossRef]
28. Zhang, Y.-G.; Ma, L.-L.; Li, J.-L.; Yu, Y. In Situ Fenton Reagent Generated from TiO_2/Cu_2O Composite Film: A New Way to Utilize TiO_2under Visible Light Irradiation. *Environ. Sci. Technol.* **2007**, *41*, 6264–6269. [CrossRef]
29. Zhao, M.; Cao, L.; Sun, Y.; Lv, J.; Shang, F.; Mao, S.; Jiang, Y.; Xu, J.; Wang, F.; Zhou, Z.; et al. Microstructure, wettability and electrical properties of n-ZnO/ZnO-SL/p-Cu_2O heterojunction. *Appl. Phys. A* **2015**, *120*, 335–340. [CrossRef]
30. Xue, M.; Wang, W.; Wang, F.; Ou, J.; Li, W. Design and understanding of superhydrophobic ZnO nanorod arrays with controllable water adhesion. *Surf. Coat. Technol.* **2014**, *258*, 200–205. [CrossRef]
31. Sawicka-Chudy, P.; Sibiński, M.; Rybak-Wilusz, E.; Cholewa, M.; Wisz, G.; Yavorskyi, R. Review of the development of copper oxides with titanium dioxide thin-film solar cells. *AIP Adv.* **2020**, *10*, 010701. [CrossRef]
32. Li, G.; Huang, J.; Chen, J.; Deng, Z.; Huang, Q.; Liu, Z.; Guo, W.; Cao, R. Highly Active Photocatalyst of Cu_2O/TiO_2 Octahedron for Hydrogen Generation. *ACS Omega* **2019**, *4*, 3392–3397. [CrossRef] [PubMed]
33. Li, Y.; Wang, B.; Liu, S.; Duan, X.; Hu, Z. Synthesis and characterization of Cu_2O/TiO_2 photocatalysts for H2 evolution from aqueous solution with different scavengers. *Appl. Surf. Sci.* **2015**, *324*, 736–744. [CrossRef]
34. Xiang, L.; Ya, J.; Hu, F.; Li, L.; Liu, Z. Fabrication of Cu_2O/TiO_2 nanotube arrays with enhanced visible-light photoelectrocatalytic activity. *Appl. Phys. A* **2017**, *123*, 160. [CrossRef]
35. Liao, Y.; Deng, P.; Wang, X.; Zhang, D.; Li, F.; Yang, Q.; Zhang, H.; Zhong, Z. A Facile Method for Preparation of Cu_2O-TiO_2 NTA Heterojunction with Visible-Photocatalytic Activity. *Nanoscale Res. Lett.* **2018**, *13*, 221. [CrossRef]
36. Lalitha, K.; Sadanandam, G.; Kumari, V.D.; Subrahmanyam, M.; Sreedhar, B.; Hebalkar, N.Y. Highly Stabilized and Finely Dispersed Cu_2O/TiO_2: A Promising Visible Sensitive Photocatalyst for Continuous Production of Hydrogen from Glycerol:Water Mixtures. *J. Phys. Chem. C* **2010**, *114*, 22181–22189. [CrossRef]

37. Aguirre, M.E.; Zhou, R.; Eugene, A.J.; Guzman, M.I.; Grela, M.A. Cu_2O/TiO_2 heterostructures for CO_2 reduction through a direct Z-scheme: Protecting Cu_2O from photocorrosion. *Appl. Catal. B Environ.* **2017**, *217*, 485–493. [CrossRef]
38. Oral, A.; Menşur, E.; Aslan, M.; Başaran, E. The preparation of copper(II) oxide thin films and the study of their microstructures and optical properties. *Mater. Chem. Phys.* **2004**, *83*, 140–144. [CrossRef]
39. Kaneva, N.V.; Dushkin, C.D.; Bojinova, A.S. ZnO thin films preparation on glass substrates by two different sol-gel meth-ods. *Bulg. Chem. Commun.* **2012**, *44*, 261–267.
40. Owens, D.K.; Wendt, R.C. Estimation of the surface free energy of polymers. *J. Appl. Polym. Sci.* **1969**, *13*, 1741–1747. [CrossRef]

Article

Investigation of the Effects of Rapid Thermal Annealing on the Electron Transport Mechanism in Nitrogen-Doped ZnO Thin Films Grown by RF Magnetron Sputtering

Simeon Simeonov [1], Anna Szekeres [1,*], Dencho Spassov [1], Mihai Anastasescu [2], Ioana Stanculescu [3,4], Madalina Nicolescu [2,*], Elias Aperathitis [5], Mircea Modreanu [6] and Mariuca Gartner [2]

1. Institute of Solid State Physics, Bulgarian Academy of Sciences, 72 Tsarigradsko Chaussee, 1784 Sofia, Bulgaria; simeon@issp.bas.bg (S.S.); d_spassov@abv.bg (D.S.)
2. Institute of Physical Chemistry "Ilie Murgulescu", Romanian Academy, 202 Splaiul Independentei, 060021 Bucharest, Romania; manastasescu@icf.ro (M.A.); mgartner@icf.ro (M.G.)
3. Horia Hulubei National Institute of Research and Development for Physics and Nuclear Engineering, 30 Aleea Reactorului, 077125 Magurele, Romania; istanculescu@nipne.ro
4. Department of Physical Chemistry, Faculty of Chemistry, University of Bucharest, 4-12 Regina Elisabeta Bd., 030018 Bucharest, Romania
5. Microelectronics Research Group, Institute of Electronic Structure and Laser, Foundation for Research and Technology (FORTH-Hellas), P.O. Box 1385, 70013 Heraklion, Crete, Greece; eaper@physics.uoc.gr
6. Tyndall National Institute, University College Cork, Lee Maltings, Dyke Parade, T12 R5CP Cork, Ireland; mircea.modreanu@tyndall.ie
* Correspondence: szekeres@issp.bas.bg (A.S.); mnicolescu@icf.ro (M.N.); Tel.: +40-21-316-79-12 (ext. 588) (M.N.)

Abstract: Nitrogen-doped ZnO (ZnO:N) thin films, deposited on Si(100) substrates by RF magnetron sputtering in a gas mixture of argon, oxygen, and nitrogen at different ratios followed by Rapid Thermal Annealing (RTA) at 400 °C and 550 °C, were studied in the present work. Raman and photoluminescence spectroscopic analyses showed that introduction of N into the ZnO matrix generated defects related to oxygen and zinc vacancies and interstitials. These defects were deep levels which contributed to the electron transport properties of the ZnO:N films, studied by analyzing the current–voltage characteristics of metal–insulator–semiconductor structures with ZnO:N films, measured at 298 and 77 K. At the applied technological conditions of deposition and subsequent RTA at 400 °C n-type ZnO:N films were formed, while RTA at 550 °C transformed the n-ZnO:N films to p-ZnO:N ones. The charge transport in both types of ZnO:N films was carried out via deep levels in the ZnO energy gap. The density of the deep levels was in the order of 10^{19} cm^{-3}. In the temperature range of 77–298 K, the electron transport mechanism in the ZnO:N films was predominantly intertrap tunneling, but thermally activated hopping also took place.

Keywords: RF magnetron sputtering; ZnO:N thin films; Raman spectroscopy; photoluminescence spectroscopy; electrical characteristics; charge carrier transport properties

1. Introduction

Zinc oxide possesses remarkable optical and semiconductor properties, such as a direct wide gap around 3.3 eV at room temperature and a large exciton binding energy of about 60 meV [1]. Because of these properties, ZnO has huge prospects in applications such as optoelectronic devices [2], homojunction LEDs [3,4], solar cells [5], sensors [6,7], and other devices and structures. However, the use of ZnO films in these applications requires the ability to control the majority-carrier type and concentration. An asymmetry exists between n- and p-type doping of ZnO thin films. While it is possible to prepare stable n-type ZnO films even without the introduction of donor dopants during the ZnO film deposition, obtaining p-type ZnO film requires thermal activation of acceptor dopants incorporated into the ZnO film during its deposition [8]. In some cases, an additional problem is the subsequent transformation of the p-type ZnO films into n-type ZnO films [9]. This is why producing stable and qualitative p-type ZnO films is still a challenge facing the technologies of the above-listed devices and structures.

Much effort has been made to obtain p-type ZnO films by applying different advanced deposition methods [10–14] and introducing different elements into ZnO as additives [15]. Among the acceptor dopants, nitrogen is considered the most promising [9]. The nitrogen atom possesses three 2p valence electrons, while the oxygen atom has four 2p valence electrons. The size of the nitrogen atom and the energy of the N 2p valence electrons are closest to the corresponding values of the oxygen atom [16]. Therefore, one might expect the replacement of some O anions in the anion sublattice with N anions and the formation of shallow acceptor levels in the ZnO energy gap. However, to prepare a stable p-type ZnO film by N doping is rather difficult due to the low solubility of nitrogen, and thus the low concentration of holes in the films. Another obstacle to effective p-type doping via the replacement of O by N in the anion sublattice is the simultaneous generation of shallow and deep donor-type defects in the ZnO energy gap, such as oxygen vacancies in the O sublattice, V_O, and zinc interstitials, Zn_i [17–19], which leads to self-compensation of N acceptors in ZnO [20]. Difficulties in the preparation of p-type ZnO films have provoked interest in the studies of ZnO heterojunctions, mainly ZnO–Si heterojunctions. The purpose of these investigations is to expand the applications of ZnO thin films in photoelectron devices and structures, especially in solar cell structures.

Zinc oxide is also the subject of our research. As a pathway for ZnO thin film deposition, we applied a radio frequency (RF) magnetron sputtering technique. This deposition technique combines the relative easiness of both nitrogen doping of the growing layers by the simple introduction of N_2 into the process chamber with control of the deposition parameters (such as gas pressure, substrate temperature, scattering power, and deposition rate). The ability to precisely maintain these parameters during film deposition determines the reproducibility and efficiency of this method. Because of this, RF magnetron sputtering is one of the most widely used methods for thin-film deposition. In order to improve the properties of our ZnO:N films, we applied postdeposition rapid thermal annealing (RTA) in a nitrogen atmosphere at different temperatures. The choice of RTA was due to the fact that it is more technologically relevant than conventional thermal treatments that take much longer and are more highly energy consuming. Moreover, during the RTA process, nitrogen release from the ZnO:N layer is less probable. It has been shown that the RTA method is a particularly suitable method for improving the structural, optical, and electrical properties of ZnO films [21,22].

In order to use nitrogen-doped ZnO films in ZnO homo- and heterojunctions and/or as transparent conductive oxide films, a detailed characterization of their structural, optical, and electrical properties is required. It is particularly important to understand, and hence to tailor, the charge transfer processes in these materials. Accordingly, detailed structural studies including X-ray Diffraction (XRD), Transmission Electron Microscopy (TEM), Atomic Force Microscopy (AFM), and Scanning Electron Microscopy (SEM) were preliminarily conducted on the RF-sputtered N-doped ZnO films [23–25]. We established that RTA treatment in N_2 at moderate temperatures essentially improves the crystalline state of the ZnO films, magnetron sputtered onto crystalline silicon [23], or fused silica [24] substrates, yielding a polycrystalline columnar structure with nanocrystallites of 9–13 nm, preferentially oriented in the (002) direction. In these studies, the ZnO films were deposited in a gas mixture of Ar:O_2:N_2 = 50:40:10 and at a total pressure of 0.66 Pa (5 m Torr). Further, we collected the same measurements as mentioned above on ZnO films deposited under increased amounts of N_2 and a constant amount of Ar. It was established that variation of the Ar:O_2:N_2 mixture to 50:25:25 and 50:10:40 weakly affected the films' crystallinity and surface morphology (the root-mean-square roughness values being lower than 5 nm), but influenced the optical properties [25], and hence alteration in the electrical properties is also expected.

The main purpose of the present research was to investigate the electrical properties of RF magnetron-sputtered nitrogen-doped ZnO films and to elucidate the charge transport mechanism in these films. There are few data in the literature concerning the charge carrier transport through N-doped ZnO films and, peculiarly, the charge transport parameters. For this purpose, we prepared Al–ZnO:N–Si–Al structures and their electrical properties were studied in detail. In addition, Raman and Photoluminescence (PL) measurements were carried out, aiming to establish the origin and nature of the deep levels associated with inherent defects in the polycrystalline ZnO:N films. This could shed more light on electron transitions and charge carrier transport involving N-induced defective centers. The final goal of the research was elucidation of the electron transport properties of zinc-oxide-based films, which would expand their application area in optoelectronics, for example as transparent ZnO:N n-p homojunctions.

2. Materials and Methods

2.1. Film Preparation

The nitrogen-doped ZnO thin films were deposited by RF magnetron sputtering employing Nordiko RFG-2500 equipment (Nordiko Technical Service Ltd., Havant, Hampshire, United Kingdom). Commercially available zinc nitride target (Testbourne Ltd., Basingstoke, Hampshire, UK), Zn:N = 1:1, purity 99.9%, 6 in diameter ×0.25 in thick) was sputtered in a gas mixture of Ar, O_2, and N_2 onto p-type Si(100) substrates with specific resistivity over 100 Ω cm. In general, the substrates never occupied an area bigger than ~12 cm^2 on the holder. The distance between the target and substrates was 11 cm.

The substrates were ultrasonically cleaned in acetone and isopropanol, rinsed in deionized water, and dried in flowing nitrogen gas. The native oxide on the Si surface was etched in HF solution before the substrates were introduced into the sputtering system. After reaching the base pressure of 1.33×10^{-5} Pa (10^{-7} m Torr) and prior to deposition, the target was presputtered for at least 15 min (Ar plasma, 0.66 Pa (5 m Torr), 100 W RF power) to remove any contaminants from the target surface and to enable equilibrium conditions to be reached. The RF power was kept at 100 W. The sputtering was carried out at a total pressure of 0.66 Pa (5 m Torr) in a gas mixture of Ar:O_2:N_2, keeping Ar constant at 50% and varying the ratio of O_2 and N_2 percentages: (i) O_2:N_2 = 40:10, (ii) O_2:N_2 = 10:40, and (iii) O_2:N_2 = 25:25. During deposition, no external heating was applied to the substrates. At the given technological conditions, the deposition rate of the ZnO films was 0.83 nm/min for the films made in the oxygen-rich plasma and 0.32 nm/min for the films made in oxygen-deficient plasma.

After deposition, and following our previous investigations concerning postdeposition annealing conditions as mentioned above [23–25], the ZnO:N samples were subjected to RTA at temperatures of 400 °C and 550 °C for 1 min in N_2 atmosphere. The annealing temperature of 550 °C was not exceeded in order to avoid any possible release of nitrogen from the structure of the films [26].

The denotations of the ZnO:N–Si samples prepared under different technological conditions are presented in Table 1. In the table, we have included the thicknesses of the corresponding films, which were determined with an accuracy of ±0.2 nm from the spectral ellipsometric measurements. The scatter of the as-deposited films' thickness resulted mainly from the low deposition rates and different sputtering durations. It should be pointed out that the thickness uniformity of the films deposited on given Si samples was ~96%, as established by the ellipsometric mapping [25].

Table 1. Technological conditions and corresponding sample number of ZnO:N films with the given layer's thickness.

ZnO:N Sample Number	Gas Mixture of Ar:O$_2$:N$_2$(%)	RTA Temperature (°C)	ZnO:N Film Thickness (nm)
1.1		as-deposited	214.8
1.2	50:40:10	400	183.8
1.3		550	185.7
2.1		as-deposited	206.6
2.2	50:10:40	400	180.7
2.3		550	202.2
3.1		as-deposited	106.4
3.2	50:25:25	400	127.0
3.3		550	124.4

The bandgap energy values of ZnO:N films deposited under the same technological condition as were used in the present work have been determined from the ellipsometric data analysis published elsewhere [25]. It was found that for the as-deposited ZnO:N films, the bandgap energy value was 3.25 ± 0.025 eV, while for the 400 °C and 550 °C RTA samples there was a slight reduction toward 3.20 ± 0.025 eV. These optical bandgap energy values are in good agreement with the literature reports [17,19,27,28].

2.2. Characterization Methods

Raman scattering spectra of the films were recorded on a complex configuration Bruker Vertex 70 FTIR/FT-Raman instrument (Bruker, Ettlingen, Germany). The spectrometer equipped with a RAM II module using a Nd:YAG laser (1064 nm) with variable power (1–500 mW) and LN2 cooled Ge detector. The Fourier-transform Raman spectra were registered between 200 and 600 cm^{-1} with 512 scans and 4 cm^{-1} spectral resolution, and a laser power of 1 mW.

The photoluminescence (PL) measurements were carried out on a Carry Eclipse (Agilent Technologies, Melbourne, Australia), fluorescence spectrometer, with the slits set at 20 nm. The photoluminescence at room temperature was excited by the 325 nm line from a Xe bulb lamp. The PL spectra were taken with a scan rate of 120 nm/min and spectral resolution of 0.5 nm.

The electrical measurements were conducted on metal–ZnO:N–silicon capacitors (further denoted as metal–insulator–semiconductor (MIS) structures) formed by vacuum thermal evaporation of Al dots with 1.96×10^{-3} cm^2 area onto the ZnO top surface through a metal mask, while continuous Al film was evaporated as a contact to the silicon backside.

The electrical properties of the MIS structures were examined from the current–voltage (I–V) characteristics, measured in two ways. In the first way, the I–V characteristics were automatically measured at room temperature in a dark chamber using a Keithley 4200 Semiconductor Characterization System with a ramp rate of 0.1 V/s. In the second way, the I–V curves were measured point by point with a cycle sequence starting from 0 V toward negative or positive voltages, with maximal amplitude applied to the top Al–dot contact followed by a voltage reversal toward zero voltage. The first measurement cycle starting from zero up to maximal applied voltage, V_a, is denoted further as the initial stage. The duration of each measurement cycle was approximately 20–25 min. This kind of I–V measurement was applied to samples 2.1 and 2.3 at temperatures of 298 and 77 K. In this way, the presence and behavior of deep levels in the ZnO:N films could be examined and their concentrations determined.

In addition, the impedance of the MIS structures with ZnO layers 2.1 and 2.3 was measured at room temperature with a Tesla BM-507 impedance meter (TESLA, Praha, Czechoslovakia) applying test voltage frequencies in the range of 0.5–500 kHz. Herein, we consider only the parallel conductance, G_m, values, calculated from the measured $|Z_m|$ and φ_m quantities using the expression $G_m = \cos(\varphi_m)/|Z_m|$.

3. Results and Discussion

The structural and morphological properties of the ZnO:N films (Table 1), as well as their chemical compositions, were reported in our previous papers [23–25]. XRD investigations revealed that all ZnO:N films were crystallized in a hexagonal ZnO wurtzite phase (JCPDS data card 36-1451), (002) oriented, and exhibited an improvement in crystallinity after RTA. There was no noticeable change in the XRD patterns induced by the variation of the nitrogen content in the sputtering reactor atmosphere during ZnO:N film deposition, but the values of the crystallite size, estimated using the Scherrer formula, increased with the RTA temperature and with N_2 content in the sputtering atmosphere, from 14.4 nm (10% N_2) to 15.7 nm (40% N_2) for the ZnO:N films after RTA at 550 °C [25]. From a topographic point of view, all ZnO:N–Si samples (Table 1) showed a homogeneous distribution of small and rounded superficial grains, with an average diameter in the range of 40–70 nm, all surfaces being smooth, with a root-mean-square (RMS) roughness of lessthan 5 nm. From TEM investigations, it was found that the morphology of the ZnO:N films was columnar, with the grain column axis oriented nearly parallel with the hexagonal <001> axis of the ZnO structure. The columnar structures had variable diameters from 10 nm at the bottom to 50 nm at the top (for the films sputtered under 40% nitrogen), so that the surface had larger grains which were well faceted, as seen by AFM. The columns of the as-deposited films contained a lot of defects and pores which diminished after RTA. The presence of the nitrogen inside the ZnO:N films was not detected by XRD analysis due to its very small amount, but was evidenced in Energy-dispersive X-ray spectroscopy (EDX) observations conducted in a ratio of 1/6 (nitrogen/oxygen) for the films deposited on Si, regardless of the RTA treatment. However, the presence of nitrogen was clearly pointed out by X-ray photoelectron spectroscopy (XPS) in the RTA samples only as zinc nitride and diluted zinc oxynitride, as reported previously [23,24]. Nevertheless, the uniformity of thickness deposition, determined by ellipsometric mapping, was found to be around 96% for the ZnO:N films deposited on silicon [25].

3.1. Analysis of FT-Raman Spectra

In Figure 1, as a representative illustration, the Raman spectra of the ZnO:N film deposited in gas mixture of $Ar:O_2:N_2$ = 50:10:40, (sample 2.1) and RT annealed at 550 °C (sample 2.3) are given. The nonsymmetrical shape of the Raman peaks is due to the multiple contributions from the polycrystalline nature of the ZnO:N [25]. The vibrational bands of Si–Si bonds are also indicated, appearing as a strong peak at 523 cm^{-1} and as a weak and broad peak in the 303–308 cm^{-1} spectral range. The kind and position of the observed vibrational modes of ZnO:N films in dependence on the oxygen/nitrogen content and RTA temperature are summarized in Table 2.

Pure, crystalline ZnO has eight sets of optical phonon modes at the Γ point in the Brillouin zone, Γ_{opt}, expressed as $\Gamma_{opt} = A_1 + E_1$ (IR, R) + $2E_2$ (R) + $2B_1$ [29]. The A_1 and E_1 modes are both Raman and infrared active and split into transverse optical (TO) and longitudinal optical (LO) phonon modes. The E_2 modes are Raman active and nonpolar at low-($E_2^{(low)}$) and high-frequency ($E_2^{(high)}$) phonon modes. The $E_2^{(low)}$ mode involves mainly Zn sublattice motion while $E_2^{(high)}$ is associated with the vibration of oxygen atoms [30]. There are also two B_1 modes, $B_1^{(low)}$ and $B_1^{(high)}$, which are inactive Raman modes but can be activated by introducing defects or by doping with other elements [29,31,32].

The recorded FT-Raman spectra exhibited the state of molecular Zn–O bonds in the ZnO:N films under changing technological conditions. The increase of the peaks' intensity after RTA treatment implies the increase of defects related to O and Zn vacancies and interstitials. The bands that appeared at 230–240 cm^{-1} and 330 cm^{-1} correspond to (2TA and $2E_2^{(low)}$) and $E_2^{(high)}$-$E_2^{(low)}$, respectively, and are assigned to the contribution of second-order vibrations, two-phonon modes which become activated in thin films. The peak around 377 cm^{-1} was identified as the A_1(TO) mode, which indicates the displacement of Zn^{2+} and O^{2-} ions parallel to the c-axis, near the center of the Brillouin zone (Γ point) [30,33,34]. In comparison to the pure crystalline ZnO, the position of the wide

$E_2^{(high)}$ peak at 410–420 cm^{-1} was significantly shifted in the doped ZnO:N films, indicating that nitrogen doping mainly affected the oxygen bonds. The peak at 550 cm^{-1} was assigned to the inactive A_1(LO) mode, activates by introducing defects or by doping with other elements [25,31–33]. The observed peak at 590 cm^{-1} was attributed to the E_1(LO) mode that appeared due to oxygen vacancies V_O, zinc interstitials Zn_i, and free carriers [35,36] existing in the analyzed ZnO films. The Raman study proved that the incorporation of nitrogen into the films caused disordering of the ZnO lattice. Considering that the spectra were taken with a resolution of 4 cm^{-1}, the close positions of the corresponding peaks indicated similar structure and chemical bonding. The exceptions were the peaks around 402–412 cm^{-1}, confirming that the change in the oxygen/nitrogen ratio in the Ar:O$_2$:N$_2$ mixture predominantly affected the oxygen bonds in the ZnO lattice.

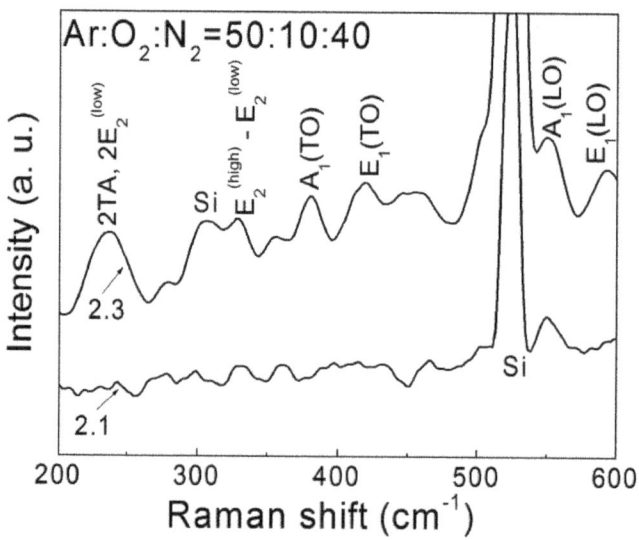

Figure 1. Representative Raman spectra of ZnO:N films, deposited in a gas mixture of Ar:O$_2$:N$_2$ = 50:10:40 (sample 2.1) and after RTA at 550 °C (sample 2.3).

Table 2. Position of the Raman phonon modes in the ZnO:N thin films.

Gas Ratio of Ar:O$_2$:N$_2$ (%)	RTA Temperature (°C)	Raman Molecular Vibration Modes (cm^{-1})					
50:40:10	as-deposited	221	332	373	411	555	589
	400	224	338	377	415	555	590
	550	234	334	378	409	557	595
50:25:25	as-deposited	237	336	376	412	553	594
	400	233	329	372	402	552	590
	550	236	330	376	418	551	593
50:10:40	as-deposited	236	330	376	402	552	593
	400	238	335	375	407	554	591
	550	239	337	376	411	550	595

3.2. Photoluminescence Analysis

In general, in the PL spectra of ZnO films, two emission bands can be observed: one is in the UV region within the 361–369 nm range and the second band is in the visible range of 450–550 nm [32,37]. As the excitation source (Xe bulb lamp) was not intense enough to excite the transitions associated with the optical band gap energy, the first characteristic band was not observed. In Figure 2, typical PL spectra in the spectral range of 400–560 nm are presented for a ZnO:N film deposited in gas mixture of Ar:O$_2$:N$_2$ = 50:10:40 (sample 2.1) and after RTA annealing at 550 °C (sample 2.3).

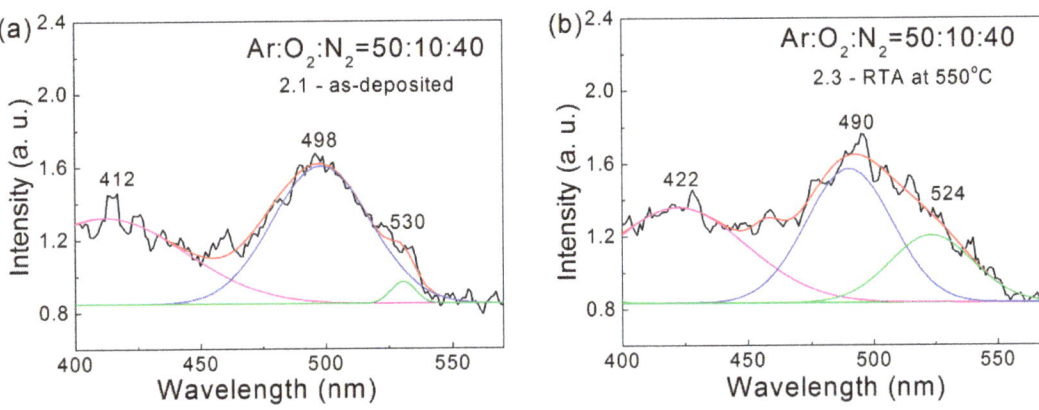

Figure 2. Representative PL spectra of the ZnO:N films deposited in a gas mixture of Ar:O$_2$:N$_2$ = 50:10:40: (**a**) as-deposited (sample 2.1) and (**b**) after RTA at 550 °C (sample 2.3).

The representative PL spectra of the studied ZnO:N films were characterized by a weak band around 455–460 nm and more strong and broad bands around 412 nm and 500 nm. Due to the broad PL peaks, their locations were difficult to resolve accurately. For this reason, we tried to find the Gaussian components until the best fit was achieved. The dominant 500 nm band was deconvoluted into two Gaussian components as the peaks were situated around 490 and 524 nm.

The emission peak from the band-to-band recombinations was expected to appear around 382 nm, the wavelength corresponding to the bandgap energy (3.25 ± 0.025 eV) of our ZnO films. However, in the measured PL spectra, the emission peak was centered within the 410–420 nm region, and hence it originated most probably from defect states close to the ZnO bandgap edges. Previous research suggests that zinc interstitial defects, Zn$_i$, creating shallow states underneath the conduction band edge, are a possible reason for the appearance of this PL peak [38]. The emission peak at ~490 nm (2.53 eV) was a typical blue green emission of ZnO, which may be attributed to a high density of point defects such as zinc vacancy, V$_{Zn}$, oxygen vacancy, V$_O$, and interstitial oxygen, O$_i$, in the polycrystalline structure of ZnO [39]. The green emission peak at ~525 nm (2.36 eV) resulted from radiative recombination of holes with electrons at the singly ionized intrinsic oxygen vacancies [40,41]. These kinds of defects are deep levels, which contribute to the electron transport properties and were considered further at the data analysis of electrical measurements.

The positions of emission peaks obtained by Gaussian deconvolution of the PL spectra of ZnO:N films are summarized in Table 3. It can be observed that for the as-deposited films, the increase of the nitrogen content from 10 to 40% resulted in a shift of the Gaussian peaks' position from 493 nm and 524 nm to 498 nm and 530 nm, respectively. This could be attributed to the increased defect generation and disordering of the ZnO lattice by nitrogen doping, as detected by the Raman measurements above. During heat treatment, these defects were partially annealed and, as a consequence, the Gaussian peaks moved to smaller wavelengths with increasing RTA temperature.

Table 3. The positions of the PL Gaussian emission peaks of the ZnO:N films.

Gas Ratio of Ar:O_2:N_2 (%)	RTA Temperature (°C)	Blue Green Spectral Range (nm)	Green Spectral Range (nm)
50:40:10	as-deposited	493	524
	400	497	522
	550	497	525
50:25:25	as-deposited	495	527
	400	495	523
	550	495	522
50:10:40	as-deposited	498	530
	400	496	525
	550	490	524

The above Raman and PL measurements show that inherent structural defects and N-induced defect centers were present in the ZnO films. These can be associated with localized states in the energy gap of ZnO, which are traps for electric charge carriers. For example, it has been established that the dominant defect centers of oxygen vacancy (PL peak at ~ 492 nm) develop deep energetic levels of 1.4 eV for electrons and 1.6 eV for holes in the ZnO bandgap [42].

3.3. Electric Charge Transport Properties

In order to prove the role of deep levels in the conduction mechanism and to study the current via these deep levels, we conducted a detailed study of the I–V characteristics of the MIS structures formed with the ZnO:N films. The I–V characteristics of the MIS structures with the ZnO:N films, automatically recorded at room temperature, are summarized in Figure 3. A schematic representation of the studied MIS structures is shown as the lower left inset in Figure 3a.

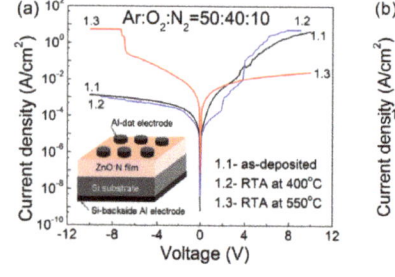
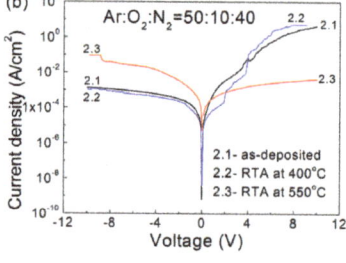
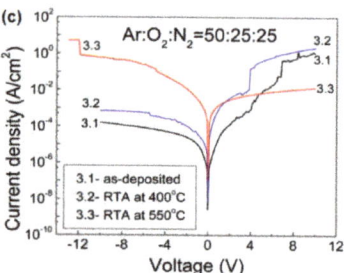

Figure 3. Automatically recorded I–V characteristics of MIS structures with ZnO:N films, deposited at different gas ratios: Ar:O_2:N_2 = 50:40:10 (a), Ar:O_2:N_2 = 50:10:40 (b), and Ar:O_2:N_2 = 50:25:25 (c). The corresponding sample numbers are given in the insets. The lower left inset in (a) shows the schematic structure of the measured samples.

In the case of as-deposited ZnO:N films (samples 1.1, 2.1, and 3.1) and those after RTA treatment at 400 °C (samples 1.2, 2.2 and 3.2), the current at positive voltages applied to the top Al-dot electrode was higher than the current in the case of negative voltage applied to the same contact. These MIS structures had diode-like properties. At absolute voltages of 8–10 V, the rectification ratios were ~4600, 7055, and 2500 for the as-deposited films obtained at 10%, 25%, and 40% N_2 in the gas mixture, respectively. This ratio decreased after 400 °C RTA but still remained in the order of 10^3. The strongest rectification effect was observed in ZnO:N films grown at 25% N_2 in the gas mixture. The existence of asymmetry between the currents at positive and negative voltage values confirms the formation of n-ZnO–p-Si heterojunction. From these observations, it follows that the nitrogen-doped ZnO films were initially n-type and remained n-type after 400 °C annealing. In the case of the films deposited at gas ratios of Ar:O$_2$:N$_2$ = 50:40:10 and Ar:O$_2$:N$_2$ = 50:25:25, the comparison of the measured I–V curves of the as-deposited ZnO:N films with those of the corresponding ones after RTA at 400 °C showed close current values, which suggests similar concentrations of the N dopants.

For the samples after RTA at 550 °C (Figure 3, samples 1.3, 2.3, and 3.3), the current through the ZnO film was higher when negative voltage was applied to the top Al-dot electrode. The asymmetry between the currents at positive and negative voltage values was less pronounced, which was expressed in considerably smaller rectification ratios. At absolute voltages of 8–10 V, their values were 230, 534, and 24 for the films deposited at 10%, 25%, and 40% N$_2$ in the gas mixture, respectively. These results imply that RTA at 550 °C transformed the nitrogen-doped ZnO films to p-type ZnO ones. In accordance with the Anderson heterojunction rule, the ZnO valence band offset at the p-ZnO–p-Si was close to 2.6 eV [17]. Because of this valence band offset, the contribution of holes in the p-Si to the forward current in the p-ZnO–p-Si MIS structure was negligible. Therefore, the forward current in these films was accompanied by electron-hole generation at the p-ZnO–p-Si interface.

It is known that acceptor dopants introduced into ZnO films are self-compensated by donor localized states generated in the ZnO lattice during the dopants' incorporation in ZnO [20,43]. From the results in Figure 3, we can conclude that during RTA at 550 °C, the concentration of donor-like defects decreased in the ZnO:N films in such a way that the concentration of nitrogen-acceptor states exceeded that of the remaining donor-like defects. As a result, the n-type as-deposited ZnO:N film was transformed to a p-type one during 550 °C RTA. The I–V characteristics of the same MIS structures, measured after a year, showed only minor changes and still exhibited p-type conductivity in these ZnO:N films. Therefore, the transformation of n-type ZnO film into p-type by 550 °C RTA treatment is a stable process.

The transformation of ZnO:N film from n-type to p-type by 550 °C RTA is similar to the appearance of p-type regions in N-implanted ZnO single crystals after annealing at 500 °C, observed by deep-level transient spectroscopy (DLTS) measurements [44]. Similar transformations of N-implanted n-type ZnO films to p-type ones have been observed after RTA treatment at 900 °C [45].

The current densities in the ZnO:N films deposited with nitrogen contents of 10 and 25% in the vacuum chamber were similar (Figure 3). Deposition of ZnO:N films at the highest nitrogen content of 40% resulted in lower current density, and it remained lower in comparison to the other ZnO:N films even after RTA treatment at 550 °C. This result is a consequence of already mentioned self-compensation of acceptor dopants by the simultaneously created donor-like deep levels during the acceptor doping of ZnO.

In ZnO films doped with nitrogen, the n-type conduction can be attributed to donors formed by interstitial zinc, Zn_i, and oxygen vacancies, V_O, whereas the p-type conduction can be related to the acceptors formed by zinc vacancies, V_{Zn}, oxygen interstitial, O_i, and nitrogen substituting oxygen in the O sublattice [18,46,47]. We have confirmed the presence of these kinds of defect states in ZnO:N films by the analysis of IRSE and FTIR spectra [25] and also by the Raman and PL spectroscopic results presented and discussed above in Sections 3.1 and 3.2.

The specific resistivity, ρ(V), values of the ZnO films were calculated from the I–V characteristics shown in Figure 3, in the forward direction with accumulation conditions at the ZnO–p-Si interface, using the relations $R_{dif} = dV/dI$ and $ρ(V) = R_{dif}S/d_f$, where R_{dif} is the differential resistance, d_f is the film thickness, and S is the Al-dot contact area. The obtained ρ(V) values for the studied ZnO:N films are summarized in Figure 4.

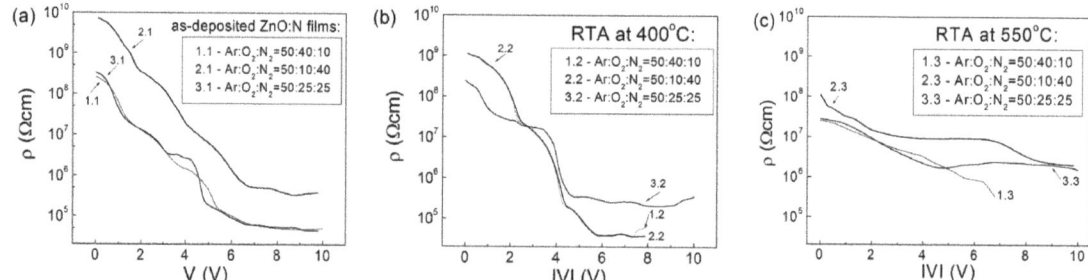

Figure 4. Specific resistivity ρ(V) versus applied forward voltage of ZnO:N films (deposited at different gas ratios (**a**) and after RTA at 400 °C (**b**) and 550 °C (**c**). The ρ values were calculated from the forward I–V curves shown in Figure 3.

For the as-deposited and 400 °C RTA n-ZnO:N films (Figure 4a,b), the specific resistivity decreased sharply by more than four orders of magnitude upon increasing the applied voltage from 1 to 8 V. After 550 °C RTA (Figure 4c), the ρ(V) values of the p-ZnO:N films decreased more gradually, as the reduction was approximately two orders of magnitude in the same voltage interval of 1–8 V, but as a whole, they remained above the range of 10^6 Ω cm. In comparison with the as-deposited ZnO:N films, the smaller change of the specific resistivity in the 550 °C RTA-treated films as a function of the applied voltage is connected with the smaller increase of the current density in the 550 °C RTA films.

In order to elucidate the role of deep levels in the ZnO:N energy gap, we further measured manually, pointbypoint, the I–V characteristics of the MIS structures with ZnO:N films. For these experiments, we chose the sample series 2.1–2.3 with maximal concentration of nitrogen in the ZnO film, expecting a larger effect of N doping on the electrical characteristics.

The room-temperature I–V characteristics of the MIS structure with ZnO:N films deposited at Ar:O_2:N_2 = 50:10:40 (sample 2.1) (a) and after RTA at 550 °C (sample 2.3) (b) are presented in Figure 5. To reveal the participation of deep levels in the ZnO energy gap, two-stage I–V measurements were performed. For the as-deposited n-type ZnO films of sample 2.1 (Figure 5a), the initial stage was measured in 20–25min increments starting from zero up to the maximal applied voltage, V_a, and the second, return stage was carried out immediately after the initial one, down from V_a to zero also in 20–25min increments. The corresponding I–V curves in the reverse direction are given as an inset in Figure 5a. The considerable asymmetry between the currents at positive and negative voltage values confirmed the formation of the n-ZnO:N–p-Si heterojunction. The forward current in the return stage was higher than the forward current in the initial stage, resulting in a counterclockwise hysteresis in the measured I–V curves. Some of the electrons injected into the ZnO film during the initial stage were captured at deep levels with time constants higher than the time needed for measuring the initial stage, and their release during the return stage led to the observed excess, higher forward current in the return stage. Thus, the hysteresis in the I–V characteristics indicates that deep levels in the ZnO energy gap take part in the charge transport mechanism through the n-ZnO:N films.

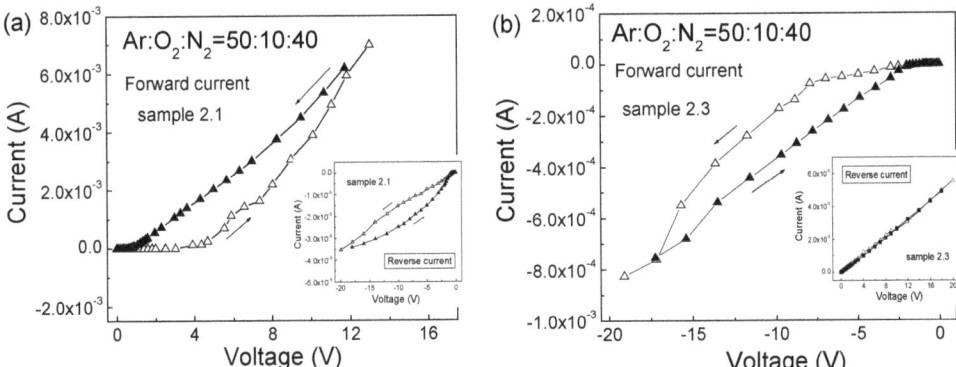

Figure 5. Room temperature I–V characteristics of MIS structures with ZnO:N films, deposited at Ar:O$_2$:N$_2$ = 50:10:40 (sample 2.1) (**a**) and after RTA at 550 °C (sample 2.3) (**b**). The initial and return stages of the I–V measurements are denoted with empty and full triangles, respectively. The corresponding I–V curves in the reverse direction are given as insets.

The forward I–V characteristics of the MIS structures with ZnO:N films annealed at 550 °C (sample 2.3) are shown in Figure 5b, with the corresponding I–V curves in reverse direction are given as inset. As expected, the forward current was with negative voltages applied to the Al-dot contact on the ZnO:N surface. As shown in Figure 5b, the forward current in the return stage was also higher than the forward current in the initial stage, leading to the appearance of counterclockwise hysteresis in the forward I–V characteristics. This counterclockwise hysteresis shown in Figure 5b, as the one in Figure 5a, also confirms that the deep levels in the ZnO energy gap take part in the charge transport mechanism in the p-ZnO:N films.

The room-temperature dependences of ln(J) versus ln|V| in the forward direction and in the initial stage are presented in Figure 6. For the MIS structures with as-deposited ZnO:N films, the slope of the plots is about 8 in the dominant parts of the plots and, around 2 for the other parts. This means that the charge transport in these ZnO:N films is carried out via deep levels in the ZnO energy gap. This transport mechanism is described as trap charge limited current (TCLC) (see for example [48]) or trap-assisted space charge limited current (see for example [49]). For the MIS structures with 550 °CRT-treated ZnO:N films, the slope of the lnJ versus ln|V| plot (Figure 6) is around 2.5 in the range of 2–18 V. Since this slope is also higher than 2, this means that the charge transport in the p-type ZnO:N is also carried out via deep levels in the ZnO energy gap.

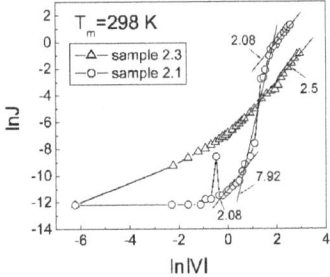

Figure 6. Logarithm of room-temperature forward current density as a function of logarithm absolute voltage for the MIS structures with ZnO:N films, deposited at Ar:O$_2$:N$_2$ = 50:10:40 (sample 2.1) and after RTA at 550 °C (sample 2.3).

The AC conductance, G_m, of these ZnO:N films, measured in the 0.5–500 kHz test frequency range, is plotted in Figure 7 as $\ln G_m$ against $\ln \omega$, where ω is the angular frequency. The slope of the plot for the as-deposited n-ZnO:N film (sample 2.1) is equal to 0.66. This value is close to the most widespread value 0.7 of the power exponent in the Jonscher universal power law for AC conductance in the case of hopping or tunneling of charge carriers via deep levels near to the Fermi level in the semiconductor energy gap [50]. For the annealed p-ZnO:N films (sample 2.3), the slope of this plot is equal to 0.24.

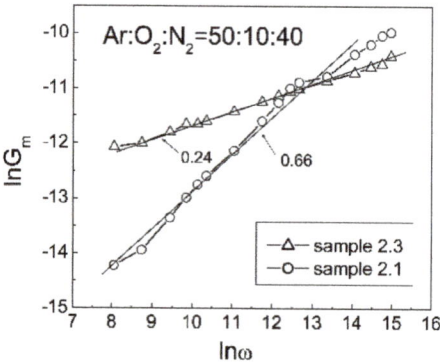

Figure 7. Logarithm of AC conductance, G_m, versus logarithm of angular frequency, ω, of MIS structures with ZnO:N films deposited at Ar:O$_2$:N$_2$ = 50:10:40 (sample 2.1) and after RTA at 550 °C (sample 2.3).

In order to reveal the character of the charge transport mechanism through the ZnO:N film, current–voltage measurements must be done at different temperatures. In Figure 8a, the I–V characteristics of the MIS structure with as-deposited n-ZnO:N, measured at 298 and 77 K, are given as the logarithm of the forward current density versus applied voltage. The averaged current density measured at 298 K under high applied voltages was 2.53 times higher than the current density measured at 77 K. Using the expression $\ln(J_{298}/J_{77}) = (q\varphi_a/k)(1/77 - 1/298)$, the effective activation energy, $q\varphi_a$, of these current densities was estimated. For the as-deposited n-ZnO:N film, it was equal to $q\varphi_a = 8.3$ meV. This small $q\varphi_a$ value indicates that the current in this film was carried out predominantly by electron tunneling from the occupied deep level to the nearest unoccupied one in the ZnO energy gap. The current at 77 K was carried out by intertrap tunneling mechanism [43]. The measured excess current at 298 K is a consequence of the appearance of thermally activated carrier hopping in these ZnO:N films at temperatures higher than 77 K [51].

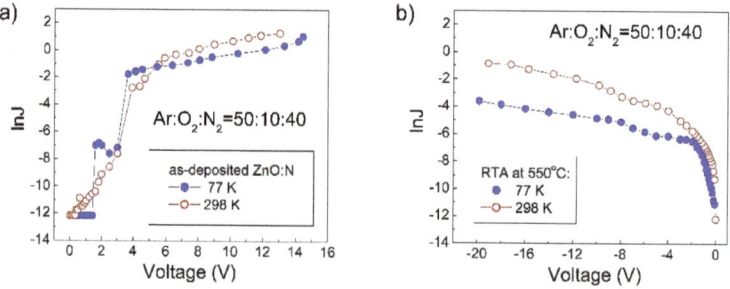

Figure 8. lnJ versus V characteristics, measured at 298 and 77 K, of MIS structures with ZnO:N films deposited at Ar:O$_2$:N$_2$ = 50:10:40 (sample 2.1) (**a**) and after RTA at 550 °C (sample 2.3) (**b**).

In the case of intertrap tunneling, the current density J is given by

$$J = J_0 \sinh[B(V - V_{fb})] \qquad (1)$$

where $J_0 = 2qv\{\exp[-2(2m^*q)^{1/2}\varphi_t^{1/2}w/\hbar]\}/w^2$ and $B = (2m^*q)^{1/2}w^2/\hbar\varphi_t^{1/2}d_f$ [52]; w is the mean distance between deep levels situated around the quasi-Fermi level; ν is the electron attempt to escape frequency from deep levels; and $q\varphi_t$ is the energy position of deep levels in the ZnO energy gap. The electron effective mass, m_n^* for n-type ZnO is taken as equal to 0.23 m_e [53]. The electron attempts to escape frequency, $\nu = 8.54 \times 10^{12}$ s^{-1}, is estimated from the relation $h\nu = kT_D$ [54], where the Debye temperature, T_D, for wurtzite ZnO is 410 K [55], and all other symbols have their common meaning. From Equation (1) it follows that the values of w and $q\varphi_t$ can be determined from the slope of the plot lnJ versus V and the intersection of the extension of this plot toward the lnJ axis at V = 0. These values of w and $q\varphi_t$ determined from the lnJ-V plot in Figure 8a at 77 K were 4.71×10^{-7} cm and 1.48 eV, respectively. The density of deep levels N_t in the ZnO energy gap was estimated by the expression $N_t = 1/w^3$ and it was equal to $N_t = 9.62 \times 10^{18}$ cm^{-3}. Because of the energy position of deep levels is $q\varphi_t = 1.48$ eV, the position of electron quasi-Fermi level in these n-ZnO MIS structures was 0.12 eV above the middle of the ZnO energy gap.

When the concentration of deep levels is known, one may calculate the mobility μ_t for this intertrap tunneling charge transport by the relation $\mu_t = J/(qN_tE)$, where J is the current density at the applied voltage V and E is the electric field across the ZnO film. In the accumulation conditions at the ZnO/Si interface, E is given by V/d_f, where d_f is the ZnO film thickness. The calculated low value of $\mu_t = 1.04 \times 10^{-6}$ cm^{2}V^{-1}s^{-1} is characteristic for charge transport via deep levels in the semiconductor band gap.

The excess of the current at 298 K over that at 77 K (Figure 8a) is evidence for thermally activated carrier hopping in these n-ZnO MIS structures. N.F. Mott [51] has proposed that in the case of variable range hopping (VRH), the temperature-dependent conductivity can be expressed by $\sigma(T) = \sigma_0 \exp\{-(T_0/T)^{1/4}\}$, where σ is the specific conductivity and T_0 is the characteristic temperature. This dependence is appropriate for the Mott VRH in ZnO [56–58]. One may calculate the value of characteristic temperature T_0 by using the expression $\ln(J_{298}/J_{77}) = T_0^{1/4}[1/(77)^{1/4} - 1/(298)^{1/4}]$, where the current density is averaged over high applied voltages. For the as-deposited n-ZnO MIS structures (Figure 8a) this value was $T_0 = 8.45 \times 10^3$ K.

The localization length, Bohr radius, $a_{B,ZnO}$, of the localized states in the ZnO energy gap is given by the expression $a_{B,ZnO} = (4\pi\varepsilon_{ZnO}\hbar^2)/(m^*q^2)$, where ε_{ZnO} is the dielectric constant of ZnO, m^* is the effective electron mass in ZnO, and other symbols have their usual meanings. In the corresponding 1 MHz C-V curve of sample 2.1 (not shown in this paper), the measured maximal capacitance value in the accumulation regime was equal to $C_{max} = 98$ pF. As C_{max} is expressed by $C_{max} = \varepsilon\varepsilon_0 S/d_f$, the dielectric constant ε_{ZnO} of the ZnO:N film was determined as $\varepsilon_{ZnO} = \varepsilon\varepsilon_0 = 1.03 \times 10^{-12}$ F/cm. Knowing ε_{ZnO} and $m_n^* = 0.23$ m_e [53] quantities, the calculated value of $a_{B,ZnO}$ was equal to $a_{B,ZnO} = 2.674 \times 10^{-7}$ cm.

The density of localized states, $N(\varepsilon)$, in the case of hopping mechanism is given by the expression $N(\varepsilon) = 16\alpha^3/kT_0$, where the decay constant α is the inverse of the Bohr radius ($\alpha = 1/a_{B,ZnO}$). The numerical factor 16 is taken from Ambegaokar et al. [59]. With $\alpha = 3.74 \times 10^6$ cm^{-1} and $T_0 = 8.449 \times 10^3$ K the calculated value of $N(\varepsilon)$ was equal to 1.15×10^{21} cm^{-3}.eV^{-1}. When $N(\varepsilon)$ is known, the most probable hopping distance R_m can be estimated by the expression $R_m = (9/8\pi\alpha N(\varepsilon)kT)^{1/4}$ [60,61]. At the temperature of 187.5 K, taken as the middle value of temperature range of 77–298 K, R_m was calculated, and it was equal to $R_m = 2.677 \times 10^{-7}$ cm. This R_m value is practically equal to the localization radius $a_{B,ZnO} = 2.67 \times 10^{-7}$ cm, and it is 14 % higher than the half-distance $w = 2.35 \times 10^{-7}$ cm between adjacent deep levels obtained from the I–V measurement at 77 K. The most probable energy difference of deep levels, ΔW, taking part in the electron hopping is given by $\Delta W = (3/4\pi R_m^3 N(\varepsilon))$ [60,61]. The calculated value of $\Delta W = 10.78$ meV for this hopping in the n-ZnO–p-Si MIS structures was close to the effective thermally activation energy $q\varphi_a$ = 8.3 meV in the as-deposited n-ZnO films. Therefore, in addition to temperature-independent intertrap tunneling, thermally activated hopping also took place in these n-type ZnO films in the temperature range of 77–298 K.

In Figure 8b, the forward I–V characteristics, measured at 298 and 77 K, of the MIS structures with ZnO:N films after RTA at 550 °C (sample 2.3) are given as the logarithm of the current density versus applied voltage. The averaged current density measured at 298 K under high applied voltage was 17.21 times higher than the current density measured at 77 K. The effective activation energy of the current density of this p-ZnO:N film, obtained from the expression $\ln(J_{298}/J_{77}) = (q\varphi_a/k)(1/77 - 1/298)$ was equal to $q\varphi_a$ = 25.46 meV. This value shows that the current in these p-ZnO:N films was also carried out predominantly by tunneling of holes from the occupied deep level to the nearest unoccupied one in the ZnO:N energy gap. Because of that, at 77 K the current density J is also given by the equation (1), where the electron effective mass, m_n^* = 0.23 m_e, is replaced with the hole effective mass, m_h^* = 0.59 m_e [62].

The values of $q\varphi_t$ and w, calculated from the slope of the plot lnJ versus V in Figure 8b and the intersection of the extension of this plot toward lnJ axis at V = 0, were $q\varphi_t$ = 1.27 eV and w = 3.724 $\times$ 10^{-7} cm, respectively. Under high forward bias the concentration of injected holes in the annealed at 550 °C p-ZnO film was equal to $N_t = 1/w^3$ = 1.94 $\times$ 10^{19} cm^{-3}. Taking into account that our previous EDX studies have shown a nitrogen/oxygen ratio of 1/8 in both RF-sputtered and annealed ZnO:N films [25], the total concentration of nitrogen in the 550 °C RTA p-ZnO film (sample 2.3) was 1.82 $\times$ 10^{21} cm^{-3}. The comparison between the concentration of acceptor deep levels (N_t = 1.94 $\times$ 10^{19} cm^{-3}) with the total concentration of N atoms incorporated into ZnO:N films (1.82 $\times$ 10^{21} cm^{-3}) shows that less than 1% of acceptor levels were not compensated. Because of this strong charge compensation, it is difficult to achieve an increase in the conductivity of ZnO by introducing a larger amount of nitrogen into a ZnO:N film.

Assuming that temperature-dependent current in these p-type ZnO:N films is also governed by the Mott VRH [51], the characteristic temperature T_0 obtained from the relation $\ln(J_{298}/J_{77}) = T_0^{1/4}[1/(77)^{1/4} - 1/(298)^{1/4}]$ was equal to T_0 = 7.44 $\times$ 10^5 K. The values of the Bohr radius $a_{B,ZnO}$, the decay constant α, the density of localized states, $N(\varepsilon)$, the most probable hopping distance, R_m, and the most probable hopping energy difference, ΔW, were calculated in the same way as the corresponding parameters for the n-type ZnO:N films. The obtained values were, respectively, $a_{B,ZnO}$ = 1.042 $\times$ 10^{-7} cm, α = 9.59 $\times$ 10^6 cm^{-1}, and $N(\varepsilon)$ = 2.2 $\times$ 10^{20} cm^{-3}·eV^{-1}, and at 187.5 K the R_m and ΔW values were 3.2 $\times$ 10^{-7} cm and 33.07 meV, respectively. These results reveal that in the p-type ZnO films, the most probable hopping distance R_m was about three times larger than the Bohr radius $a_{B,ZnO}$ and this relation confirms that Mott variable range hopping of holes occurred in these p-type ZnO:N films. Therefore, the additional current at 298 K compared to that at 77 K in these p-type ZnO:N films was carried out by charge carrier hopping, close to the Fermi or the quasi-Fermi level in the ZnO:N energy gap.

For better insight, the above calculated charge transport parameters of the intertrap tunneling and variable range hopping mechanism in ZnO:N films deposited at a gas ratio of Ar:O$_2$:N$_2$ = 50:10:40 (sample 2.1) and treated by RTA at 550 °C (sample 2.1) are summarized in Table 4.

Table 4. Intertrap tunneling and variable range hopping parameters of ZnO:N films deposited at gas ratio of Ar:O$_2$:N$_2$ = 50:10:40 (sample 2.1) and treated by RTA at 550 °C (sample 2.1): qφ_a—effective thermally activation energy; w—mean distance between deep levels; qφ_t and N$_t$—energy position and density of deep levels, respectively, in the ZnO energy gap; T$_0$—characteristic temperature; a$_{B,ZnO}$—Bohr radius of the localized states in the ZnO energy gap; N(ε)—density of localized states; R$_m$ and ΔW—most probable hopping distance and energy difference of deep levels, respectively.

Parameters	As-Deposited ZnO:N	After RTA at 550 °C
Intertrap tunneling		
qφ_a (meV)	8.3	25.46
w (cm) at 77 K	4.70×10^{-7}	3.724×10^{-7}
qφ_t(eV) at 77 K	1.48	1.27
N$_t$(cm^{-3}) at 77 K	9.62×10^{18}	1.94×10^{19}
Variable range hopping		
T$_0$(K)	8.45×10^3	7.44×10^5
a$_{B,ZnO}$(cm)	2.67×10^{-7}	1.04×10^{-7}
N(ε)(cm^{-3}.eV^{-1})	1.15×10^{21}	2.2×10^{20}
R$_m$ (cm) at 187.5 K	2.677×10^{-7}	3.2×10^{-7}
ΔW (meV) at 187.5 K	10.78	33.07

The decrease of the ZnO band gap density of localized states from N(ε) = 1.15×10^{21} cm^{-3}eV^{-1} for the as-deposited ZnO:N (sample 2.1) to N(ε) = 2.2×10^{20} cm^{-3}eV^{-1} for the p-ZnO:N after 550 °C RTA (sample 2.3) correlates well with the observed decrease in the slope values of lnG$_m$ versus lnω plots (Figure 7) from 0.66 to 0.24, respectively. The decrease of the density of localized states in the ZnO bandgap by the 550 °C RTA is evidence that the concentration of donor-like defects decreased during this treatment. The densities of localized states N(ε) given in Table 4 are within the range of 9.93×10^{19}–2.08×10^{22} cm^{-3}eV^{-1} reported for VRH conduction in n-type polycrystalline ZnO films [63]. Values of N(ε) = 2×10^{20} cm^{-3}eV^{-1} and ΔW = 29 meV, close to ours, for the densities of localized states and corresponding characteristic energies, respectively, have been obtained from the current–voltage characteristics of amorphous ZnON thin film transistors [64].

The established transformation of n-type conduction to p-type in the studied ZnO:N films viaannealing the samples at 550 °C suggests that if the deposition could be accomplished at such elevated temperatures as 550 °C, the density of localized states could be reduced below N(ε) = 2.2×10^{20} cm^{-3}.eV^{-1}, as was obtained in the 550 °C RTA-treated ZnO:N films. In this way, it would be possible to obtain nitrogen-doped ZnO films with weaker N-acceptor self-compensation and a smaller density of localized states in the p-type ZnO band gap than that we observed herein.

4. Conclusions

Nitrogen-doped ZnO thin films were deposited onto Si substrates by RF magnetron sputtering in nitrogen-containing ambient gas, followed by rapid thermal annealing at 400 °C and 550 °C in nitrogen. Raman and PL spectroscopic analyses revealed defects due to O and Zn vacancies and interstitials, which are related to carrier traps in the ZnO:N films.

Detailed analyses of the I–V characteristics of the MIS structures with the ZnO:N films revealedn-type conduction in both as-deposited and annealed at 400 °C ZnO:N films. The reason for n-type conductivity in these ZnO:N films is a result of the self-compensation of the N-acceptor levels by donor-like defects generated during N doping. RTA at 550 °C decrease the concentration of donor-like defects below the concentration of N acceptors and, as a consequence, the n-type ZnO:N films transformed into p-type ZnO:N ones. Repeated measurements after a year proved that the resulting conduction-type transformation at the annealing stage at 550 °C is a stable process.

The observed counterclockwise hysteresis in the measured point-by-point I–V characteristics of both n- and p type ZnO:N films confirms the presence of deep levels in the ZnO energy gap which participate in the charge transport through these ZnO:N films. The slope of the lnJ versus lnV plots reveals that the current in both n- and p-type ZnO:N films is trap charge limited current (TCLC), while the slope of the lnG_m versus $ln\omega$ plots confirms the tunneling or hopping of charge carriers via deep levels in the ZnO energy gap.

At 77 K, the current in both n- and p-type ZnO:N films is carried out by intertrap tunneling via deep levels in the ZnO energy gap. The higher forward averaged current density at 298 K than that measured at 77 K reveals additional thermally activated variable range carrier hopping in these ZnO:N films.

Author Contributions: Formal analysis, investigation, writing-original draft preparation, S.S. and M.N.; conceptualization, investigation, writing—original draft preparation, writing-review and editing, A.S.; formal analysis, investigation, D.S. and I.S., formal analysis, writing—review and editing M.A.; visualization, E.A. and M.M.; conceptualization, writing-review and editing, data curation, project administration, funding acquisition, M.G. All authors have read and agreed to the published version of the manuscript.

Funding: This research was funded by the Romanian National Authority for Scientific Research on Innovation, grant number PN-III-P2-2.1-PED-2019-2073. Part of this work was supported by the project "Materials and Processes for Energy and Environment Applications-AENAO" (MIS 5002556) co-financed by Greece and the European Regional Development Fund.

Institutional Review Board Statement: Not applicable.

Informed Consent Statement: Not applicable.

Data Availability Statement: The data presented in this study are available on request from the corresponding author.

Acknowledgments: The paper was carried out within the research program "Science of Surfaces and Thin Layers" of the "Ilie Murgulescu" Institute of Physical Chemistry. The support of the Romanian Government that allowed for the acquisition of the research infrastructure under POS-CCE O 2.2.1 project INFRANANOCHEM—No. 19/01.03.2009 and the support of inter-academic exchange program between the Romanian and Bulgarian Academies Collaboration Agreements are gratefully acknowledged.

Conflicts of Interest: The authors declare that they have no conflict of interest.

References

1. Jellison, G.E.; Boatner, L.A. Optical functions of uniaxial ZnO determined by generalized ellipsometry. *Phys. Rev. B* **1998**, *58*, 3586–3589. [CrossRef]
2. Han, C.; Duan, L.; Zhao, X.; Hu, Z.; Niu, Y.; Geng, W. Effect of Fe doping on structural and optical properties of ZnO films and nanorods. *J. Alloy. Compd.* **2019**, *770*, 854–863. [CrossRef]
3. Przezdziecka, E.; Guziewicz, E.; Witkowski, B.S. Photoluminescence investigation of the carrier recombination processes in N-doped and undoped ZnO ALD films grown at low temperature. *J. Lumin.* **2018**, *198*, 68–76. [CrossRef]
4. Xu, W.Z.; Ye, Z.Z.; Zeng, Y.J.; Zhu, L.P.; Zhao, B.H.; Jiang, L.; Lu, J.G.; He, H.P.; Zhang, S.B. ZnO light-emitting diode grown by plasma-assisted metal organic chemical vapor deposition. *Appl. Phys. Lett.* **2006**, *88*, 173506. [CrossRef]
5. Jee, S.W.; Park, S.J.; Kim, J.; Park, Y.C.; Choi, J.H.; Jeong, J.H.; Lee, J.H. Efficient three-dimensional nanostructured photoelectric device by Al-ZnO coating on lithography-free patterned Si nanopillars. *Appl. Phys. Lett.* **2011**, *99*, 053118. [CrossRef]
6. Socol, G.; Axente, E.; Ristoscu, C.; Sima, F.; Popescu, A.; Stefan, N.; Mihailescu, I.N.; Escoubas, L.; Ferreira, J.; Bakalova, S.; et al. Enhanced gas sensing of Au nanocluster –doped or –coated zinc oxide thin films. *J. Appl. Phys.* **2007**, *102*, 083103. [CrossRef]
7. Kumar, R.; Al-Dossary, O.; Kumar, G.; Umar, A. Zinc Oxide Nanostructures for NO2 Gas–Sensor Applications: A Review. *Nano-Micro Lett.* **2015**, *7*, 97–120. [CrossRef]
8. Mosca, M.; Macaluso, R.; Caruso, F.; Lo Muzzo, V.; Calì, C. *The P-Type Doping of ZnO: Mirage or Reality Advances in Semiconductor Research: Physics of Nanosystems, Spintronics and Technological Applications*; Nova Science Publishers: New York, NY, USA, 2015; Chapter 12; pp. 245–282.
9. Barnes, T.M.; Olson, K.; Wolden, C.A. On the formation and stability of p-type conductivity in nitrogen-doped zinc oxide. *Appl. Phys. Lett.* **2005**, *86*, 112112.
10. Kennedy, O.W.; Coke, M.L.; White, E.R.; Shaffer, M.S.P.; Warburton, P.A. MBE growth and morphology control of ZnO nanobelts with polar axis perpendicular to growth direction. *Mater. Lett.* **2018**, *212*, 51–53. [CrossRef]

11. Li, J.; Wang, J.; Pei, Y.; Wang, G. Research and optimization of ZnO-MOCVD process parameters using CFD and genetic algorithm. *Ceram. Int.* 2020, *46*, 685–695. [CrossRef]
12. Coutancier, D.; Zhang, S.T.; Bernardini, S.; Fournier, O.; Mathieu-Pennober, T.; Donsanti, F.; Tchernycheva, M.; Foldyna, M.; Schneide, N. ALD of ZnO:Ti: Growth Mechanism and Application as an Efficient Transparent Conductive Oxide in Silicon Nanowire Solar Cells. *ACS Appl. Mater. Interfaces* 2020, *12*, 21036–21044. [CrossRef] [PubMed]
13. Dave, P.Y.; Patel, K.H.; Chauhan, K.V.; Chawla, A.K.; Rawal, S.K. Examination of zinc oxide films prepared by magnetron Sputtering. *Proc. Technol.* 2016, *23*, 328–335. [CrossRef]
14. Manikandan, B.; Endo, T.; Kaneko, S.; Murali, K.R.; John, R. Properties of sol gel synthesized ZnO nanoparticles. *J. Mater. Sci.* 2018, *29*, 9474–9485. [CrossRef]
15. Fan, J.C.; Sreekanth, K.M.; Xie, Z.; Chang, S.L.; Rao, K.V. p-Type ZnO materials: Theory, growth, properties and devices". *Prog. Mater. Sci.* 2013, *58*, 874–985. [CrossRef]
16. Wang, Z.; Li, Q.; Yuan, Y.; Yang, L.; Zhang, H.; Liu, Z.; Ouyang, J.; Chen, Q. N doped ZnO (N:ZnO) film prepared by reactive HiPIMS deposition technique. *AIP Adv.* 2020, *10*, 035122. [CrossRef]
17. Hussain, B.; Aslam, A.; Khan, T.M.; Creighton, M.; Zohuri, B. Electron Affinity and Bandgap Optimization of Zinc Oxide for Improved Performance of ZnO/Si Heterojunction Solar Cell Using PC1D Simulations. *Electronics* 2019, *8*, 238. [CrossRef]
18. Czternastek, H. ZnO thin films prepared by high pressure magnetron sputtering. *Opto-Electron. Rev.* 2004, *12*, 49–52.
19. Tsay, C.-Y.; Chiu, W.-Y. Enhanced Electrical Properties and Stability of P-Type Conduction in ZnO Transparent Semiconductor Thin Films by Co-Doping Ga and N. *Coatings* 2020, *10*, 1069.
20. Lee, E.-C.; Kim, Y.-S.; Jin, Y.-G.; Chang, K.J. Compensation mechanism for N acceptors in ZnO. *Phys. Rev. B* 2001, *64*, 085120. [CrossRef]
21. Kim, J.; Ji, J.H.; Min, S.W.; Jo, G.H.; Jung, M.W.; Koh, J.H. Enhanced conductance properties of UV laser/RTA annealed Al-doped ZnO thin films. *Ceram. Int.* 2017, *43*, 3900–3904. [CrossRef]
22. Watanabe, F.; Shirai, H.; Fujii, Y.; Hanajiri, T. Rapid thermal annealing of sputter-deposited ZnO/ZnO:N/ZnO multilayered structures. *Thin Solid Film.* 2012, *520*, 3729–3735. [CrossRef]
23. Nicolescu, M.; Anastasescu, M.; Preda, S.; Calderon-Moreno, J.M.; Osiceanu, P.; Gartner, M.; Teodorescu, V.S.; Maraloiu, A.V.; Kampylafka, V.; Aperathitis, E.; et al. Investigation of microstructural properties of nitrogen doped ZnO thin films formed by magnetron sputtering on silicon substrate. *J. Optoelectron. Adv. Mater.* 2010, *12*, 1045–1051.
24. Nicolescu, M.; Anastasescu, M.; Preda, S.; Calderon-Moreno, J.M.; Osiceanu, P.; Gartner, M.; Teodorescu, V.S.; Maraloiu, A.V.; Kampylafka, V.; Aperathitis, E.; et al. Surface topography and optical properties of nitrogen doped ZnO thin films formed by radio frequency magnetron sputtering on fused silica substrates. *J. Optoelectron. Adv. Mater.* 2010, *12*, 1343–1349.
25. Nicolescu, M.; Anastasescu, M.; Preda, S.; Stroescu, H.; Stoica, M.; Teodorescu, V.S.; Aperathitis, E.; Kampylafka, V.; Modreanu, M.; Zaharescu, M.; et al. Influence of the substrate and nitrogen amount on the microstructural and optical properties of thin r.f.-sputtered ZnO films treated by rapid thermal annealing. *Appl. Surf. Sci.* 2012, *261*, 815–823. [CrossRef]
26. Himmlich, M.; Koufaki, M.; Ecke, G.; Mauder, C.; Cimalla, V.; Schaefer, J.A.; Kondilis, A.; Pelekanos, N.T.; Modreanu, M.; Krischok, S.; et al. Effect of Annealing on the Properties of Indium-Tin-Oxynitride Films as Ohmic Contacts for GaN-Based Optoelectronic Devices. *ACS Appl. Mater. Interfaces* 2009, *1*, 1451–1456. [CrossRef]
27. Chang, H.T.; Chen, G.J. Influence of nitrogen doping on the properties of ZnO films prepared by radio-frequency magnetron sputtering. *Thin Solid Film.* 2016, *618*, 84–89. [CrossRef]
28. Kumar, R.R.; Raja Sekhar, M.; Raghvendra, R.L.; Kumar Pandey, S. Comparative studies of ZnO thin films grown by electron beam evaporation, pulsed laser and RF sputtering technique for optoelectronics applications. *Appl. Phys. A* 2020, *126*, 859. [CrossRef]
29. Panda, J.; Sasmal, I.; Nath, T.K. Magnetic and optical properties of Mn-doped ZnO vertically aligned nanorods synthesized by hydrothermal technique. *AIP Adv.* 2016, *6*, 035118. [CrossRef]
30. Cusco, R.; Alarcon-Llado, E.; Ibanez, J.; Artus, L.; Jimenez, J.; Wang, B.; Callahan, M.J. Temperature dependence of Raman scattering in ZnO. *Phys. Rev. B Condens. Matter. Mater. Phys.* 2007, *75*, 165202. [CrossRef]
31. Manjon, F.J.; Mari, B.; Serrano, J.; Romero, A.H. Silent Raman modes in zinc oxide and related nitrides. *J. Appl. Phys.* 2005, *97*, 053516. [CrossRef]
32. Raji, R.; Gopchandran, K.G. ZnO nanostructures with tunable visible luminescence: Effects of kinetics of chemical reduction and annealing. *J. Sci. Adv. Mater. Dev.* 2017, *2*, 51–58. [CrossRef]
33. Musa, I.; Qamhieh, N.; Mahmoud, S.T. Synthesis and dependent photoluminescence property of zinc oxide nanorods. *Results Phys.* 2017, *7*, 3552–3556. [CrossRef]
34. Zerdali, M.; Hamzaoui, S.; Teherani, F.H.; Rogers, D. Growth of ZnO thin film on SiO2/Si substrate by pulsed laser deposition and study of their physical properties. *Mater. Lett.* 2006, *60*, 504–508. [CrossRef]
35. Song, Y.; Zhang, S.; Zhang, C.; Yang, Y.; Lv, K. Raman Spectra and Microstructure of Zinc Oxide irradiated with Swift Heavy Ion. *Crystals* 2019, *9*, 395. [CrossRef]
36. Decremps, F.; Pellicer-Porres, J.; Saitta, A.M.; Chervin, J.C.; Polian, A. High-pressure Raman spectroscopy study of wurtzite ZnO. *Phys. Rev. B* 2002, *65*, 092101–092105. [CrossRef]
37. Ng, Z.N.; Chan, K.Y.; Muslimin, S.; Knipp, D. P-Type Characteristic of Nitrogen-Doped ZnO Films. *J. Electron. Mater.* 2018, *47*, 5607–5613. [CrossRef]

38. Kegel, J.; Laffir, F.R.; Povey, I.M.; Pemble, M.E. Defect-promoted photo-electrochemical performance enhancement of orange-luminescent ZnO nanorod-arrays. *Phys. Chem. Chem. Phys.* **2017**, *19*, 12255–12268. [CrossRef] [PubMed]
39. Fatima, A.A.; Devadason, S.; Mahalingam, T. Structural, luminescence and magnetic properties of Mn doped ZnO thin films using spin coating technique. *J. Mater. Sci.-Mater. Electron.* **2014**, *25*, 3466–3472. [CrossRef]
40. Ismail, A.; Abdullah, M.J. The structural and optical properties of ZnO thin films prepared at different RF sputtering power. *J. King Saud Univ. Sci.* **2013**, *25*, 209–215. [CrossRef]
41. Kaur, G.; Mitra, A.; Yadav, K.L. Pulsed laser deposited Al-doped ZnO thin films for optical applications. *Prog. Nat. Sci.-Mater. Int.* **2015**, *25*, 12–21. [CrossRef]
42. Schmidt, M.; von Wenckstern, H.; Pickenhain, R.; Grundmann, M. On the investigation of electronic defect states in ZnO thin films by space charge spectroscopy with optical excitation. *Solid-State Electron.* **2012**, *75*, 48–54. [CrossRef]
43. Look, D.C.; Leedy, K.D.; Vines, L.; Svensson, B.G.; Zubiaga, A.; Tuomisto, F.; Doutt, D.R.; Brillson, L.J. Self-compensation in semiconductors: The Zn vacancy in Ga-doped ZnO. *Phys. Rev. B* **2011**, *84*, 115202. [CrossRef]
44. von Wenckstern, H.; Pickenhain, R.; Schmidt, H.; Brandt, M.; Biehne, G.; Lorenz, M.; Grundmann, M. Deep acceptor states in ZnO single crystals. *Appl. Phys. Lett.* **2006**, *89*, 092122. [CrossRef]
45. Huang, Z.; Ruan, H.; Zhang, H.; Shi, D.; Li, W.; Qin, G.; Wu, F.; Fang, L.; Kong, C. Investigation of the p-type formation mechanism of nitrogen ion implanted ZnO thin films induced by rapid thermal annealing. *Opt. Mater. Express* **2019**, *9*, 3098. [CrossRef]
46. Zhang, H.; Kong, C.; Li, W.; Qin, G.; Ruan, H.; Tan, M. The formation mechanism and stability of p-type N-doped Zn-rich ZnO films. *J. Mater. Sci. Mater. Electron.* **2016**, *27*, 5251–5258. [CrossRef]
47. Kampylafka, V.; Kostopoulos, A.; Modreanu, M.; Schmidt, M.; Gagaoudakis, E.; Tsagaraki, K.; Kontomitrou, V.; Konstantinidis, G.; Deligeorgis, G.; Kiriakidis, G. Long-term stability of transparent n/p ZnO homojunctions grown by rf-sputtering at room-temperature. *J. Mater.* **2019**, *5*, 428–435. [CrossRef]
48. Aksoy, S.; Caglar, Y. Structural transformations of TiO$_2$ films with deposition temperature and electrical properties of nanostructure n-TiO$_2$/p-Si heterojunction diode. *J. Alloy. Compd.* **2014**, *613*, 330–337. [CrossRef]
49. Ghatak, S.; Ghosh, A. Observation of trap-assisted space charge limited conductivity in short channel MoS2 transistor. *Appl. Phys. Lett.* **2013**, *103*, 122103. [CrossRef]
50. Jonscher, A.K. The 'Universal' Dielectric Response. *Nature* **1977**, *267*, 673–679. [CrossRef]
51. Mott, N.F. Conduction in glasses containing transition metal ions. *J. Non-Cryst. Solids* **1968**, *1*, 1–17. [CrossRef]
52. Simeonov, S.; Yourukov, I.; Kafedjiiska, E.; Szekeres, A. Inter-trap tunnelling in thin SiO$_2$ films. *Phys. Status Solidi* **2004**, *201*, 2966–2979. [CrossRef]
53. Oshikiri, M.; Imanaka, Y.; Aryasetiawan, F.; Kido, G. Comparison of the electron effective mass of the n-type ZnO in the wurtzite structure measured by cyclotron resonance and calculated from first principle theory. *Phys. B Cond. Matter* **2001**, *298*, 472–476. [CrossRef]
54. Mott, N.F.; Algaier, R.S. Localized States in Disordered Lattices. *Phys. Status Solidi* **1967**, *21*, 343–356. [CrossRef]
55. Choudhary, K.K. Analysis of temperature-dependent electrical resistivity of ZnO nano-structures. *J. Phys. Chem. Solids* **2012**, *73*, 460–464. [CrossRef]
56. Tiwari, A.; Jin, C.; Narayan, J.; Park, M. Electrical transport in ZnO1-δ films: Transition from band-gap insulator to Anderson localized insulator. *J. Appl. Phys.* **2004**, *96*, 3827–3830. [CrossRef]
57. Heluani, S.P.; Braunstein, G.; Villafuerte, M.; Simonelli, G.; Duhalde, S. Electrical conductivity mechanisms in zinc oxide thin films deposited by pulsed laser deposition using different growth environments. *Thin Solid Film.* **2006**, *515*, 2379–2386. [CrossRef]
58. Kumar, R.; Khare, N. Temperature dependence of conduction mechanism of ZnO and Co-doped ZnO thin films. *Thin Solid Film.* **2008**, *516*, 1302–1307. [CrossRef]
59. Ambegaokar, V.; Halperin, B.I.; Langer, I.S. Hopping Conductivity in Disordered Systems. *Phys. Rev. B* **1971**, *4*, 2612–2620. [CrossRef]
60. Ziqan, A.M.; Qasrawi, A.F.; Mohammad, A.H.; Gasanly, N.M. Thermally assisted variable range hopping in Tl$_4$S$_3$Se crystal. *Bull. Mater. Sci.* **2015**, *38*, 593–598. [CrossRef]
61. Mott, N.F.; Davis, E.A. *Electronic Processes in Non-Crystalline Materials*; Clarendon Press: Oxford, UK, 1971; p. 437.
62. Norton, D.P.; Heo, Y.W.; Ivill, M.P.; Ip, K.; Pearton, S.J.; Chisholm, M.F.; Steiner, T. ZnO: Growth, doping & processing. *Mater. Today* **2004**, *7*, 34–40.
63. Huang, Y.-L.; Chiu, S.-P.; Zhu, Z.-X.; Li, Z.-Q.; Lin, J.-J. Variable-range-hopping conduction processes in oxygen deficient polycrystalline ZnO films. *J. Appl. Phys.* **2010**, *107*, 063715. [CrossRef]
64. Lee, S.; Nathan, A.; Ye, Y.; Guo, Y.; Robertson, J. Localized Tail States and Electron Mobility in Amorphous ZnON Thin Film Transistors. *Sci. Rep.* **2015**, *5*, 13467. [CrossRef] [PubMed]

Article

Silica@zirconia Core@shell Nanoparticles for Nucleic Acid Building Block Sorption

Livia Naszályi Nagy [1], Evert Dhaene [2], Matthias Van Zele [2], Judith Mihály [3], Szilvia Klébert [3], Zoltán Varga [3], Katalin E. Kövér [4], Klaartje De Buysser [2], Isabel Van Driessche [2], José C. Martins [1] and Krisztina Fehér [5,*]

[1] NMR and Structure Analysis Research Group, Department of Organic and Macromolecular Chemistry, Ghent University, Krijgslaan 281 S4, B-9000 Ghent, Belgium; lnaszalyi@gmail.com (L.N.N.); jose.martins@ugent.be (J.C.M.)

[2] Sol-Gel Centre for Research on Inorganic Powders and Thin Films Synthesis, Department of Chemistry, Ghent University, Krijgslaan 281 S3, B-9000 Ghent, Belgium; Evert.Dhaene@UGent.be (E.D.); Matthias.VanZele@ugent.be (M.V.Z.); Klaartje.DeBuysser@ugent.be (K.D.B.); Isabel.VanDriessche@ugent.be (I.V.D.)

[3] Institute of Materials and Environmental Chemistry, Research Centre for Natural Sciences, Eötvös Loránd Research Network (IMEC RCNS ELKH), Magyar Tudósok Körútja 2, H-1117 Budapest, Hungary; mihaly.judith@ttk.mta.hu (J.M.); klebert.szilvia@ttk.mta.hu (S.K.); varga.zoltan@ttk.mta.hu (Z.V.)

[4] Department of Inorganic and Analytical Chemistry, University of Debrecen, Egyetem tér 1, H-4032 Debrecen, Hungary; kover@science.unideb.hu

[5] Molecular Recognition and Interaction Research Group, Hungarian Academy of Sciences-Eötvös Loránd Research Network at University of Debrecen, Egyetem tér 1, H-4032 Debrecen, Hungary

* Correspondence: feher.krisztina@science.unideb.hu; Tel.: +36-52-512-900

Citation: Naszályi Nagy, L.; Dhaene, E.; Van Zele, M.; Mihály, J.; Klébert, S.; Varga, Z.; Kövér, K.E.; De Buysser, K.; Van Driessche, I.; Martins, J.C.; et al. Silica@zirconia Core@shell Nanoparticles for Nucleic Acid Building Block Sorption. *Nanomaterials* **2021**, *11*, 2166. https://doi.org/10.3390/nano11092166

Academic Editor: Yuanbing Mao

Received: 1 July 2021
Accepted: 13 August 2021
Published: 25 August 2021

Publisher's Note: MDPI stays neutral with regard to jurisdictional claims in published maps and institutional affiliations.

Copyright: © 2021 by the authors. Licensee MDPI, Basel, Switzerland. This article is an open access article distributed under the terms and conditions of the Creative Commons Attribution (CC BY) license (https://creativecommons.org/licenses/by/4.0/).

Abstract: The development of delivery systems for the immobilization of nucleic acid cargo molecules is of prime importance due to the need for safe administration of DNA or RNA type of antigens and adjuvants in vaccines. Nanoparticles (NP) in the size range of 20–200 nm have attractive properties as vaccine carriers because they achieve passive targeting of immune cells and can enhance the immune response of a weakly immunogenic antigen via their size. We prepared high capacity 50 nm diameter silica@zirconia NPs with monoclinic/cubic zirconia shell by a green, cheap and up-scalable sol–gel method. We studied the behavior of the particles upon water dialysis and found that the ageing of the zirconia shell is a major determinant of the colloidal stability after transfer into the water due to physisorption of the zirconia starting material on the surface. We determined the optimum conditions for adsorption of DNA building blocks, deoxynucleoside monophosphates (dNMP), the colloidal stability of the resulting NPs and its time dependence. The ligand adsorption was favored by acidic pH, while colloidal stability required neutral-alkaline pH; thus, the optimal pH for the preparation of nucleic acid-modified particles is between 7.0–7.5. The developed silica@zirconia NPs bind as high as 207 mg dNMPs on 1 g of nanocarrier at neutral-physiological pH while maintaining good colloidal stability. We studied the influence of biological buffers and found that while phosphate buffers decrease the loading dramatically, other commonly used buffers, such as HEPES, are compatible with the nanoplatform. We propose the prepared silica@zirconia NPs as promising carriers for nucleic acid-type drug cargos.

Keywords: nanocarrier; nanoparticle; age-dependent adsorption; Langmuir isotherm; deoxynucleoside monophosphate; silica@zirconia; core@shell; solution NMR; buffer interference with adsorption

1. Introduction

With the rapid emergence of third-generation vaccines, nucleic acids are increasingly deployed both as coding for antigenic proteins as well as adjuvants in immunity-inducing preparations. This strategy has been particularly successful in fighting the worldwide epidemic caused by SARS-CoV-2, for which several nucleic acid-based vaccines have been developed in a spectacularly short amount of time, with multiple further preparations in the pipeline [1]. In third-generation vaccines, antigens are encoded by genetically engineered plasmid DNA in DNA vaccines [2] or by messenger RNA in mRNA vaccines [3], both of which induce expression of the target protein in the cell in situ. Nucleic acid type adjuvants [4–6], such as CpG motif-containing oligodeoxynucleotides (CpG ODNs) [7], are sequences with immune-modulating properties and are needed to induce and direct the immune response of the vaccine. However, nucleic acid delivery is not only relevant in prophylactic vaccines against viral diseases but also in the development of therapeutic cancer vaccines [8] and in gene therapies [9].

Due to their polyanionic nature, nucleic acids cannot easily permeate the cell membrane, and they are susceptible to enzymatic degradation by nucleases. For these reasons, therapeutic nucleic acid molecules can vastly benefit from delivery systems. Indeed, one of the major stumbling blocks for the mass deployment of SARS-CoV-2 mRNA vaccines is the requirement of ultra-low temperature storage needed for their stability in cationic lipid nanoparticle formations [1]. Since DNA and RNA are negatively charged, delivery systems based on electrostatic interactions for immobilization can be used for both types of molecules. Particle-mediated in vivo delivery options [10] includes cationic lipids and polymers, cell-penetrating peptides and biological and inorganic particles. [11] The development of suitable nanocarriers for the immobilization of nucleic acids requires cheap, non-toxic, biodegradable materials available in large quantities that are able to adsorb nucleoside–phosphate-containing molecules with high capacity. Furthermore, the process leading to the nanocarriers should be controllable, reproducible and up-scalable to be of industrial interest. The size of the nanocarrier is of importance as the 20–200 nm size range was shown to enhance the immunogenic effect of the vaccine formulation [12].

Zirconia is a material studied extensively as a dental and orthopedic implant coating because of its biocompatibility and favorable osseointegration [13]. Some research groups studied its potential as a drug delivery nanocarrier [14–16]. However, as the size and shape control of sol–gel-derived pure and colloidally stable zirconia particles has only been achieved for diameters <10 nm [17–20] and >200 nm [21], researchers started to use organic (such as bacterial or polymeric) or inorganic templates for zirconia deposition [15,22–27]. In the last step of the template removal, these procedures all apply a thermal treatment to the particles, which was, however, shown to lower the amount of ligand-adsorbing sites on the surface of zirconia due to recrystallization into a tetragonal form and decrease the number of surface defects [16,28]. Moreover, as we were aiming at the preparation of core@shell particles in the 20–100 nm diameter range, which is smaller than those formerly described in the literature, it was crucial to avoid the sintering effect. Thus, here in our wet chemical synthetic route, very mild conditions were used to avoid recrystallization and sintering, and extreme care was taken to control the colloidal stability of the resulting materials.

As a proof of concept, we aimed at the elucidation of the optimal conditions for DNA building blocks using deoxynucleoside monophosphate (dNMP) adsorption at the surface of the silica@zirconia core@shell NPs. The steps of the chemical synthesis are visualized in Scheme 1a. The strategy used for the optimization of dNMP sorption is shown in Scheme 1b.

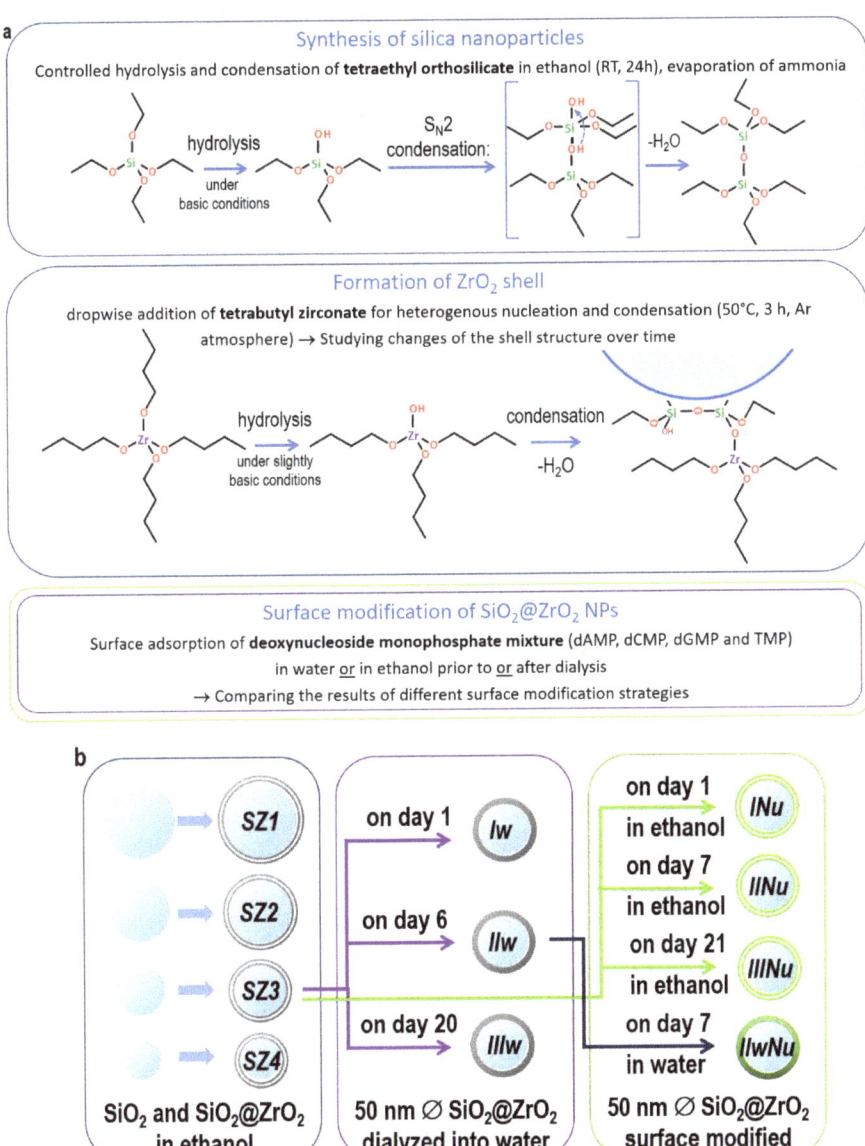

Scheme 1. A summary of the chemical synthesis steps (**a**) and sorption optimization strategy (**b**). According to our strategy, the optimal nanocarrier size was chosen first to achieve the highest possible specific surface area (**left panel**), then the optimal aging time before transfer into the water was defined to get maximal stability of the shell (**middle panel**); finally, the best surface modification strategy, leading to the highest cargo was chosen by the characterization of the four surface-modified samples (**right panel**). The silica core is depicted in blue; the deposited zirconia shell is in grey and the adsorbed dNMPs are shown in green. Batches *SZ1*, *SZ2*, *SZ3* and *SZ4* are native silica@zirconia NPs with increasing sizes in the 10-100 nm range; samples *Iw*, *IIw* and *IIIw* are NPs transferred into water by dialysis at day 1, 7 and 21, respectively; samples *INu*, *IINu* and *IIINu* are NPs modified using the ethanolic surface modification procedure at the age of 0, 6 and 20 days after dialysis, and finally, the *IIwNu* sample is treated by the aqueous surface modification procedure 7 days after synthesis.

First, we optimized the synthetic conditions of the core@shell nanoparticles for the different size ranges (left panel in Scheme 1b) and characterized their morphology and size by transmission electronmicroscopy (TEM) and dynamic light scattering (DLS). We studied the time-dependent structural changes of the native NPs during the condensation step of the zirconia deposition. TEM and nuclear magnetic resonance (NMR) revealed details of the evolution of the zirconia shell composition showing a physisorbed layer of zirconia starting material on the surface in addition to a solid zirconia shell. X-ray diffraction (XRD) also indicated changes over time in the composition of the crystalline phases. Fourier-transform infrared (FTIR) gave insight into the number and the state of the potential surface binding sites via monitoring the surface attached carbonate species [16,29] by quantifying the number of hydrogenocarbonate, monodentate and bridged carbonate molecules in the solution of the synthesis.

In the following, we monitored the behavior of the particles during and after dialysis into the water (the middle panel in Scheme 1b). DLS measurements and TEM pictures were used to monitor aggregation upon dialysis into the water for samples with different aging times of the zirconia shell. FTIR and NMR spectroscopies were indispensable for characterizing the surface chemistry at the solid-solution interface. We used the NMR toolbox [30] to confirm the presence of a physisorbed layer of zirconia starting material on the surface, which rapidly hydrolyzed upon dialysis into water and was found to be a major determinant of colloidal stability versus aggregation during this process. Zeta potential versus pH curves of the native NPs established the pH ranges, in which the particles are electrostatically stabilized by surface charges. FTIR revealed that the smallest size NPs did not retain their zirconia shell by monitoring the silica vibrational bands and also provided further insight into the overall adsorption capacity of the surface and the acid-base nature of the adsorption sites after transfer into the water.

We intended to understand how and under which conditions the adsorption takes place (right panel of Scheme 1b) and how it affects the colloidal stability of the particles. A shift of the isoelectric point (IEP) of the NPs under the effect of dNMP chemisorption in combination with the determination of the free ligand content in the solution revealed that the adsorption of ligands is favored at acidic pH. This enabled us to pinpoint the pH range optimal for both colloidal stability and ligand binding. We compared different surface modification methods by DLS, TEM, FTIR and zeta potential measurements and also studied the effect of the aging time of the zirconia shell onto the adsorption.

In view of the desired biological application, we inspected the effect of several buffers, such as phosphate, HEPES, PIPES, MES, MOPS and MOPSO—usually referred to as non-interfering buffers [31]—on the adsorption phenomenon. We established the adsorption isotherms of dNMP equimolar mixtures at the surface of the silica@zirconia NPs in four of the chosen buffers and derived the maximum adsorption capacity at pH 7.4 under equilibrium conditions. In parallel, we measured the dNMP content of 50 nm diameter nanocarriers surface modified using two different procedures by thermogravimetric analysis (TGA). In this way, we could define the optimal procedure for the surface modification of the nanocarriers whereby the dNMP load was found to be very close to the theoretical maximum obtained by Langmuir isotherms. The zirconia surface showed microporosity as determined by N_2 adsorption-desorption isotherms, which were also used to determine specific surface areas.

As a result of this development process, we were able to establish the optimum conditions for a procedure that yielded colloidally stable silica@zirconia NPs with the ability to absorb DNA-type molecules with high capacity.

2. Experimental

2.1. Materials

Tetraethyl orthosilicate (TEOS, 99%, abcr), ammonia solution (fresh, 32%, EMPLURA, Merck, Darmstadt, Germany), absolute ethanol (LiChrosolv®, gradient grade for liquid chromatography, water content <0.1%, Merck, Darmstadt, Germany), zirconium(IV) butoxide solution (TBOZ, 80% in n-butanol, Sigma-Aldrich, St. Louis, MO, USA, kept under argon,) were used for the NP synthesis. The biomolecules adsorbed at the surface of the NPs were: 2′-deoxyadenosine-5′ monophosphate (dAMP, 98–100%, Sigma grade, 2′-deoxycytidine-5′ monophosphate (dCMP, ≥95.0%, Sigma grade), 2′-deoxyguanosine-5′ monophosphate disodium salt hydrate (dGMP, ≥98.0%, Aldrich), thymidine-5′ monophosphate disodium salt hydrate (TMP, ≥99.0%, Sigma) and 2′-deoxyadenosine monohydrate (≥99%, Sigma-Aldrich, St. Louis, MO, USA).

High-purity deionized water (Sartorius, Göttingen, Germany, Arium 611, 18.2 MΩ·cm) basified with 1 M KOH solution (Titripur® Reag. Ph Eur, Reag. USP, EMD Millipore, Burlington, MA, USA) was used for the transfer of NPs into the water by dialysis. Buffers were prepared using 3-(N-morpholino) propane sulfonic acid (MOPS, BioPerformance certified, >99.5%, Sigma-Aldrich, St. Louis, MO, USA), β-hydroxy-4-morpholinepropanesulfonic acid (MOPSO, Pharma grade, Sigma-Aldrich, St. Louis, MO, USA), 1,4-piperazinediethanesulfonic acid (PIPES, Bioperformence certified, Sigma-Aldrich, St. Louis, MO, USA), 2-(N-morpholino) ethane sulfonic acid hydrate (MES, 99.5%, Sigma-Aldrich, St. Louis, MO, USA), 4-(2-hydroxyethyl)piperazine-1-ethanesulfonic acid (HEPES, 99.5%, Sigma-Aldrich, St. Louis, MO, USA), potassium phosphate dibasic trihydrate (99+%, Acros Organics, Waltham, MA, USA) and potassium phosphate monobasic (p.a., Acros Organics, Waltham, MA, USA). Deuterium oxide was added to samples prior to NMR measurements (99.96%, Euriso-top, Tewksbury, MA, USA).

2.2. Preparation of Silica@zirconia NPs

Silica sols of different particle sizes were prepared by the Stöber method [32]. Briefly, 125 mL abs. ethanol was poured in a tall form beaker, and 7.0, 6.0, 5.0 or 3.5 mL ammonia solution was added. The mixture was stirred at 330 rpm covered with parafilm. 5.0 mL TEOS was quickly added, and the reaction mixture was being stirred at room temperature for 24 h. Thereafter, the parafilm was discarded, and additional ethanol (~70 mL) added to the sample prior to the removal of ammonia at 70 °C under vigorous stirring in a ventilated fume hood. The pH of the ethanolic mixture was controlled with a wet pH paper, and the evaporation of ammonia was stopped at pH 8. The solid content of the sample was determined on 3 × 1 mL sol dried out at 85 °C, 10 h. The sol was stored at 4 °C.

The deposition of zirconia at the surface of 100 nm diameter silica cores was carried out by a controlled hydrolysis-condensation of the metal organic precursor. The process was based on Kim and coworkers' method [33,34]. We applied 8-fold dilution to 50 mL of the silica sol and put the ethanolic reaction mixture under argon bubbling and stirring at 50 °C. Based on TEM mean diameters, 0.11 mmol TBOZ/m^2 silica surface was added to 25 mL ethanol in the dropping funnel, and dropwise addition was performed. After 2.0 h, the reaction mixture was cooled down and stored at 4 °C. In our optimized procedure for 50 nm diameter NPs, 10-fold dilution was found to be more suitable, and 4–7 days of aging was applied prior to the surface modification or the dialysis into water.

2.3. Solvent Exchange and Surface Modification of the Silica@zirconia NPs

In the case of dAMP, 2 mg/mL ligand solutions were prepared in water using heating to reach complete dissolution of the solid material. The four-component mixture of dNMPs was obtained by adding equal volumes of dAMP, dCMP, dGMP and TMP. The native pH of dNMP solutions were: dAMP (3.1), dCMP (3.4), dGMP (8.2), TMP (7.8) and dNMP mixture (4.9). For controlled adsorption conditions, the pH of all the ligand solutions and mixtures were set to 6.0 ± 0.1.

2.3.1. Surface Modification in Ethanol

The ligand was added to an aliquot of the freshly prepared silica@zirconia core@shell sol at a ligand-to-NP surface ratio of 0.8 µmol/m^2 to 22.7 µmol/m^2 (calculation based on the TEM mean particle size). The excess of ligands was removed by dialysis against pure or neutralized water (CelluSep dialysis tubing, MWCO 10.000 Da, 50 mm flat width).

2.3.2. Surface Modification in Water

An aliquot of the ethanolic core@shell sol was filled into a dialysis bag and was dialyzed against cold (4–8 °C), basified pure water (pH 8.5–9.5). The ligand solution was added in a ratio of 50 µmol/m^2, and further dialysis was done against pure water.

2.4. Characterization

Morphological investigations of the NPs were carried out on a JEOL JEM-2200FS transmission electron microscope operated at 200 kV with a Cs corrector. Diluted samples were dropped and dried on holey carbon-coated copper grids (200 mesh).

DLS and zeta potential measurements were performed at 25 °C on a Malvern Zetasizer Nano ZS (Malvern, Worcs, UK) equipped with He-Ne laser (λ = 633 nm) and backscatter detector at a fixed angle of 173°. The pH of the samples was measured using an extended length pH electrode with micro-bulb together with a 9126 pH/ORP meter (HANNA Instruments Ltd., Leighton Buzzard, UK). The parameters used for the measurement were: RI_{SiO2} = 1.475, $RI_{SiO2@ZrO2}$ = 2.152, absorption for both oxides: 0.001, $RI_{ethanol}$ = 1.364, RI_{water} = 1.330, Viscosity$_{ethanol}$ = 1.0740 mPa.s, Viscosity$_{water}$ = 0.8872 mPa.s and dielectric constant$_{water}$ = 78.5.

The solid content of the samples was assessed on 3 × 1 mL aliquots dried out at 85 °C (for samples in ethanol) or 95 °C for samples in water in a Binder ED53 drying oven.

FTIR absorption spectrum was recorded using a Perkin-Elmer Frontier MIR Spectrometer equipped with a deuterated triglicine sulfate (DTGS) detector and a single reflection attenuated total reflection (UATR) unit (SPECAC "Golden Gate") with diamond ATR element. Records of 32 scans and a 4 cm^{-1} resolution were applied.

One-dimensional and two-dimensional NMR spectra were recorded on a Bruker Avance II 700 MHz (frequency for protons) NMR spectrometer equipped with a Prodigy cryoprobe. During sample preparation, we applied samples with an 8-fold ligand to NP ratio using an Eppendorf centrifuge 5415R (14.000 rpm, 5 min), followed by the addition of 10% D$_2$O and filling into 5 mm diameter quartz tubes or Shigemi tubes.

Powder X-ray diffraction patters of the adsorbents were collected on a Thermo Scientific ARL XTra diffractometer, operated at 40 kV, 40 mA using Cu K$_\alpha$ radiation of 0.15418 nm wavelength. Pas Academic V4.1 software was used for Rietveld refinement.

The Brunauer–Emmett–Teller (BET) specific surface area and micropore volume were evaluated from the N$_2$ adsorption-desorption isotherm. Nitrogen physisorption measurement was performed at 77 K (−196 °C) using a static volumetric apparatus (Quantachrome Autosorb 1C analyzer). The samples were previously degassed at 373 K (100 °C) for 24 h. Nitrogen adsorption data were obtained using ca. 0.05 g of sample and successive doses of nitrogen until p/p$_0$ = 1 relative pressure was reached. Only the nitrogen adsorption volumes up to a relative pressure of 0.1 were considered in the micropore size distribution. The micropore volume was obtained from the DR (Dubinin–Radushkevich) plot.

TGA was performed in a NETZSCH STA 449 F3 Jupiter instrument. Powder samples were heated to 800 °C at a rate of 10°/min in air flow.

The adsorption of biomolecules to the silica@zirconia NP surface was monitored by zeta potential measurements of the particulate samples complemented by the UV-visible study of their supernatant as a function of the pH. First, a large batch of native NP suspension was prepared in water with a solid content set to 1 mg/mL. For each run, 10 mL aliquot of this suspension was added 2 or 3 µmol/m^2 ligand solution (and eventually 10 mM buffer) in a stirred vessel. The pH of the suspension was controlled in the range of pH 9–4 using 1 M NaOH and 1 M HCl, and at every step, a 730 µL sample was placed in a disposable folded capillary cell. After recording the zeta potential in 3 measurements, the sample was transferred into an Eppendorf tube for centrifugation at 12.000 rpm, 3 min. Finally, the UV spectrum of the supernatant was recorded in 5-fold dilution in quartz cuvette in the wavelength range of 200–400 nm (Perkin-Elmer, Waltham, MA, USA, Lambda 950 spectrometer, Deuterium lamp, slit width: 2 nm, resolution: 2 nm).

The adsorption isotherms were determined on the supernatants of 1 mL samples containing 0.45 mg/mL NP, 20 mM buffer and 4, 12, 16, 20, 24, 28, 32, 40, 80, 120, 160 and 200 µg/mL dNMP mixture. The equilibration was done at room temperature for 24 h. The NPs were centrifuged at 12.400 rpm for 10 min, and the supernatants were collected. Finally, 2-, 5- and 10-fold dilutions were applied according to the concentration of the dNMP in the solution. The same dilutions were used for the buffer as background.

3. Results

3.1. Native Silica@zirconia Core@shell NPs

In order to obtain high adsorption capacity nanocarriers, we gradually increased the specific surface area of our samples by reducing the size of the Stöber silica cores from 81 ± 12 nm to 13 ± 2 nm resulting in batches *SZ1* to *SZ4* (Table 1). We adapted the synthetic procedure of the zirconia deposition accordingly. Keeping the original n_{TBOZ}/SiO_2 surface ratio was optimal for smaller particle sizes of batches *SZ1* and *SZ2*, and on the other hand, the optimal dilutions for *SZ3* and *SZ4* were found to be 10 and 12, respectively.

Table 1. The mean size and size distribution of NPs.

Sample	SiO_2 Core TEM Mean Ø * (nm)	SiO_2 Core DLS Volume Mean ** (nm)	SiO_2@ZrO_2 TEM Mean Ø * (nm)	SiO_2@ZrO_2 DLS Volume Mean ** (nm)	SiO_2@ZrO_2 DLS PdI
SZ1	81 ± 12	123 ± 34	95 ± 12	109 ± 29	0.021
SZ2	55 ± 8	75 ± 25	63 ± 6	134 ± 53	0.133
SZ3	40 ± 6	35 ± 13	48 ± 5	47 ± 13	0.043
SZ4	13 ± 2	6 ± 3	20 ± 3	25 ± 6	0.021

* TEM diameter evaluated on 50–100 NPs, ** hydrodynamic diameter, PdI: polydispersity index.

We summarized the mean size and size distribution properties of the core and core@shell NPs as evaluated by DLS and TEM investigations in Table 1. The NPs were spherical, and the samples showed a relatively narrow size distribution. TEM pictures and DLS size distribution functions are shown in Figures S1 and S2 in Supplementary Materials.

According to our experience, in the second reaction step, it was crucial to stop the deposition of TBOZ after 2.0 h and cool it to 4 °C. The formation of the zirconia shell was, however, not complete at this time point. As Arnal et al. also observed [22], the "aging" of the reaction mixture was important: they aged their particles for 3 days prior to the removal of the silica core but provided no explanation for this treatment. We decided to study the structural changes occurring in this time period using TEM, FTIR and XRD measurements.

TEM pictures of the one-day-old core@shell NPs revealed that in addition to a very thin shell of crystalline ZrO_2 (high contrast contour) surrounding the silica sphere, there is a thicker amorphous or polymer-like deposit on the surface of the NPs (Figure 1, left). The surface of the deposit is smooth, and its contrast is much lower than expected for crystalline zirconia. We suggest that this is a physisorbed multilayer of TBOZ that condensates slowly to give ZrO_2. For a two-week-old sol, only the high-contrast thin shell can be seen on the surface of the NPs (Figure 1, middle). At this time point, the ripening of the sample has begun leading to dissolution-redeposition of the oxides next to the silica@zirconia core@shell NPs (see the advanced stage for a seemingly stable 1-year-old sample in Figure 1, right).

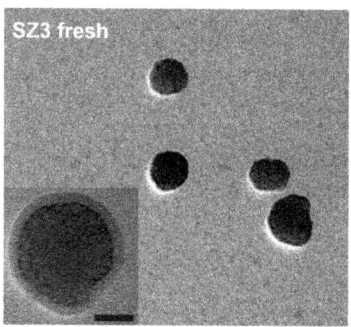 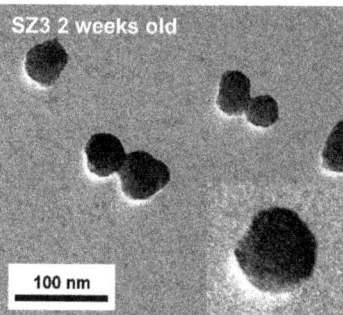 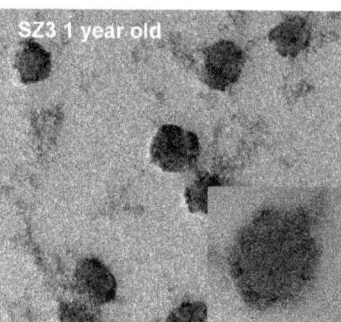

Figure 1. TEM pictures of batch *SZ3* in ethanol showing an incomplete condensation of the zirconia precursor on the day after addition to the silica cores (**left**), completed ZrO_2 shell formation and beginning of the ripening at the age of 2 weeks (**middle**) and the advanced ripening of the 1-year-old core@shell NPs leading to the dissolution-redeposition of the material next to the NPs (**right**).

The broad bands in the XRD spectra (see the example XRD spectrum of a 2-week-old *SZ3* sample shown in Figure S3 in Supplementary Materials) of differently aged *SZ3* batches evidenced a small crystallite size of ca. 5 nm. The Rietveld analysis showed that freshly prepared *SZ3* presented purely monoclinic zirconia, while 36% of zirconia adopted a cubic crystalline phase one week later (Table 2). This does not appear to be in full agreement with the phase diagram of pure zirconia, according to which the cubic crystalline phase is not stable at room temperature [35]. While metallic dopants (Ca^{2+}, Y^{3+}, etc.) were found to lower the temperature for the formation of cubic ZrO_2, high annealing temperatures are, in general, required for its formation (800–1000 °C). "Low-temperature" syntheses of pure, cubic ZrO_2 were described by Prakashbabu [36] and Salavati-Niazari [37], using organic ligands to fix the precursor's symmetry during the thermal degradation process at 400 °C and 245 °C. Our hypothesis is that due to the high purity of the system, the Stöber silica surface serves as a scaffold that induces and stabilizes the formation of cubic zirconia. This was supported by the observation of decreasing amount of cubic zirconia during the ripening: when zirconia left the surface, it redeposited under monoclinic crystalline phase (or amorphous form) next to the NPs.

Table 2. The proportion of zirconia crystalline phases in core@shell NPs as a function of age.

Sample	Fm-3m (Cubic)	P121/c1 (Monoclinic)
SZ3 fresh	0.04%	99.96%
SZ3 1 week	36.08%	63.92%
SZ3 2 weeks	22.02%	77.98%
SZ3 1 year	14.33%	85.67%

Although FTIR spectroscopy cannot give direct information on ZrO_2 as its vibrational bands appear at 500 cm^{-1}, it allows monitoring surface attached carbonate species that occupied potential ligand-binding sites. It has been previously shown [16,29] that soon after deposition, the most basic sites on the surface of zirconia are immediately binding CO and CO_2 from the air, thus covering accessible adsorption sites in the form of hydrogenocarbonate, alkaline monodentate carbonate and bridged carbonate. We monitored the ratio of hydrogenocarbonate and carbonate species adsorbed at the particles' surface, which changed slightly over the observed time. In order to view these changes, we normalized the raw spectra to the Si-O-Si vibrational band at 1090 cm^{-1} (Figure S4 in Supplementary Materials). The intensity of the hydrogenocarbonate (HCO_3^-) vibrational band at 1621 cm^{-1} decreased after the first day, while monodentate carbonate (m-CO_3^{2-}) vibrations (1568 cm^{-1}) showed a slight increase in intensity starting with day 2. The vibrational bands of bridged carbonate (br-CO_3^{2-}) became more prominent on day 3, with a shoulder appearing at 1549 cm^{-1} and 1393 cm^{-1}. These changes evidenced an evolution of the zirconia surface structure during the days following the deposition reaction. The observed carbonate species appear to reflect the acidity of the zirconia surface: at first, acidic hydrogenocarbonate is prominent, which is then converted into more alkaline monodentate and bridged carbonate species as the zirconia layer is ripening.

As a summary of the above TEM, XRD and FTIR analyses, we propose that the transformation of physisorbed TBOZ to zirconia takes several days (4–7), which is deposited first as a mixture of amorphous and monoclinic crystalline states, and then the silica surface induces the formation of cubic zirconia. Furthermore, at this stage, the adsorption sites of the deposited zirconia are occupied by various carbonate species, which are evolving from acidic to more alkaline sites over time.

3.2. Transfer of Native Silica@zirconia Core@shell NPs into Water by Dialysis

In view of the biological application, we transferred the NPs of different sizes into the water to study their structural changes and colloidal stability. We have chosen dialysis as a gentle solvent exchange technique to avoid the adverse effects of centrifugation on particle size distribution. Even with dialysis, we observed an increase in DLS particle sizes after solvent exchange, indicating aggregation, particularly when the age of the sol, the temperature and the pH of the dialysis water were not controlled. Regarding the temperature of the dialyzing water, using cold water was found to be crucial for the maintenance of the original NP structure: we used 4–8 °C water changed every 2 h. As to the pH of the dialyzing water and the age of the sol, we carried out dialysis of the ethanolic sol after different aging times, and as expected, both parameters had a significant effect on the properties of the resulting material as follows.

Freshly prepared native NPs (*Iw*) underwent immediate, irreversible aggregation when dialysis against water started, independently of the pH of the water. This was evidenced by the whitening and sedimentation of the sample. Since the agglomerates were no longer in suspension, evaluation of the size distribution of these samples by DLS was not possible. Our interpretation based on the above hypothesis of the zirconia shell formation was that upon addition of water to *Iw* during dialysis, a large amount of the remaining physisorbed TBOZ suddenly converts to zirconia, which crystallizes on the surface and connects colliding particles by generating chemical bonds between them.

Using 4–7-day-old native NPs, the sol was still whitened during dialysis (*IIw*), but this process was slow, and—what is more interesting—depending on the pH, a spontaneous redispersion of the NPs occurred in 1–2-weeks' time at neutral-basic pH. Setting the pH of the dialyzing water to 8.5–9.5 reduced significantly the aggregation resulting in slight or no whitening of the sample. DLS showed predominantly primary particles after 4 days of aging time, while 7-day old particles increased in size with a broader distribution (Figure S5 in Supplementary Materials). Our interpretation in the case of *IIw* is that the water added to the sample for dialysis suddenly converts all remaining TBOZ into zirconia, just as with *Iw*. However, since there remains only a low amount of TBOZ after 4–7 days of condensation time in ethanol, and it is mostly chemisorbed at the surface, this process does not connect the particles chemically, and the aggregation is not significant. The basicity of the dialyzing water may influence the relative rates of the hydrolysis and condensation of TBOZ, which determines whether the deposition of TBOZ is mostly taking place on the surface or next to the NPs. The basic pH might be slowing down hydrolysis and speeding up condensation, as a result of which zirconia is deposited on the existing surface and not between the NPs, thus reducing aggregation of the NPs. The spontaneous redispersion means that the formed aggregates are in a metastable state that can be solubilized again by Brownian motion even at 4–8 °C.

Colloidal stability also depended on the size of the NPs: smaller NPs (*SZ3-4*) were more prone to aggregation than larger ones (*SZ1-2*) (Figure S5 in Supplementary Materials). Due to their higher surface-to-volume ratio, smaller NPs can adsorb a larger amount of TBOZ, which more easily connects the particles upon dialysis when it suddenly is converted into zirconia.

It is also important to note here that the pH outside of the dialysis membrane always remained higher than the pH inside the membrane, and both values were continuously decreasing in time. Using ultra-pure water without pH control resulted in a sample pH of 5.5–5.8. When basified water was used, the pH of the final sol varied between 6.9–7.7. The reason for the aqueous sol becoming more acidic with increasing aging time may either be the continued dissolution of the silica core through the porous zirconia shell (porosity investigated in paragraph 3.5) or processes altering the composition of the surface attached carbonate species. In any case, it was not possible to completely prevent the aggregation of the native NPs by further increasing the pH of the dialyzing water (10.4), at which point the dissolution of silica would become dominant.

When dialysis was carried out on native NPs aged for 3 weeks after deposition (*IIIw*), the sol did not whiten during dialysis. However, fewer primary particles were observed by DLS (Figure S5 in Supplementary Materials) than for the samples dialyzed between 4–7 days. According to these results, it was disadvantageous to wait 3 weeks before transferring the NPs into water.

Besides the effect on sol dispersity, we studied the structural changes induced by the transfer of the NPs from ethanol into the water by TEM, XRD, NMR, FTIR and zeta-potential measurements.

Structural changes were seen in TEM pictures of the native NPs upon transfer into the water at either 1 or 4 days of age (Figure 2): the surface of the NPs became rough, being covered by zirconia crystallites, also presenting spurs. For *Iw* and *IIw*, we observed zirconia crystallites exclusively at the surface of silica cores, not next to them.

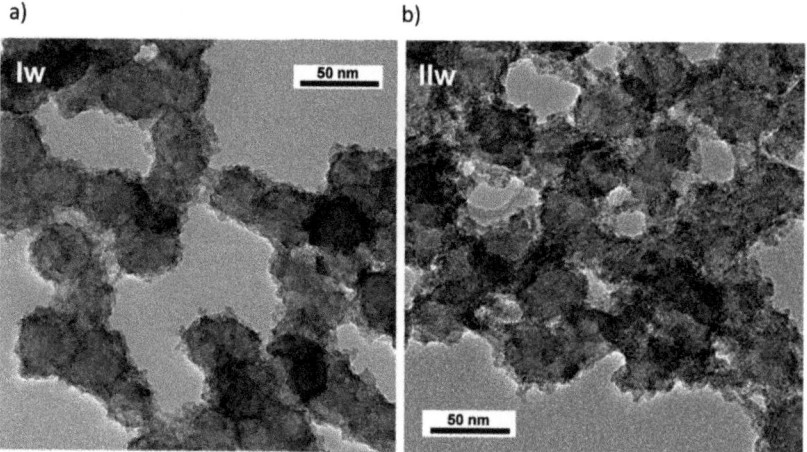

Figure 2. TEM pictures of *SZ3* (50 nm diameter) NPs after dialysis into water at different age: (**a**) *Iw* at 1. day) and (**b**) *IIw* at 7. days. Both samples presented rough NP surface, showing a structural change of zirconia when transferred from ethanol to water.

The X-ray diffractogram of *Iw* indicated a small crystallite size similar to that of the ethanolic sol. According to the Rietveld refinement, more than half of the formed crystallites were in the cubic phase (55.1%), accompanied by the monoclinic phase (41.2%) and the newly arisen tetragonal phase (3.7%) (Figure S6 in Supplementary Materials). This was the highest cubic phase ratio observed in our study.

To confirm our hypothesis on physisorbed TBOZ driving the aggregation behavior of the NPs transferred into the water, we investigated residual TBOZ signals in the NMR spectra of the dialyzed samples. Free and surface-bound TBOZ species were observed in NMR spectra at any age of the sample after dialysis into water. ^{1}H NMR spectra of the *IIIw* featured multiplets in the aliphatic region (Figure 3a), which were assigned to residual methyl groups of tetrabutyl moieties of TBOZ. The 2D nuclear Overhauser effect spectroscopy (NOESY) spectra showed negative NOE cross-peaks between these signals (Figure 3b). The narrow resonances in the TBOZ multiplets became broad after a month spent in water (Figure 3c) while still showing negative NOE cross-peaks between them (Figure 3d). The initial narrow resonances with negative NOEs indicate the attachment of loosely associated TBOZ on the NP surface with a fast exchange of the TBOZ molecules between surface-bound and solution-state free forms. Broadening of the signals later suggests stronger attachment to the zirconia surface, indicating chemisorption of the rest of TBOZ to the zirconia surface in the aqueous environment.

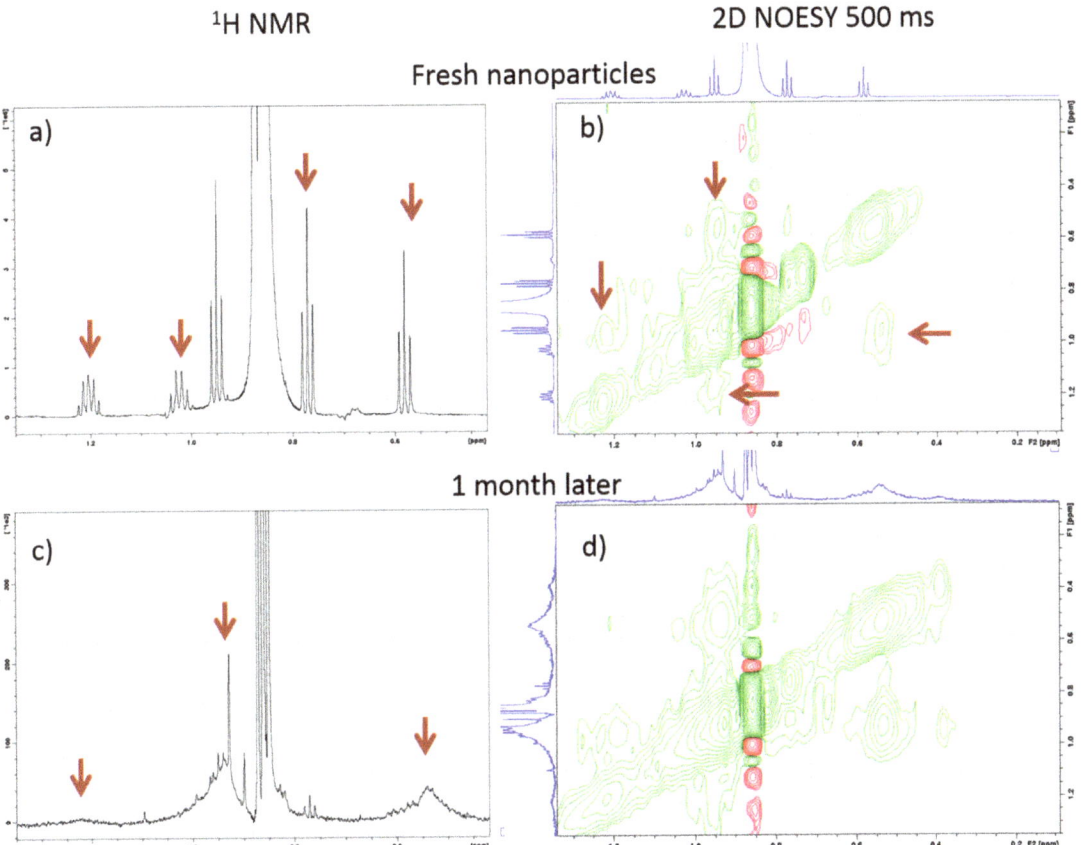

Figure 3. One-dimensional ^{1}H NMR spectrum of fresh NPs of *IIIw* showing narrow signals of TBOZ (**a**), between which negative NOE cross peaks are observed in 2D NOESY spectra (**b**). One month later broad signals of residual TBOZ are detected (**c**), which also showed negative NOE cross peaks (**d**).

The IEP of the NPs as determined by zeta potential vs. pH curves shown in Figure 4a was monotonously decreasing with the age of the transferred sol from 7.4 (*Iw*) through 7.2 (*IIw*) to 6.1 (*IIIw*). Thus, depending on condensation time, the native NPs are stabilized by electrostatic repulsion between negative charges at higher, alkaline pH, prone to aggregation at intermediate values and could, in theory, be again stable at acidic pH by repulsion between positive surface charges. However, increased silica dissolution prevents the application of acidic pH for practical purposes.

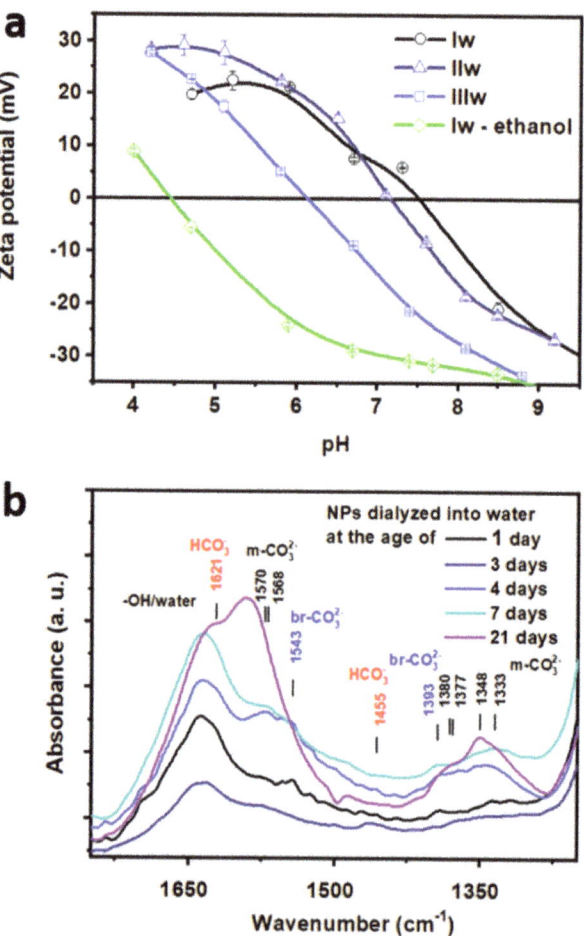

Figure 4. Zeta potential vs. pH curves (**a**) and FTIR spectra (**b**) and of *SZ3* (50 nm diameter) NPs after dialysis into the water at different ages: *Iw* (1 day), *IIw* (7 days), *IIIw* (21 days) and *Iw-ethanol* (freshly dialyzed against ethanol and then against water). The zeta potential measurements show that the IEP of zirconia is continuously decreasing with time, which will affect the colloidal stability and the adsorption properties of the nanocarrier. The FTIR results reveal that the difference in the surface chemistry of samples transferred into water at different ages is significant.

As a final proof for our hypothesis of TBOZ physisorption on the zirconia surface after deposition, we carried out an experiment whereby the freshly prepared *SZ3* was first dialyzed against ethanol, and then the dialysis was continued against basic water (*Iw-ethanol*, green line in Figure 4b). In this case, the IEP (4.4) was strikingly lower compared to all the other samples described above. It was closer to that of native silica surface (~2) [38] and was much lower than the IEP of native pure zirconia surface (8.2–observed on similarly prepared pure zirconia particles [39]). We think that the physisorbed TBOZ was removed during the ethanolic dialysis leaving a thin layer of chemisorbed zirconia on the surface, which only partially covers the silica core reflected by the IEP closer to that of silica instead of zirconia. This finding indicated the reversibility of the surface modification on the day of the deposition reaction and supports our hypotheses on the temporary formation of a physisorbed TBOZ multilayer.

The FTIR study of the dialyzed native NPs revealed an alteration of the silica nanoparticulate structure for samples dialyzed without temperature or pH control. As an example of native NPs dialyzed into the water without temperature control (Figure S7 in Supplementary Materials) and an example of surface-modified samples dialyzed without pH control (Figure S8 in Supplementary Materials). The figures show that the intensity of the νSi-O-Si vibrational band at 1090 cm^{-1} is drastically reduced. This is probably a result of silica dissolution that has been reported and studied by several research groups for mesoporous silica in simulated body fluid [40–42] but has also been reported to occur in neutral-alkaline water [43]. For *SZ4*, with the smallest particle size, the zirconia shell did not remain attached to the core after dialysis even under controlled conditions, as indicated by the reappearance of vibrational bands of native silica surface in the FTIR spectrum. Thus, *SZ3* was found to be the smallest suitable particle size, and we conducted all further studies on this sample.

We found significant differences in the surface chemistry of the aqueous samples according to the age of *SZ3* at the start of the dialysis using FTIR spectroscopy (Figure 4b). If the dialysis was carried out in the first 3 days after zirconia deposition (*Iw*), the hydrogenocarbonate species (1621 and 1445 cm^{-1}) were predominant next to the vibrational bands of monodentate carbonate (1570, 1568, 1380, 1377, 1348 and 1333 cm^{-1}). Bridged carbonate was visible first on the surface of the sample dialyzed on the third day (1543 and 1393 cm^{-1}). The samples transferred at the age of 4–7 days (*IIw*) showed a surface with a similarly high amount of monodentate and bridged carbonates. For samples transferred at the age of 20 days or more (*IIIw*), the monodentate form was predominant. The trend observed for the ripening of the zirconia shell in ethanolic sol, as described with respect to Figure S4, are thus conserved after transferring the NPs into water: the more acidic hydrogenocarbonate species are gradually replaced by the alkaline monodentate and bridged carbonate as zirconia is further aged in the ethanolic solution. This is in agreement with observation of increasing acidification of the sols with time. The carbonate species could contribute to this process by two consecutive processes that also cause acidification: (a) protons could be removed from surface in the process of hydrogenocarbonate to monodentate carbonate conversion, and (b) monodentate carbonates could become bridged carbonates releasing a CO_2 molecule into the solution, which can further acidify the medium. These differences in the surface chemistry of the zirconia show the importance of the careful timing during the synthesis, as the resulting material surface is highly dependent on it.

3.3. Aging of Native Silica@zirconia Core@shell NPs in Water Following Dialysis

As the NMR study suggested, the transformation of the NP surface did not stop after dialyzing the NPs into water, which is why we carried out a follow-up study using FTIR on samples *Iw*, *IIw* and *IIIw* in the weeks following the dialysis.

The enlarged region of hydrogenocarbonate/carbonate vibrational bands showed that the surface structure did not change in the first week in the *Iw* sample, and then the carbonate vibrational bands (1544, 1465 and 1389 cm^{-1}) slowly disappeared (Figure S9 in Supplementary Materials). For *IIw*, the sample aged for 7 days before transfer into the water, we observed a high amount of hydrogenocarbonate and m-CO_3^{2-}/br-CO_3^{2-} on the surface, which were still partly present 5 weeks after dialysis with an increased ratio of bridged carbonate. *IIIw* had a lower amount of binding carbonate species after 3 weeks of time spent in water than *IIw* after 5 weeks in water, though the composition of surface species was similar. *IIw* and *IIIw* samples kept their colloidal stability for at least two months. The longest persistence of surface carbonate species that indicates the highest binding capacity of the zirconia surface was observed for *IIw*.

In conclusion, we determined that the optimum time for transferring the NPs into the water was 4–7 days of ripening after synthesis because not only the NP surface is best covered by accessible sites, but also the dispersity and surface properties of the samples are stable for the longest time, at least for 5 weeks. We also established that the native NPs are best stabilized depending on the aging time at neutral-alkaline pH, whereby electrostatic repulsion between negatively charged surface prevent aggregation.

3.4. Adsorption of Deoxynucleoside Monophosphates on Silica@zirconia NPs

As we observed a significant transformation of the zirconia shell during the transfer of the core@shell NPs into water, we expected to see differences in their adsorption properties as well. We undertook investigations to identify the experimental design that leads to the maximum surface loading and stability of the vaccine nanocarriers. Two experimental designs were used for the surface modification of the NPs differing in the order of steps: (i) NP synthesis-dNMP functionalization-dialysis, and (ii) NP synthesis-dialysis-dNMP functionalization-dialysis. In the first method, an aqueous solution of dNMPs was added to the NPs of different ages (0, 6 and 21 days of condensation time) in ethanol, followed by the dialysis of the NPs into the water. In the second design, the dialysis of the NPs took place first (after 4 days of condensation time), and then the aqueous dNMP solution was added and followed by further dialysis. We refer to these two methods as the ethanolic and the aqueous surface modification methods.

In the first study, we identified the optimal pH range for the adsorption of single dNMP solutions and four-component mixtures. For this, we used the aqueous surface modification method, adding 63 mg ligands to 1 mg of NPs (without the last dialysis step) and performed a pH-dependent zeta potential study. We recorded the free ligand content of the supernatant by UV detection at the same time after centrifugation of the NPs at each pH (Figure 5).

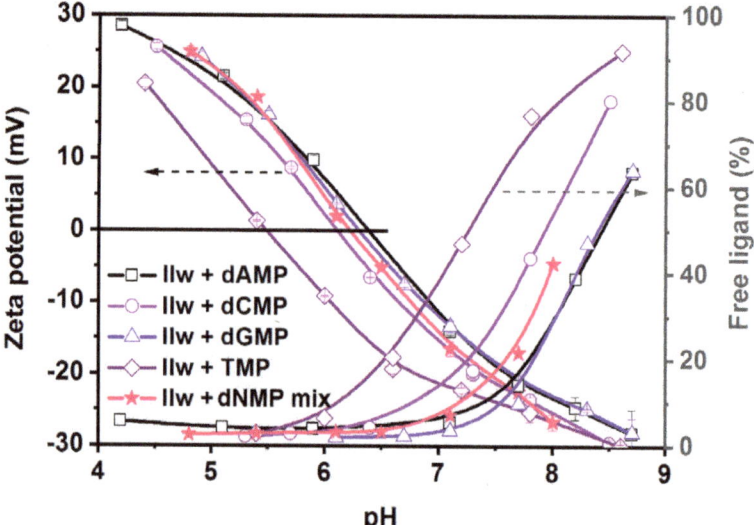

Figure 5. Variation of zeta potential (**left axes**) and the percentage of free ligands (**right axes**) in the function of pH for single-component dNMP solutions and a four-component dNMP mixture. The resulting IEP was 6.3 ± 0.2 for all, except for TMP ligand, whose adsorption took place at lower pH, below pH 7. For the three other single components and the dNMP mixture, the adsorption was almost complete at pH below 7.5.

Specific adsorption took place in the acidic-neutral pH range in all of the examined cases of single- and four-component mixtures, as evidenced by the low free ligand content in the supernatant (Figure 5). In general, in the acidic pH range, the surface of NPs was somewhat positive upon surface modification, which was gradually turning to neutral around the IEP (pH 6.3 ± 0.2) and further to strong negative charges as the pH was increased accompanied by the desorption of the ligands. As the electrostatic repulsion, either between positive or negative charges, is a strong determinant of colloidal stability, it was extremely important to control the pH of the surface-modified NPs. We demonstrated that at neutral-acidic pH, the silica cores started to dissolve (especially at room temperature), and the present results indicate that in this region, they are also prone to aggregation due to low surface charge. On the other hand, high surface loading can only be reached at a pH lower than 7.5. This delimits a narrow useful pH range for the efficient surface modification and the use of silica@zirconia NPs with dNMPs to pH 7.0–7.5. This is not completely in line with the observations of Wu et al., who reported a pH range of 5–7 as optimal for DNA adsorption on titania/zirconia NPs [44]. Although the stability of the nanocarriers is just as good at pH 5 (due to a positively charged surface) as at pH 7, we found that the addition of single components or a dNMP mixture solution at the native pH (pH 4.9) yielded aggregated samples. Since the target nucleic acid sequences contain all nucleotides, we continued the optimization of the surface modification using the dNMP mixture as the ligand solution in the pH range of 7.0–7.5.

3.5. Surface Modification in Ethanol vs. in Water and the Effect of Zirconia Shell Ageing

We compared the ethanolic and aqueous surface modification methods using a dNMP mixture ligand solution at a controlled pH and *SZ3* NPs of 50 nm mean diameter (Table 3). Since the NPs are colloidally more stable above pH 7 due to electrostatic repulsion between negative surface charges, while the ligand adsorption is favored at acidic pH, the pH used in this study for the NPs and the dNMP mixture was chosen to be pH 6.0 to aid swift ligand adsorption, and the resulting solution was dialyzed against alkaline water to prevent aggregation. Three samples were surface modified using the ethanolic procedure at the age of 0, 6 and 20 days after dialysis (*INu*, *IINu* and *IIINu*, respectively), and one sample was surface-modified using the aqueous procedure (*IIwNu*), which was dialyzed into the water 7 days after synthesis, which was found optimal in the previous investigations described in paragraph 3.2. We investigated the pH of the samples, the dispersity with DLS, the structure and composition of the surface with FTIR and the morphology of the NPs by TEM, and we determined the loading capacity of the particles by quantitative analysis with TGA and desorptional titration.

Table 3. A comparison of stability and dNMP content of NPs surface modified in ethanol and in water. *INu*, *IINu* and *IIINu* are samples modified using the ethanolic procedure at the age of 0, 6 and 20 days after dialysis, and the *IIwNu* sample is treated by the aqueous procedure, which is dialyzed into water 7 days after synthesis.

Sample	pH at the End of Dialysis	Zeta Potential (mV)	Z-Average (nm) and PdI from DLS	dNMP Mix Content from Titration		dNMP Mix Content from TGA	
				mg/g	µmol/m^2 *	mg/g	µmol/m^2 *
INu	8.2	−36 ± 2	106 (0.586)	196	1.26	124	0.74
IINu	7.4	−39 ± 1	107 (0.501)	108	0.63	120	0.71
IIINu	5.7	−3 ± 1	high	53	0.29	63	0.33
IIwNu	6.9	−30 ± 1	247 (0.179)	n.a.	n.a.	207	1.34

* Calculated using the volume average mean diameter of the sample. PdI: polydispersity Index.

Similarly, to the case of native silica@zirconia NPs, we observed that independently of the medium of the surface modification and despite the careful pH control of the dialyzing water, the pH of the resulting samples was gradually decreasing with age, e.g., 6-week-old *INu* had a pH of 6.7 in contrast to pH 8.2 when freshly made (Table 3). We ascribe this phenomenon to the continuous dissolution of the silica core due to the microporous structure of the zirconia layer (see below), which was not inhibited by the surface modification. Thus, the pH of the sample ought to be controlled for longer-term stability.

Despite the highly charged surface for samples *INu*, *IINu* and *IIwNu* (Table 3), the dispersity of the resulting samples could not be completely restored after solvent exchange according to DLS. While large aggregates were present in very low amounts for the samples surface modified using the ethanolic procedure at 0 or 6 days of shell condensation (*INu*, *IINu*), there were smaller clusters of primary particles in a higher amount for samples surface modified with the aqueous procedure (*IIwNu*) (Figure 6a). The sample surface modified using the ethanolic procedure at 20 days of shell condensation (*IIINu*) was found to be strongly aggregated.

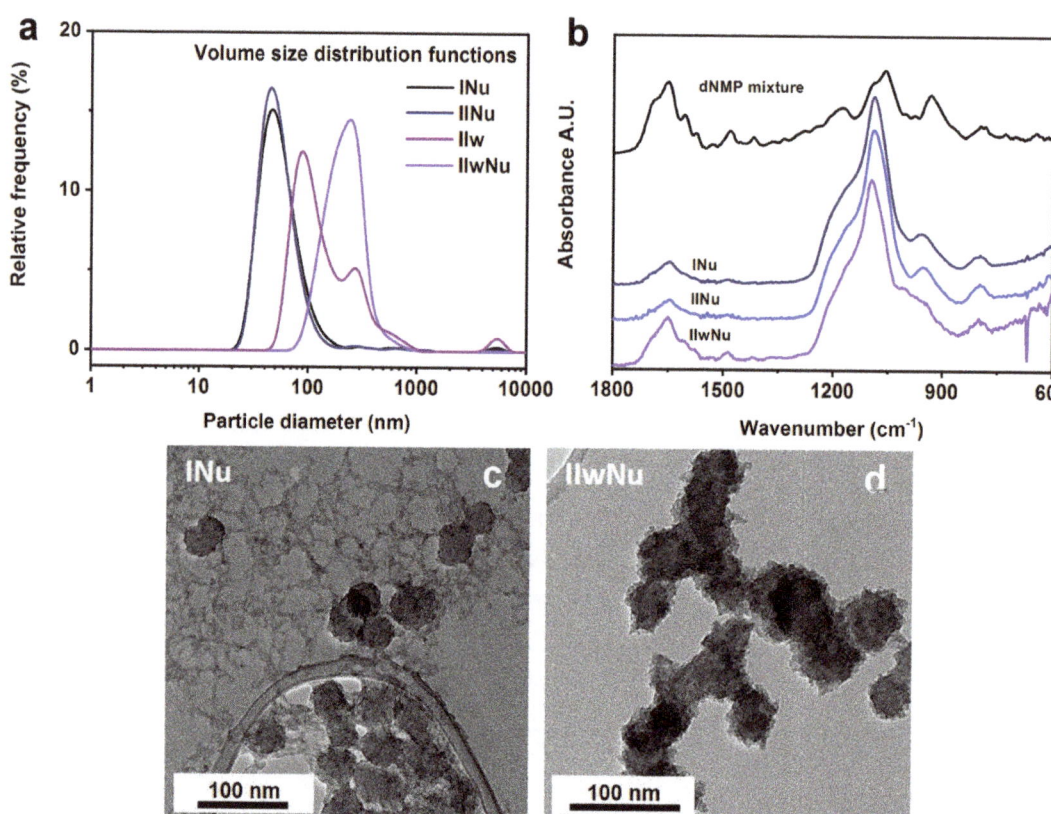

Figure 6. A comparison of samples surface modified in ethanol and in water by (**a**) DLS, (**b**) FTIR spectra and (**c–d**) TEM pictures. The dispersity of the sol is affected by the ethanolic and the aqueous surface modification methods (**a**). A higher surface load can be achieved by surface modification in water (**b**). The addition of aqueous ligand solutions always results in the deposition of zirconia next to the NPs using the ethanolic surface modification method, as shown for dNMP mixture in (**c**) for sample *INu*, while this is not observed for surface modification after transfer into water (**d**) for sample *IIwNu*.

The silica nanoparticulate structure was preserved during both surface modification processes according to the FTIR spectra. When spectra were normalized for the νSi-O-Si band at 1080 cm^{-1}, it became clear that the aqueous surface modification led to higher dNMP cargo load, as indicated by the higher intensity of vibrational bands at 1650 cm^{-1}, 1607 cm^{-1}, 1486 cm^{-1}, and 1575 cm^{-1} (Figure 6b).

TEM pictures revealed that during the surface modification in ethanol, part of the zirconia deposited next to the NPs, giving rise to an inhomogeneous system (Figure 6c). When the surface modification was carried out after the transfer of NPs into the water, zirconia was only deposited on the silica surface and did not form stand-alone crystallites (see Figure 6d). This analysis showed that the addition of the aqueous ligand solution to the NPs in ethanol also provoked a sudden crystallization of the unreacted TBOZ, but the TBOZ deposited next to the particles, unlike during the dialysis of native NPs when the condensation of the unreacted TBOZ caused the aggregation. This observation was identical in all cases of single-component ligand solutions and the four-component mixture. We hypothesize that the ligands quickly exchange the surface-adsorbed TBOZ; thus, the organometallic compound cannot deposit at the surface of NPs anymore but reacts readily with the injected water forming stand-alone ZrO_2 crystallites. However, in the case of the aqueous surface modification, the unreacted TBOZ physisorbed on the NPs are already removed by the preceding dialysis step when the ligand solution is added to it, so no zirconia aggregates are formed. Altogether, these results demonstrated the clear superiority of the aqueous surface modification method.

Quantitative analysis of the bound dNMP mixture was performed by desorption titration on sol samples as well as by TGA on dried powders (Table 3). In order to evaluate the quantity of dNMP mixture desorbed from the surface of NPs at pH 8.8, the UV-visible absorption of the supernatant was evaluated at this pH based on a calibration curve and the solid content of the samples (Figure S10 in Supplementary Materials). There was a good agreement between the quantities obtained by the two methods (except for *INu*) and a clear trend was found: although the adsorption of ligands was observed to become complete at an acidic pH, below 6.5 or lower, the capacity of NPs to bind ligands was decreasing with growing age despite the decreasing pH. According to XRD and FTIR investigations on the transformation of crystalline phases on the NP surface, as described in Section 3.1, the decreasing binding capacity is probably due to the ripening of the zirconia shell accompanied by a reduction of the number of crystal lattice defects. The surface load was found to be the highest (207 mg/g) for sample *IIwNu* surface modified in water following dialysis on day 6 of the shell condensation. This capacity is much higher than the value reported by Wu et al. [44], capturing nucleosides from urine with titania-zirconia NPs (35 mg/g). The material synthesized by Wu et al. is similar in composition to our system, but they used surfactant-assisted deposition of mesoporous titania-zirconia coating on the surface of mesoporous silica and had to eliminate surfactants at 400 °C. This had obviously reduced the adsorption capacity of the obtained material even though they had achieved a high surface area (350 m^2/g).

The BET-specific surface area of dried *SZ3* was assessed by the N_2 adsorption-desorption method and was found to be 315 m^2/g, which is somewhat smaller obtained by Wu et al. and higher compared to the surface load. Some mesoporosity was found in *SZ3*, but this we attributed to interparticulate spaces (Figure S11 in Supplementary Materials). The microporosity of the sample was determined to be 0.121 cm^3/g with a mean pore diameter of 1.4 nm. Thus, the silica@zirconia NPs prepared under mild wet conditions in our work presented similar surface area, with more than 5 times the adsorption capacity of the mesoporous titania-zirconia described by Wu et al. The research group of Badruddoza [45] captured nucleosides using cyclodextrin-modified magnetic NPs, whereby the binding mode was based on Van der Waals and hydrophobic interactions, which is thus different than the electrostatic interaction-based adsorption in our system. They achieved a load of 12.4 mg/g (adenosine) and 29.9 mg/g (guanosine), which is also considerably lower than obtained in our study.

To summarize, we found that the optimal way to prepare highly loaded silica@zirconia materials is to use the smallest stable size (50 nm in our study), transfer the NPs into cold, basic water after an ideal aging period of 4–7 days and carry out surface modification in water with a dNMP mixture at a controlled pH (7.0–7.5). The regulation of the pH plays a crucial role in the long-term colloidal stability of the sample; therefore, the possibility of buffering was further investigated.

3.6. Influence of Buffers on the Adsorption Equilibrium

In view of further biological applications, we selected several buffers based on the suggestions of Ferreira et al. [31] and studied the adsorption of the dNMP mixture on the surface of the NPs dialyzed into the water after 6 days of shell condensation (sample *IIw*), in the presence of 10 mM HEPES, PIPES, MES, MOPS, MOPSO and K-phosphate.

The study was conducted under identical conditions as described earlier in the investigation of the dNMP adsorption. The zeta potential of the NPs and the UV spectrum of the supernatants were recorded, and the latter was analyzed for free ligand content (Figure S12 in Supplementary Materials). The zeta potential vs. pH curves of the dNMP-modified NPs was similar in the presence of PIPES, MOPS, MOPSO and MES and was shifted towards the acidic region compared to the non-buffered system without buffer (IEP 5.8 vs. IEP 6.3). This indicates the presence of stronger negative charges on the surface at the pH range of interest, between pH 7.0–7.5, in these buffers. In the presence of HEPES, we observed a more pronounced shift of IEP (pH 5.5). The percentage of free ligands showed a similar picture: it fell below 10% at pH 6.5 and lower for all the buffered systems except for phosphate buffer, for which it remained around 70%. K-phosphate buffer behaved completely differently from the other buffers, as the charge of the NPs was -31 ± 1 mV in the entire observed pH range, indicating a chemical transformation taking place on the surface. The phosphate ions compete with dNMP molecules and occupy most of the surface sites, thereby providing a high negative zeta potential in the whole of the observed pH range. As a result, 70% of the dNMP ligands remain free and cannot bind to the surface.

We concluded that the phosphate buffer was incompatible with our system. For the rest of the buffers, we did not see interference with the adsorption phenomenon. They are, therefore, all suitable buffers to control the pH and to ensure colloidal stability of the present system.

Finally, in order to choose the best buffering medium that enables the highest amount of dNMP mixture adsorbed to the surface, we recorded adsorption isotherms in the four selected buffers (we discarded MES because of its upper buffering limit at pH 6.7) at 22 °C and pH 7.4 (Figure 7a). The isotherms followed the Langmuir model (Figure 7b); therefore, by plotting the reciprocal of the adsorbed amount of ligands against the reciprocal of the initial concentration, we obtained a linear relationship, which allowed us to evaluate the maximum adsorbing amount of ligands (q_m) in each buffer (Table S1 in Supplementary Materials) [45,46]. The highest amount of dNMP mixture adsorbed to the surface was achieved for *IIw* in HEPES buffer: 187 mg/g (pH 7.4), which is also in agreement with its largest IEP shift corresponding to the highest negative charges on the surface. Comparing this value with the surface loading for non-buffered *IIwNu* obtained by TGA (207 mg/g; pH 6.9), we can state that the maximum load in a buffered system is only slightly lower than the one obtained in non-buffered one. We would expect that the maximum load may even be identical at pH 6.9.

In summary of these steps, we evidenced that silica@zirconia core@shell NPs prepared under mild wet synthetic conditions bind a high amount of dNMP on their surface and buffering necessary to prevent their early degradation does not inhibit this process. Moreover, the pH stability range of the nanocarriers coincides with the physiological pH optimal for the administration of vaccine delivery systems. This way, we confirmed that the buffered system sustained colloid stability and is suitable for its intended biological use.

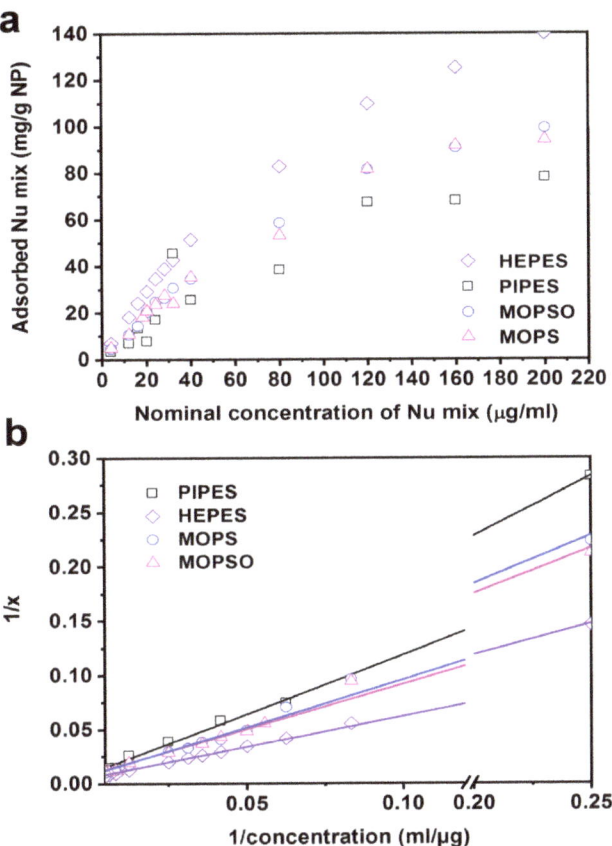

Figure 7. Adsorption isotherms of the dNMP mixture on *IIw* in 20 mM buffers at 22 °C (**a**) and a linearized plot used for fitting (**b**). The adsorption isotherms followed the Langmuir model, allowing us to derive the maximum adsorption capacity for each system.

4. Conclusions

Our paper presents the systematic progress towards the establishment of an optimized synthetic procedure of high adsorption capacity silica@zirconia nanocarrier.

We achieved high specific surface area nanocarriers through lowering particle size and using very mild synthetic conditions. We studied the time-dependent changes in the zirconia shell structure and optimized the solvent exchange and the surface modification accordingly. The aging time of the zirconia shell in the synthetic mixture largely determined the colloidal stability and the aggregation behavior of the NPs during the dialysis step due to a layer of physisorbed TBOZ zirconia starting material attached to the surface. Furthermore, the pH was found to be crucial for the maintenance of colloid stability during and following the surface modification, with the optimum being between pH 7.0–7.5. We tested several biocompatible buffers of which HEPES, PIPES, MOPS and MOPSO were proven to be suitable and compatible with our nanocarrier system. Based on these results, we anticipate stable and high-capacity adsorption of nucleic acid-type antigens or adjuvants on the surface of the optimized NPs, and in our future work, we proceed to the elaboration of vaccine adjuvant nanocarriers using these cargo molecules.

Supplementary Materials: The following are available online at https://www.mdpi.com/article/10.3390/nano11092166/s1, **Figure S1:** TEM pictures of silica core for batches *S1*, *S2*, *S3* and *S4* and silica@zirconia core@shell for batches *SZ1*, *SZ2*, *SZ3* and *SZ4* NPs, **Figure S2:** DLS size distribution functions of core and core@shell NPs, **Figure S3:** XRD spectrum of 2-week-old SZ3 NPs, line fitted using Rietveld analysis and the error to the fit, **Figure S4:** FTIR spectra of *SZ3* silica@zirconia core@shell NPs in ethanol at different aging times after deposition, **Figure S5:** DLS size distribution functions of different size NPs dialyzed into water at pH 9 at the age of 4-7 days after deposition of zirconia shell and 50 nm diameter particles dialyzed after different aging times at pH 9, **Figure S6:** X-ray diffractogram and Rietveld refinement evaluation of *Iw* powder, **Figure S7:** FTIR spectra of native NPs dialyzed into basic water without temperature control, **Figure S8:** FTIR spectra of *SZ1* NPs surface modified in ethanol without controlling the pH of ligand solutions and *SZ3* NPs surface modified freshly after synthesis in ethanol using dNMP mixtures at pH 6.0, **Figure S9:** Structural changes of *SZ3* core@shell NPs after transfer into water by dialysis at pH 9 at different ages according to FTIR spectra. *Iw* (transfer on day 1 after zirconia deposition); *IIw* (transfer on day 7); *IIIw* (transfer on day 21), **Figure S10:** Combined zeta potential and free ligand quantity of the supernatant vs. pH curves obtained during desorption titration and calibration curve used for *Nu* mix quantity assessment, **Figure S11:** N_2 adsorption-desorption isotherm and micropore size distribution of *SZ3* (dried at the age of 4 days) powder, **Figure S12:** Variation of the zeta potential (left axis) and the percentage of free ligands (right axis) in the function of the pH for the dNMP mixture in the presence of 10 mM buffers, **Table S1:** Goodness of fit and maximum absorbing dNMP mixture on *IIw* in 20 mM buffers at 22 °C and at pH 7.2.

Author Contributions: Conceptualization, L.N.N., J.C.M. and K.F.; funding acquisition, L.N.N., K.E.K., J.C.M. and K.F.; investigation, L.N.N., E.D., M.V.Z., J.M. and S.K.; methodology, L.N.N., Z.V., K.D.B., I.V.D. and K.F.; project administration, L.N.N., J.C.M. and K.F. All authors have read and agreed to the published version of the manuscript.

Funding: This project has received funding from the European Union's Horizon 2020 research and innovation programme under the Marie Sklodowska-Curie grant agreement No 703374. E.D. acknowledges FWO-Vlaanderen for a fellowship (FWO-SB fellowship). I.V.D. acknowledges Ghent University (project BOF2015/GOA/007). K.F. acknowledges the support of the Marie Curie Career Integration Grant (303917 PGN-INNATE) and the Research Grant from the Research Foundation-Flanders (1525517N), the János Bolyai Research Scholarship of the Hungarian Academy of Sciences (BO/004333/18/7) and the New National Excellence Program of Debrecen University (ÚNKP-20-4-DE-165 Bolyai+). This research was supported by the National Research, Development and Innovation Office of Hungary (grant NKFI/OTKA NN 128368).

Conflicts of Interest: There are no conflict of interest to declare.

References

1. Park, K.S.; Sun, X.; Aikins, M.E.; Moon, J.J. Non-viral COVID-19 vaccine delivery systems. *Adv. Drug Deliv. Rev.* **2020**, *169*, 137–151. [CrossRef] [PubMed]
2. Xu, Z.; Patel, A.; Tursi, N.J.; Zhu, X.; Muthumani, K.; Kulp, D.W.; Weiner, D.B. Harnessing Recent Advances in Synthetic DNA and Electroporation Technologies for Rapid Vaccine Development Against COVID-19 and Other Emerging Infectious Diseases. *Front. Med. Technol.* **2020**, *2*. [CrossRef]
3. Pardi, N.; Hogan, M.J.; Porter, F.W.; Weissman, D. mRNA vaccines—A new era in vaccinology. *Nat. Rev. Drug Discov.* **2018**, *17*, 261–279. [CrossRef]
4. Van den Boorn, J.G.; Barchet, W.; Hartmann, G. Nucleic Acid Adjuvants: Toward an Educated Vaccine. In *Advances in Immunology*; Melief, C.J.M., Ed.; Academic Press: Cambridge, MA, USA, 2012; Chapter 1; Volume 114, pp. 1–32.
5. Brown, K.; Puig, M.; Haile, L.; Ireland, D.; Martucci, J.; Verthelyi, D. Nucleic Acids as Adjuvants. In *Oligonucleotide-Based Drugs and Therapeutics*; John Wiley & Sons: Hoboken, NJ, USA, 2018; pp. 411–444.
6. Temizoz, B.; Kuroda, E.; Ishii, K.J. Combination and inducible adjuvants targeting nucleic acid sensors. *Curr. Opin. Pharmacol.* **2018**, *41*, 104–113. [CrossRef] [PubMed]
7. Klinman, D.M.; Currie, D.; Gursel, I.; Verthelyi, D. Use of CpG oligodeoxynucleotides as immune adjuvants. *Immunol. Rev.* **2004**, *199*, 201–216. [CrossRef] [PubMed]
8. Melief, C.J.M.; van Hall, T.; Arens, R.; Ossendorp, F.; van der Burg, S.H. Therapeutic cancer vaccines. *J. Clin. Investig.* **2015**, *125*, 3401–3412. [CrossRef] [PubMed]
9. High, K.A.; Roncarolo, M.G. Gene Therapy. *N. Engl. J. Med.* **2019**, *381*, 455–464. [CrossRef] [PubMed]

10. Slivac, I.; Guay, D.; Mangion, M.; Champeil, J.; Gaillet, B. Non-viral nucleic acid delivery methods. *Expert Opin. Biol. Ther.* **2017**, *17*, 105–118. [CrossRef] [PubMed]
11. Luther, D.C.; Huang, R.; Jeon, T.; Zhang, X.; Lee, Y.-W.; Nagaraj, H.; Rotello, V.M. Delivery of drugs, proteins, and nucleic acids using inorganic nanoparticles. *Adv. Drug Deliv. Rev.* **2020**, *156*, 188–213. [CrossRef]
12. Mamo, T.; Poland, G.A. Nanovaccinology: The next generation of vaccines meets 21st century materials science and engineering. *Vaccine* **2012**, *30*, 6609–6611. [CrossRef]
13. Apratim, A.; Eachempati, P.; Krishnappa Salian, K.; Singh, V.; Chhabra, S.; Shah, S. Zirconia in dental implantology: A review. *J. Int. Soc. Prev. Community Dent.* **2015**, *5*, 147–156. [CrossRef] [PubMed]
14. Srinivas, M.; Buvaneswari, G. A Study of in Vitro Drug Release from Zirconia Ceramics. *Trends Biomater. Artif. Organs* **2006**, *20*, 24–30.
15. Catauro, M.; Raucci, M.G.; Ausanio, G. Sol–gel processing of drug delivery zirconia/polycaprolactone hybrid materials. *J. Mater. Sci. Mater. Med.* **2007**, *19*, 531–540. [CrossRef] [PubMed]
16. Nagy, L.N.; Mihály, J.; Polyak, A.; Debreczeni, B.; Csaszar, B.; Szigyártó, I.C.; Wacha, A.; Czégény, Z.; Jakab, E.; Klébert, S.; et al. Inherently fluorescent and porous zirconia colloids: Preparation, characterization and drug adsorption studies. *J. Mater. Chem. B* **2015**, *3*, 7529–7537. [CrossRef]
17. De Keukeleere, K.; De Roo, J.; Lommens, P.; Martins, J.C.; Van Der Voort, P.; Van Driessche, I. Fast and Tunable Synthesis of ZrO_2 Nanocrystals: Mechanistic Insights into Precursor Dependence. *Inorg. Chem.* **2015**, *54*, 3469–3476. [CrossRef] [PubMed]
18. Qin, W.; Zhu, L. Anisotropic morphology, formation mechanisms, and fluorescence properties of zirconia nanocrystals. *Sci. Rep.* **2020**, *10*, 1–10. [CrossRef]
19. Xia, Y.; Shi, J.; Sun, Q.; Wang, D.; Zeng, X.-F.; Wang, J.-X.; Chen, J.-F. Controllable synthesis and evolution mechanism of monodispersed Sub-10 nm ZrO_2 nanocrystals. *Chem. Eng. J.* **2020**, *394*, 124843. [CrossRef]
20. Wang, Y.; Bouchneb, M.; Mighri, R.; Alauzun, J.G.; Mutin, P.H. Water Formation in Non-Hydrolytic Sol–Gel Routes: Selective Synthesis of Tetragonal and Monoclinic Mesoporous Zirconia as a Case Study. *Chem. A Eur. J.* **2020**, *27*, 2670–2682. [CrossRef]
21. Widoniak, J.; Maret, G.; Eiden-Assmann, S. Synthesis and Characterisation of Monodisperse Zirconia Particles. *Eur. J. Inorg. Chem.* **2005**, *2005*, 3149–3155. [CrossRef]
22. Arnal, P.M.; Weidenthaler, C.; Schuth, F. Highly Monodisperse Zirconia-Coated Silica Spheres and Zirconia/Silica Hollow Spheres with Remarkable Textural Properties. *Chem. Mater.* **2006**, *18*, 2733–2739. [CrossRef]
23. Tang, S.; Huang, X.; Chen, X.; Zheng, N. Hollow Mesoporous Zirconia Nanocapsules for Drug Delivery. *Adv. Funct. Mater.* **2010**, *20*, 2442–2447. [CrossRef]
24. Nomura, T.; Tanii, I.; Ishikawa, M.; Tokumoto, H.; Konishi, Y. Synthesis of hollow zirconia particles using wet bacterial templates. *Adv. Powder Technol.* **2013**, *24*, 1013–1016. [CrossRef]
25. Chenan, A.; Ramya, S.; George, R.; Mudali, U.K. Hollow mesoporous zirconia nanocontainers for storing and controlled releasing of corrosion inhibitors. *Ceram. Int.* **2014**, *40*, 10457–10463. [CrossRef]
26. Huang, X.-Q.; Yang, H.-Y.; Luo, T.; Huang, C.; Tay, F.R.; Niu, L.-N. Hollow mesoporous zirconia delivery system for biomineralization precursors. *Acta Biomater.* **2018**, *67*, 366–377. [CrossRef]
27. Wysokowski, M.; Motylenko, M.; Bazhenov, V.V.; Stawski, D.; Petrenko, I.; Ehrlich, A.; Behm, T.; Kljajic, Z.; Stelling, A.L.; Jesionowski, T.; et al. Poriferan chitin as a template for hydrothermal zirconia deposition. *Front. Mater. Sci.* **2013**, *7*, 248–260. [CrossRef]
28. Pokrovski, K.; Jung, K.T.; Bell, A. Investigation of CO and CO_2 Adsorption on Tetragonal and Monoclinic Zirconia. *Langmuir* **2001**, *17*, 4297–4303. [CrossRef]
29. Köck, E.M.; Kogler, M.; Bielz, T.; Klötzer, B.; Penner, S. In Situ FT-IR Spectroscopic Study of CO_2 and CO Adsorption on Y_2O_3, ZrO_2, and Yttria-Stabilized ZrO_2. *J. Phys. Chem. C Nanomater. Interfaces* **2013**, *117*, 17666–17673. [CrossRef] [PubMed]
30. Hens, Z.; Martins, J.C. A Solution NMR Toolbox for Characterizing the Surface Chemistry of Colloidal Nanocrystals. *Chem. Mater.* **2013**, *25*, 1211–1221. [CrossRef]
31. Ferreira, C.; Pinto, I.; Soares, E.; Soares, H. (Un)suitability of the use of pH buffers in biological, biochemical and environmental studies and its interaction with metal ions—A review. *RSC Adv.* **2015**, *5*, 30989–31003. [CrossRef]
32. Stöber, W.; Fink, A.; Bohn, E. Controlled growth of monodisperse silica spheres in the micron size range. *J. Colloid Interface Sci.* **1968**, *26*, 62–69. [CrossRef]
33. Kim, J.; Chang, S.; Kim, S.; Kim, K.-S.; Kim, J.; Kim, W.-S. Design of SiO_2/ZrO_2 core–shell particles using the sol–gel process. *Ceram. Int.* **2009**, *35*, 1243–1247. [CrossRef]
34. Nagy, L.N.; Polyak, A.; Mihály, J.; Szécsényi, Á.; Szigyártó, I.C.; Czégény, Z.; Jakab, E.; Németh, P.; Magda, B.; Szabó, P.; et al. Silica@zirconia@poly(malic acid) nanoparticles: Promising nanocarriers for theranostic applications. *J. Mater. Chem. B* **2016**, *4*, 4420–4429. [CrossRef]
35. Kurapova, O.; Konakov, V.G. Phase evolution in zirconia based systems. *Rev. Adv. Mater. Sci.* **2014**, *36*, 177–190.
36. Prakashbabu, D.; Krishna, R.H.; Nagabhushana, B.; Nagabhushana, H.; Shivakumara, C.; Chakradar, R.; Ramalingam, H.; Sharma, S.; Chandramohan, R. Low temperature synthesis of pure cubic ZrO_2 nanopowder: Structural and luminescence studies. *Spectrochim. Acta Part A Mol. Biomol. Spectrosc.* **2014**, *122*, 216–222. [CrossRef]
37. Salavati-Niasari, M.; Dadkhah, M.; Nourani, M.R.; Fazl, A.A. Synthesis and Characterization of Single-Phase Cubic ZrO_2 Spherical Nanocrystals by Decomposition Route. *J. Clust. Sci.* **2012**, *23*, 1011–1017. [CrossRef]

38. Iler, R.K. *The chemistry of Silica: Solubility, Polymerization, Colloid and Surface Properties, and Biochemistry*; Wiley: New York, NY, USA, 1979.
39. Nagy, L.N.; Dhaeneb, E.; Szigyártó, I.C.; Mihályc, J.; May, Z.; Varga, Z.; Van Driessche, I.; Martins, J.C.; Fehéred, K. An unsought and expensive way to make gold nanoparticles on the way to the development of $SiO_2@ZrO_2$ nanocarriers for cancer vaccination. *J. Mol. Liq.* **2020**, *311*, 113307. [CrossRef]
40. Braun, K.; Pochert, A.; Beck, M.; Fiedler, R.; Gruber, J.; Lindén, M. Dissolution kinetics of mesoporous silica nanoparticles in different simulated body fluids. *J. Sol Gel Sci. Technol.* **2016**, *79*, 319–327. [CrossRef]
41. Paris, J.L.; Colilla, M.; Izquierdo-Barba, I.; Manzano, M.; Vallet-Regí, M. Tuning mesoporous silica dissolution in physiological environments: A review. *J. Mater. Sci.* **2017**, *52*, 8761–8771. [CrossRef]
42. Quignard, S.; Coradin, T.; Powell, J.J.; Jugdaohsingh, R. Silica nanoparticles as sources of silicic acid favoring wound healing in vitro. *Colloids Surf. B Biointerfaces* **2017**, *155*, 530–537. [CrossRef] [PubMed]
43. Pham, A.L.-T.; Sedlak, D.L.; Doyle, F.M. Dissolution of mesoporous silica supports in aqueous solutions: Implications for mesoporous silica-based water treatment processes. *Appl. Catal. B Environ.* **2012**, *126*, 258–264. [CrossRef] [PubMed]
44. Wu, Q.; Wu, D.; Guan, Y. Hybrid Titania–Zirconia Nanoparticles Coated Adsorbent for Highly Selective Capture of Nucleosides from Human Urine in Physiological Condition. *Anal. Chem.* **2014**, *86*, 10122–10130. [CrossRef] [PubMed]
45. Badruddoza, A.Z.; Junwen, L.; Hidajat, K.; Uddin, M.S. Selective recognition and separation of nucleosides using carboxymethyl-beta-cyclodextrin functionalized hybrid magnetic nanoparticles. *Colloids Surf. B Biointerfaces* **2012**, *92*, 223–231. [CrossRef] [PubMed]
46. Duff, D.G.; Ross, S.M.C.; Vaughan, D.H. Adsorption from solution: An experiment to illustrate the Langmuir adsorption isotherm. *J. Chem. Educ.* **1988**, *65*. [CrossRef]

Review

The Combination of Two-Dimensional Nanomaterials with Metal Oxide Nanoparticles for Gas Sensors: A Review

Tao Li [1], Wen Yin [1], Shouwu Gao [2], Yaning Sun [1], Peilong Xu [2], Shaohua Wu [1], Hao Kong [3], Guozheng Yang [3] and Gang Wei [3,*]

1. College of Textile & Clothing, Qingdao University, No. 308 Ningxia Road, Qingdao 266071, China; qdlitao2008@126.com (T.L.); yiwin@126.com (W.Y.); qdfzxsyn@163.com (Y.S.); shaohua.wu@qdu.edu.cn (S.W.)
2. State Key Laboratory, Qingdao University, No. 308 Ningxia Road, Qingdao 266071, China; qdgsw@126.com (S.G.); xpl@qdu.edu.cn (P.X.)
3. College of Chemistry and Chemical Engineering, Qingdao University, No. 308 Ningxia Road, Qingdao 266071, China; konghao5@outlook.com (H.K.); yangguozheng123@outlook.com (G.Y.)
* Correspondence: weigroup@qdu.edu.cn; Tel.: +86-1506-6242-101

Abstract: Metal oxide nanoparticles have been widely utilized for the fabrication of functional gas sensors to determine various flammable, explosive, toxic, and harmful gases due to their advantages of low cost, fast response, and high sensitivity. However, metal oxide-based gas sensors reveal the shortcomings of high operating temperature, high power requirement, and low selectivity, which limited their rapid development in the fabrication of high-performance gas sensors. The combination of metal oxides with two-dimensional (2D) nanomaterials to construct a heterostructure can hybridize the advantages of each other and overcome their respective shortcomings, thereby improving the sensing performance of the fabricated gas sensors. In this review, we present recent advances in the fabrication of metal oxide-, 2D nanomaterials-, as well as 2D material/metal oxide composite-based gas sensors with highly sensitive and selective functions. To achieve this aim, we firstly introduce the working principles of various gas sensors, and then discuss the factors that could affect the sensitivity of gas sensors. After that, a lot of cases on the fabrication of gas sensors by using metal oxides, 2D materials, and 2D material/metal oxide composites are demonstrated. Finally, we summarize the current development and discuss potential research directions in this promising topic. We believe in this work is helpful for the readers in multidiscipline research fields like materials science, nanotechnology, chemical engineering, environmental science, and other related aspects.

Keywords: two-dimensional materials; metal oxide; nanoparticles; composite materials; gas sensors

1. Introduction

In actual production and life, flammable, explosive, toxic, and harmful gases pose a serious threat to environmental safety and human health. Therefore, devices with high performance are urgently needed to detect those flammable, explosive, toxic, and harmful gases. The gas sensors play great importance in determining various gases as they can convert a certain gas volume fraction into electrical signals. They have the advantages of low cost, fast response, high sensitivity, and high selectivity. In addition, in some cases, the gas sensor device can be directly used in electronic interfaces. Therefore, gas sensors have been widely used in environmental monitoring, air quality monitoring, vehicle exhaust monitoring, medical diagnosis, food/cosmetics monitoring, and many other fields [1–8].

The fabrication of nanomaterial-based gas sensors has been the focus of research over the past few decades. According to the working principle of the sensors, gas sensors can be divided into several types, such as semiconductor type, polymer type, contact combustion type, and solid electrolyte [9–12]. Among these gas sensors, the semiconductor gas sensors have developed into one of the largest and most widely used sensors in the world due to their large types of gases, high sensitivity, low price, and simple fabrication [13]. According to the different gas detection methods, semiconductor gas sensors can be divided into two types: resistive and non-resistive, in which the resistive semiconductor gas sensor detects gas concentration according to the change of the resistance value of the semiconductor when it comes into contact with the gas [13]. Currently, gas-sensitive materials such as semiconductor metal oxides, conductive polymers, and carbon materials have been used for the fabrication of resistive semiconductor-type gas sensors [14–17].

Among these sensing materials for the fabrication of semiconductor gas sensors, metal oxides, including zinc oxide (ZnO), indium oxide (In_2O_3), tin oxide (SnO_2), and tungsten oxide (WO_3) have been proved to be the best candidates for the fabrication for making resistive gas sensors due to their advantages of simple fabrication, low cost, easy portability, and high sensitivity [18–21]. However, the high operating temperature, high power, and low selectivity limited their rapid development [22,23]. Two-dimensional (2D) nanomaterials, including graphene, transition metal chalcogenides, layered metal oxides, black phosphorus, and others, have shown great potential in gas sensors due to their unique single-atom-layer structure, high specific surface area, and many surface-active sites [24,25].

2D material-based gas sensors have the advantages of high sensitivity, fast response speed, low energy consumption, and room temperature operation [26–30]. However, since 2D nanomaterials tend to form a dense stack structure during the formation of the conductive network, it is not conducive to the full contact between the flakes inside the conductive network and the gas molecules, making the sensitivity and response recovery speed relatively low at room temperature. Combining metal oxides with 2D nanomaterials to construct a heterostructure can combine the advantages of each other and overcome their respective shortcomings, thereby improving the sensing performance of the fabricated gas sensors. The combination of the metal oxide and 2D materials can drive transformations in the design and performance of 2D nanoelectronics devices, such as the graphene/2D indium oxide/SiC heterostructure [31,32]. Currently, the combination of metal oxides with graphene, transition metal chalcogenide, and other 2D materials to form heterojunction nanostructures for gas sensors have been studied widely, which have exhibited significantly enhanced sensing performance at room temperature [33,34].

In this review, we present the advances in the fabrication and sensing mechanisms of 2D material- and metal oxide nanoparticle-based gas sensors. For this aim, we first introduce the detection mechanism of the resistive semiconductor gas sensors and the factors that can affect the sensitivity of the gas sensors. Then, various types of gas sensors based on metal oxides, 2D materials, and 2D materials/metal oxides composites are introduced. Special emphasis is placed on the recent progress of the combination of metal oxides and 2D nanomaterials for gas sensors. We believe that this review will be helpful for readers to understand the synthesis of functional 2D material-based composites and promote the fabrication of 2D material-based sensors for the high-performance determination of gases.

2. Working Principles of Gas Sensors

2.1. Mechanism of Oxygen Ion Adsorption on the Surface of Metal Oxide Nanoparticles

The sensing mechanism of traditional metal oxide-based gas sensors is related to the resistance change of the sensing materials caused by the adsorption of oxygen ions on the material surface [35,36]. When metal oxides are exposed to air, O_2 in the air is adsorbed onto the surface of metal oxides, which acts as electron acceptors to extract electrons from the conduction band of oxides and dissociates into different forms of negative oxygen ions (O_2^-, O^-, O^{2-}). Since the electrons in the conduction band of the material are captured by oxygen anions, a hole accumulation layer (also called electron depletion layer) rich in hole carriers is formed on the surface of the material, thereby increasing the resistance of the gas-sensing material. Various gases are adsorbed on the surface of metal oxides and interact with oxygen anions to change the electrical conductivity of metal oxides. Taking NiO as an example, when exposed to a reducing gas, the adsorbed oxygen undergoes a redox reaction with the gas, and the captured electrons are released into the conduction band of the semiconductor, and these electrons combine with the holes present in the hole accumulation layer of the sensor material [37]. Therefore, the number of carriers is reduced, and the resistance value is increased, so as to achieve the purpose of gas detection, as shown in Figure 1.

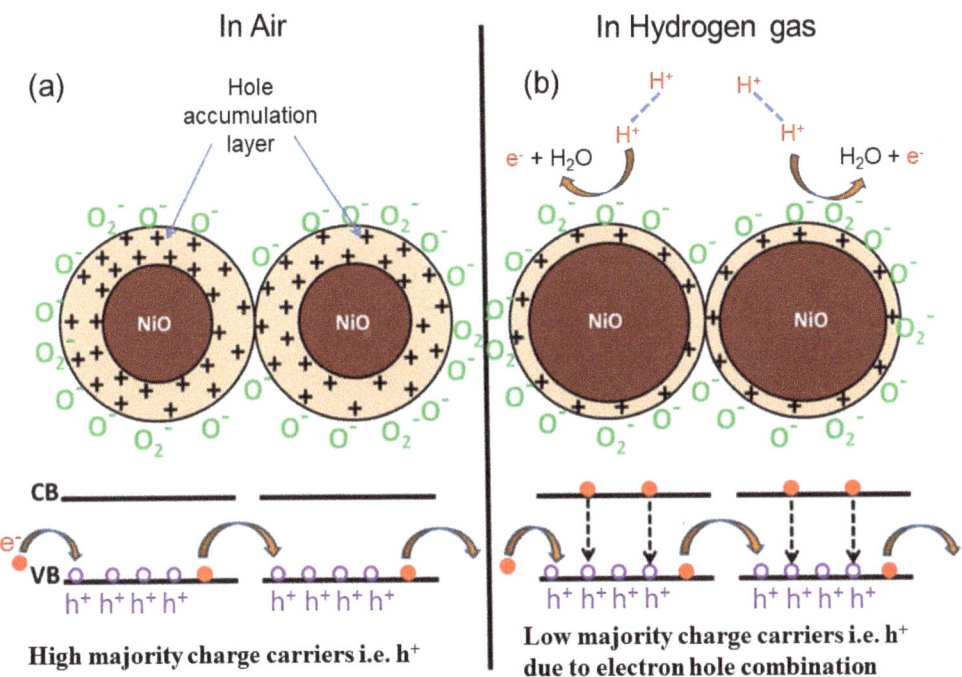

Figure 1. Schematics of H_2 sensing mechanism for NiO sensor. Hole accumulation of the NiO sensor exposed in air (**a**) and hydrogen (**b**), respectively. Reprinted with permission from Ref. [37]. Copyright 2018 Elsevier.

2.2. Charge Transfer Mechanism of 2D Material-Based Gas Sensors

The sensing mechanism of gas sensors based on graphene and related 2D layered materials is mainly related to the charge transfer process, in which the sensing material acts as a charge acceptor or donor [38]. When exposed to different gases, a charge transfer reaction occurs between the sensing material and the adsorbed gas, and the direction and amount of charge transfer are different, resulting in different changes in the material resistance. Taking layered MoS_2 as an example, the charge transfer between different gas molecules (including O_2, H_2O, NH_3, NO, NO_2, CO) and monolayer MoS_2 is different [39], as shown in Figure 2.

Before gas adsorption, some electrons already exist in the conduction band of the n-type MoS_2 monolayer. When MoS_2 is exposed to O_2, H_2O, NO, NO_2, and CO gases, the electron charge is transferred from MoS_2 to the gas atmosphere, resulting in a decrease in the carrier density of MoS_2 and an increase in the resistance of MoS_2. On the contrary, When MoS_2 is exposed to NH_3, the NH_3 molecules adsorbed on MoS_2 act as charge donors to transfer electrons to the MoS_2 monolayer, increasing the carrier density of the MoS_2 monolayer and reducing its resistance (Figure 2).

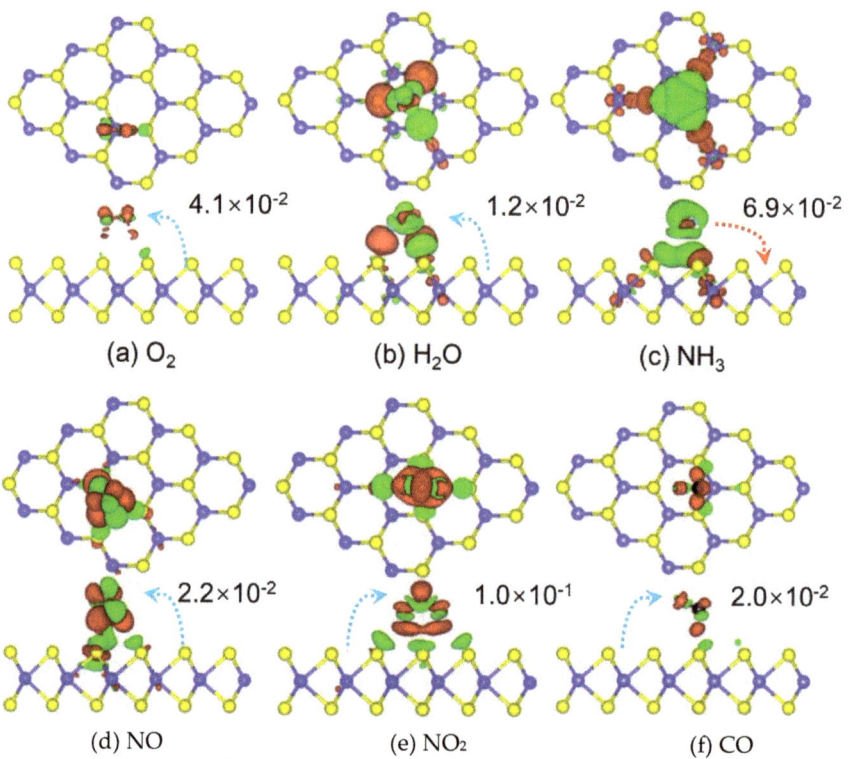

Figure 2. Charge density difference plots for (**a**) O_2, (**b**) H_2O, (**c**) NH_3, (**d**) NO, (**e**) NO_2, and (**f**) CO interacting with monolayer MoS_2. Reprinted with permission from Ref. [39]. Copyright 2013 Springer.

2.3. Gas Sensing Mechanism of 2D Material/Metal Oxide Composites

When one material is composited with another, the bonding between different materials forms p-n, n-n, p-p, and Schottky heterojunctions. Among them, the formation of p-n heterojunction is beneficial for adjusting the thickness of the electron depletion layer, thereby further improving the sensing performance of the fabricated gas sensors. For example, when SnO_2 is combined with graphene oxide (GO), the p-n heterojunction can be formed to form a new energy-level structure, as shown in Figure 3 [40]. The electron dissipation layer expands at the interface of SnO_2 and GO, resulting in increased resistance. When formaldehyde is introduced, the trapped electrons are released back into the conduction band, resulting in a reduction in the width of the dissipation layer, which reduces the resistance of the sample. The porous and ultrathin structure of the SnO_2/GO composite increases the specific surface area and active sites, facilitates the reaction with HCHO gas, and contributes to the ultrahigh response for gas sensing. It should be noted that the ultrathin nanosheet structure of GO shortens the transport path and greatly improves the response of the gas sensor. Meanwhile, the abundant pores in SnO_2 are favorable for gas diffusion and help to improve the response/recovery performance. In addition, GO can act as a spacer, which reduced the agglomeration of SnO_2 nanoparticles, and provided abundant adsorption sites for HCHO gas, thereby enhancing the gas sensing response.

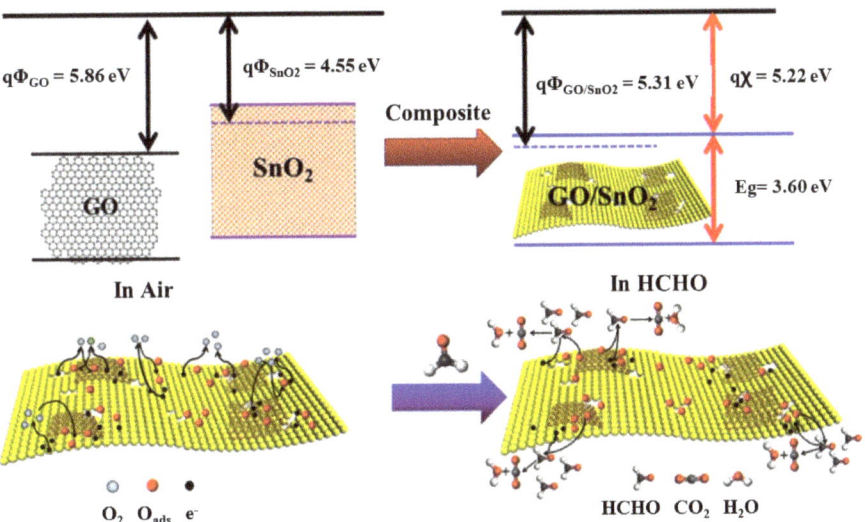

Figure 3. Schematic diagram of the sensing mechanism for GO/SnO$_2$. Reprinted with permission from Ref. [40]. Copyright 2019 ACS.

3. Factors for Affecting the Sensitivity of Gas Sensors

3.1. Size, Morphology, and Porosity

The grain size, morphology, and porosity are important factors for affecting the sensing performance of semiconductor-based gas sensors. In gas sensors, semiconductor nanoparticles are connected to adjacent particles through grain boundaries to form larger aggregates [41]. On the particle surface, the adsorbed oxygen molecules extract electrons from the conduction band and capture electrons in the form of ions on the surface, resulting in band bending and electron depletion layers, as shown in Figure 4 [42,43]. Since the transport of electrons between grains needs to pass through the electron depletion layer, the grain size has a great influence on the conductivity, which in turn affects the gas sensing performance of the material-based gas sensor. When the particle size (D) is much larger than twice the thickness (L) of the electron depletion layer (D >> 2L), there is a wide electron channel between the grains, and the gas sensitivity of the material is mainly controlled by the surface of the nanoparticles (boundary control). When D ≥ 2L, there is a constricted conduction channel. The change in electrical conductivity depends not only on the particle boundary barrier but also on the cross-sectional area of the channel, and the gas sensitivity of the material is mainly controlled by the contact neck between nanoparticles (neck control). When D < 2L, the electron depletion layer dominates, the entire nanoparticle is contained in the electron depletion layer (grain control), and the sensitivity of the material is very high. The energy bands are nearly flat throughout the interconnected grain structure, and there is no significant impediment to the inter-grain charge transport. The small amount of charge gained from the surface reaction results in a large change in the conductivity of the entire structure. Therefore, the smaller grain size is beneficial to improving the sensitivity of the gas sensor. When the grain size is small enough, the crystal becomes very sensitive to the surrounding gas molecules. For example, Min et al. used SnO$_2$ as the sensing material to prepare SnO2 films with different particle sizes (8.4–18.5 nm) and different porosity (70.8–99.5%), and found that in the gas sensors with porous structure, high porosity and low average particle size exhibited quicker gas sensing response [44].

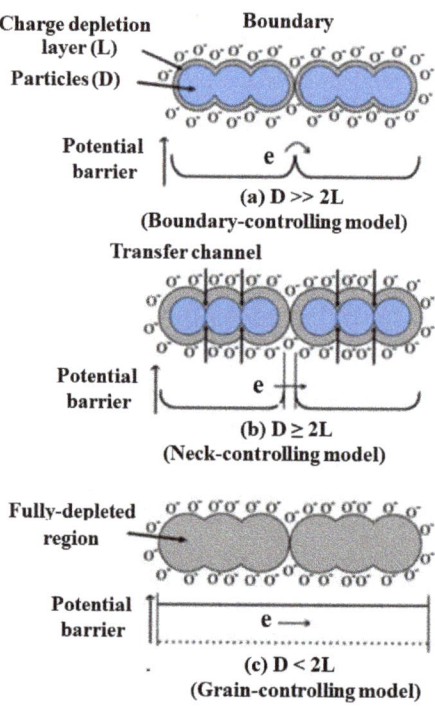

Figure 4. Schematic model of the effect of the crystallite size on the sensitivity of metal-oxide gas sensors: (**a**) D >> 2L, (**b**) D ≥ 2L, and (**c**) D < 2L. Reprinted with permission from Ref. [42]. Copyright 1991 Elsevier.

However, when the grain size is excessively reduced, the agglomeration between particles is serious. If the aggregates are relatively dense, only the particles on the surface of the aggregates could participate in the gas sensing reaction, and the internal materials are wasted because they are not in contact with the gas, resulting in a decrease in the utilization rate of the material. In addition, the agglomeration is not conducive to the diffusion of gas inside the material and will reduce the gas sensing performance [45,46]. Increasing the specific surface area not only facilitates the adsorption of oxygen molecules in the air on the surface of the material, but also increases more effective active sites and more gas transmission channels to facilitate the diffusion and absorption of the test gas. Therefore, increasing the specific surface area of gas-sensing materials is an important way to modify the sensing properties of gas sensors. The regulation of both the morphology (flower-like, sea urchin-like, etc.) and porous structure (macropores, mesopores, and micropores) of materials is an effective way to improve the specific surface area of materials. Nanomaterials with porous structures can increase the effective surface area and active sites of the material due to their special pore structure, so that the material has better permeability, making gas molecules easier to diffuse into the interior of the material, and increasing the contact between the material and the gas. It can accelerate the diffusion of gas, improve the response and recovery speed of the gas sensor, and thus improve its gas sensing performance.

For example, Boudiba et al. synthesized WO_3 materials with different morphologies by direct precipitation, ion exchange, and hydrothermal methods, and further used the as-prepared materials to fabricate gas sensors. Their results indicated that the greater the porosity, the higher the sensitivity to SO_2 gas [47]. In another case, Jia et al. prepared WO_3 semiconductor materials with different morphologies by hydrothermal method as sensitive materials. Under the same test conditions, they found that the sensitivity of WO_3 nanorods towards acetone was 19.52, while the sensitivity of WO_3 nanospheres towards acetone was 25.71. In addition, WO_3 nanoshpheres exhibited better selectivity than nanorods [48]. Lü et al. [49] successfully prepared porous materials with extremely high specific surface area (120.9 $m^2 \cdot g^{-1}$) by simple chemical transformation of Co-based metal-organic frameworks (Co-MOFs) template and controlling the appropriate calcination temperature (300 °C). The prepared Co_3O_4 concave nanocubes were systematically tested for their gas-sensing properties to volatile organic compounds (VOCs), including ethanol, acetone, toluene, and benzene. to the fabricated sensors exhibited excellent performance in gas sensing, such as high sensitivity, low detection limit (10 ppm), fast response and recovery (<10 s), and high selectivity for ethanol. Wang et al. synthesized concave Cu_2O octahedral nanoparticles with a diameter of about 400 nm and performed gas-sensing tests for benzene (C_6H_6) and NO_2 [50]. It was found that the concave Cu_2O octahedral nanoparticles exhibited better gas sensing properties than Cu_2O nanorods. Unlike conventional octahedrons, Cu_2O octahedral nanoparticles have a structure similar to icosahedral with sharp boundaries. Therefore, compared with nanorods, the synthesized Cu_2O octahedral nanoparticles have a larger specific surface area, which can provide more reactive sites and thus exhibit better gas sensing properties.

3.2. Doping of Metals

The doping of metal elements can effectively improve the gas recognition ability of gas sensing materials, and is an important method to improve gas-sensing performance. Different dopant species may lead to different types of crystallites, defects and electronic properties [51,52]. Doping or surface modification by adding metal elements (such as Ag, Au, Pt, Pd, etc.) on the surface of the gas-sensing materials can increase the number of active sites, promote the adsorption/desorption reaction on the surface of the gas-sensing materials, and reduce the reaction activation energy, and reduce the operating temperature, thereby improving the gas-sensing performance [53].

For instance, Fedorenko et al. [54] prepared Pd-doped SnO_2 semiconductor sensors by a sol–gel method. The effect of Pd additives on methane sensitivity was studied, and it was found that due to the catalytic activity of Pd, compared with undoped materials, the addition of Pd to SnO_2 significantly improved the sensor response to methane (about 6–7 times). Barbosa et al. [55] studied the sensing responses of SnO micro-sheets that modified with Ag and Pd noble metal catalysts towards the gases such as NO_2, H_2, and CO, and found that the Ag/Pd surface-modified SnO micro-sheets exhibited higher sensitivity to gases such as H_2 and CO. However, the catalyst particles reduced the sensing response to oxidizing gases such as NO_2. It is clear that the catalytic activity of Pd nanoparticles is related to chemical sensitization, while the catalytic activity of Ag nanoparticles is related to electronic sensitization. The Ag-modified samples showed high response to H_2, and Pd-modified samples showed high response and selectivity to CO. Zhang et al. synthesized Co-doped sponge-like In_2O_3 cubes by a simple and environmentally friendly hydrothermal method with the help of organic solvents, and studied their acetone gas-sensing properties [56]. It was found that Co-doped In_2O_3 has good gas-sensing performance for acetone gas, and its porous structure can create more adsorption sites for the adsorption of oxygen molecules and the diffusion of the target gas, thereby significantly improving the sensing performance. Compared with the undoped sample, the response value of the doped Co-In_2O_3 sample to acetone was increased by 3.25 times, the response recovery time was fast (1.143 s/37.5 s), the detection limit was low (5 ppm), the reproducibility was good, and the selectivity was high. In another case, Ma et al. reported a Pt-modified WO_3 mesoporous material with high sensitivity to CO [57]. Pt acted as a chemical sensitizer, and gas molecules were adsorbed and flowed into the gas-sensitive material through the spillover effect. In addition, the PtO formed on the surface of Pt further increased the electron depletion layer, and enhanced the electron sensitivity. The synergistic effects of both components further improved the gas-sensing performance.

Compound doping is another important method to improve the sensing performance of gas sensors. When a metal oxide is combined with other metal oxides, a heterojunction structure is constructed. Since the two materials each have their own Fermi energy levels, there will be a mutual transfer of carriers between the two materials to form a space charge layer, so as to achieve the purpose of enhancing the gas-sensing properties of the compound materials. For example, Ju et al. [58] prepared SnO_2 hollow spheres by a template-assisted hydrothermal method and successfully implanted p-type NiO nanoparticles onto the surface of SnO_2 hollow spheres by the pulsed laser deposition (PLD) to prepare NiO/SnO_2 p-n hollow spheres. The gas-sensing performance test indicated that its response to 10 ppm triethylamine (TEA) gas could reach 48.6, which was much higher than that of the original SnO_2 hollow spheres, and the detection limit was as low as 2 ppm. The optimal operating temperature dropped to 220 °C, which was 40 °C lower than that of the original SnO_2 hollow sphere sensor. Compared with the pristine SnO_2 sensor, the enhanced response of NiO/SnO_2 sensor to TEA is mainly attributed to the formation of a depletion layer by the p-n heterojunction interface, which makes the resistance of hybrid materials in air and TEA gas change a lot.

In addition, when the gas-sensing materials of different dimensions are compounded, the stacking of the gas-sensing materials can be prevented, the porosity can be increased, and the gas-sensing performance can be improved. For example, Kida et al. [59] introduced monodispersed SnO_2 nanoparticles (about 4 nm) into WO_3 nanosheet-based films, which could improve its porosity, prevent the aggregation of flakes, and increase the diffusion paths and adsorption sites of gas molecules. The response sensitivity of the composites to NO_2 was enhanced when the concentration in air was 20–1000 ppb, indicating the effectiveness of the microstructure control of the WO_3-based film on high-sensitivity NO_2 detection. In another case, Mishra et al. [60] prepared nanocubic In_2O_3@RGO composites by combining In_2O_3 with reduced graphene oxide (RGO), and the sensor based on nanocubic In_2O_3@RGO heterostructures exhibited high resistance to acetone (~85%) and formaldehyde (~88%) with good selectivity, long-term stability, and fast response/recovery rates

4. 2D Material-Based Gas Sensors

With the successful preparation of graphene materials, its unique structure and excellent properties have attracted widespread attention, thus setting off a research upsurge in 2D materials. 2D nanomaterials have a large specific surface area and special electrical properties due to their nanoscale thin-layer structure. After the gas is adsorbed on the surface, it will affect the conductivity of the surface, so it can be used as a gas-sensing material to adsorb and capture certain single species of gas molecules, with excellent gas-sensing properties. Gas sensors based on 2D nanomaterials exhibit many advantages, such as high sensitivity, fast response speed, low energy consumption, and the ability to work at room temperature.

4.1. 2D Graphene-Based Gas Sensors

Graphene is a honeycomb 2D carbon nanomaterial composed of a single layer of sp^2 carbon atoms. It is currently the thinnest 2D material in the world, with a thickness of only 0.35 nm. Graphene has many excellent properties due to its special structure, such as good electrical conductivity, high carrier mobility, transparency, and mechanical strength. As a typical 2D material, every atom in the graphene structure can be considered as a surface atom, so ideally every atom can interact with the gas, which makes graphene promising as a kind of gas sensor with ultrahigh sensitivity. In the process of adsorption and desorption of gas molecules, graphene nanosheets will affect the change of local carrier concentration in the material, thus showing the transition of electrical signal in the detection of electrochemical performance, and has a good development prospect in gas adsorption. In 2007, Novoselov's group first reported a graphene-based gas sensor, which confirmed that a graphene-based nanoscale gas sensor can be used to detect the adsorption or desorption of single gas molecules on the graphene surface [61]. This study opens the door to research on 2D graphene-based gas sensors.

Single-layer graphene nanosheets, RGO, chemically modified graphene, and GO have been proven to be good gas sensing materials [62–64]. Since the main advantage of graphene nanostructures is the low temperature response, this sensor can greatly reduce the energy consumption of the sensing device. Various graphene-based gas sensors have been used to detect various harmful gases such as NO_2, NH_3, CO_2, SO_2, and H_2S. For example, Ricciardella et al. developed a graphene film-based room temperature gas sensor with a sensitivity of up to 50 ppb (parts-per-billion) to NO_2 [65]. Various methods, such as mechanical exfoliation, chemical vapor deposition (CVD), and epitaxy, have been used to prepare graphene for gas sensing applications. For example, Balandin et al. [66] prepared monolayer graphene using a mechanical exfoliation method and reported a monolayer intrinsic graphene transistor, which can utilize low-frequency noise in combination with other sensing parameters to realize selective gas sensing of monolithic graphene transistors. Choi et al. [67] prepared graphene by CVD and transferred it onto flexible substrates, and demonstrated a gas sensor using graphene as a sensing material on a transparent (Tr > 90%) flexible substrate. Nomani et al. demonstrated that epitaxial growth of graphene on Si and C surfaces of semi-insulating 6H-SiC substrates can provide very high NO_2 detection sensitivity and selectivity, as well as fast response times [68]. Yang et al. directly grew multilayer graphene on various substrates through the thermal annealing process of catalytic metal encapsulation, and tested it as a gas sensor for NO_2 and NH_3 gas molecules to detect its response sensitivity to NO_2 and NH_3 [69]. The schematic diagram of the graphene sensor is shown in Figure 5a. The NO_2 molecules are electron acceptors (p-type dopant), which extract electrons from graphene, while the NH_3 molecules are electron donors (n-type dopant), which donate electrons to graphene. Therefore, when NO_2 molecules are adsorbed, the conductivity of graphene is enhanced, while when NH_3 molecules are adsorbed on the graphene surface, the conductivity decreases due to the compensation effect (Figure 5b).

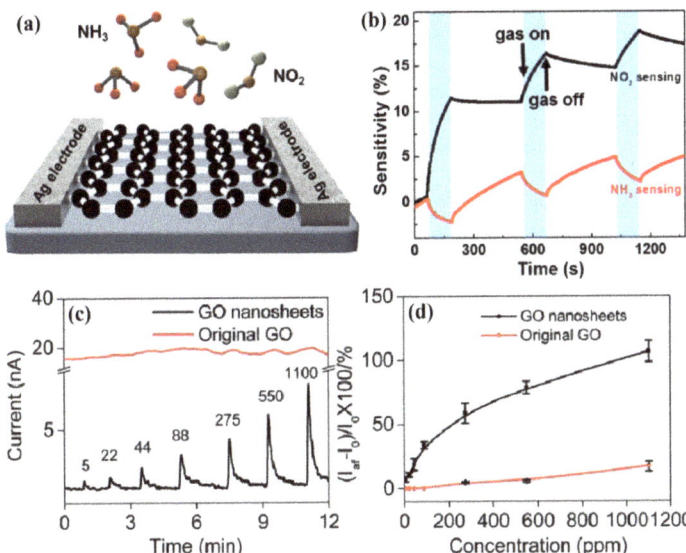

Figure 5. Schematic (a) and time-resolved sensitivity (b) of the graphene sensor toward NH_3 and NO_2 gas molecules. (c) Current vs. time curves for 5–1100 ppm of SO_2 for the original GO and edge-tailored GO nanosheets, and (d) the corresponding sensitivities of the sensors to SO_2 gas. Reprinted with permission from Ref. [69]. Copyright 2016 ACS.

GO is suitable for gas sensors due to its multiple properties, such as easy processing, high solubility in various solvents, and containing oxygen functional groups or defects. Since the defects or functional groups in GO can act as reaction sites for gas adsorption, making the gas easily adsorbed on the surface of GO and improving the selectivity and sensitivity of the GO-based sensor, the response of the GO-based sensor can be tuned by functionalization. Shen et al. [70] prepared edge-trimmed GO nanosheets by periodically acid-treating GO, and then fabricated field effect transistors (FETs) for gas sensing testing of SO_2 at room temperature (Figure 5c,d). Compared with pristine GO nanosheets, edge-clipped GO nanosheets were found to have a significant response enhancement effect to SO_2 gas, and the detection concentration range was 5–1100 ppm. Meanwhile, the edge-trimmed GO device also exhibited a fast response time, which was mainly attributed to the hygroscopic properties of the GO nanosheets, which can trap water molecules and react with SO_2 to generate sulfuric acid to facilitate their fast protonation process. By utilizing different reducing agents to remove oxygen from GO and recover aromatic double-bonded carbons, the selectivity of RGO-based gas sensors could be improved significantly. For example, Guha et al. [71] developed a gas sensor for $NaBH_4$ reduction of GO on a ceramic substrate and reported its performance for detecting NH_3 at room temperature. The response to NH_3 can be optimized by the reduction time of GO. Through chemical modification, RGO can introduce some foreign groups or atoms to change its surface properties, which can enhance its sensing performance. For example, the response of RGO reduced with p-phenylenediamine (PPD) to dimethyl methylphosphonate (DMMP) was 4.7 times higher than that of RGO reduced by ordinary methods [72]. The RGO-based gas sensor reduced by ascorbic acid has high selectivity for corrosive NO_2 and Cl_2, and the detection limit can reach 100 and 500 ppb, respectively [73].

4.2. 2D Transition Metal Sulfide-Based Gas Sensors

As a typical p-type inorganic 2D material with a hexagonal filled layered structure of TMDs, it has received extensive attention in energy conversion and storage, especially in room temperature gas sensors, which have unique advantages and are widely used in various gas detection. Similar to graphene, MoS_2 consists of vertically stacked layers, each formed by covalently bonded Mo-S atoms, with adjacent layers connected by relatively weak van der Waals forces. The weak van der Waals interactions allow gas molecules to permeate and diffuse freely between the layers, so the resistance of MoS_2 can change dramatically with the adsorption and diffusion of gas molecules within the layers. Various methods for gas sensing using few-layer MoS_2 have been reported in the literature, including detectors for many kinds of chemical vapors such as H_2, NO_2, and ethanol [74–76]. For example, Li et al. [77] found for the first time that mechanically exfoliated multilayer MoS_2 exhibited high sensitivity to NO gas, while monolayer MoS_2 had an unstable response to NO gas. They used a mechanical lift-off technique to deposit monolayer and multilayer MoS_2 films on Si/SiO_2 surfaces for the fabrication of FETs. The FET acted as a gas sensor, which realized gas detection by monitoring the change of the conductance of the FET channel during the adsorption of target gas molecules. Since the mechanically cut MoS_2 sheet is an n-type semiconductor, when the MoS_2 channel was exposed to NO gas, it would result in p-doping of the channel, resulting in an increase in channel resistance and a decrease in current flow. It was found that although the single-layer MoS_2 device exhibited a fast response after exposure to NO, the current was unstable; the two-layered, three-layered, and four-layered MoS_2 devices all exhibited stable and sensitive responses to NO at a concentration of 0.8 ppm. Late et al. [78] systematically studied the relationship between the number of MoS_2 layers and gas sensing performance, and found that the sensitivity and recovery time of 5-layered MoS_2 to NH_3 and NO_2 gases were better than those of double-layered MoS_2 (Figure 6a–d). These findings suggest that a small amount of layered MoS_2 has great potential to detect various polar gas molecules.

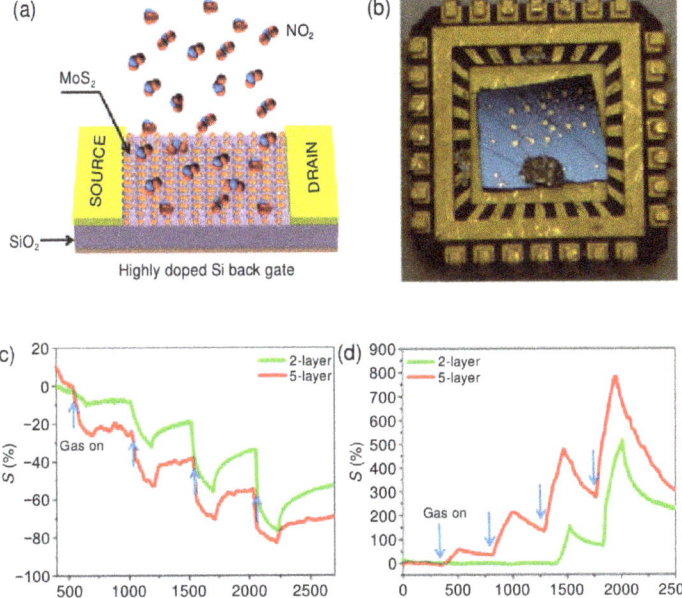

Figure 6. Sensing behavior of atomically thin-layered MoS$_2$ transistors. (**a**) Schematic of the MoS$_2$ transistor-based NO$_2$ gas-sensing device. (**b**) Optical photograph of the MoS$_2$ sensing device mounted on the chip. Comparative two- and five-layer MoS$_2$ cyclic sensing performances with NH$_3$. (**c**) and NO$_2$ (for 100, 200, 500, 1000 ppm) (**d**). Reprinted with permission from Ref. [78]. Copyright 2013 ACS.

At present, 2D Sn-based sulfide materials (SnS and SnS$_2$) are also used in the field of gas sensors due to their unique performance advantages. For example, Wang et al. [79] successfully synthesized free-standing large-scale ultrathin SnS crystalline materials by utilizing the 2D directional attachment growth of colloidal quantum dots in a high-pressure solvothermal reaction. The SnS ultrathin crystals were rectangular with uniform shape, the lateral dimension was between 20 and 30 µm, and the thickness was less than 10 nm (Figure 7a,b). The obtained material was used to fabricate a gas sensor, which exhibited excellent sensitivity and selectivity for NO$_2$ at room temperature with a detection limit of 100 ppb (Figure 7c–e). Xiong et al. [80] synthesized 3D flower-like SnS$_2$ nanomaterials assembled from nanosheets and fabricated them into gas sensors by a simple solvothermal method, As shown in Figure 7f–h. When 100 ppm NH$_3$ was detected at 200 °C, the response value was 7.4, the response time was 40.6 s, and the recovery time was 624 s. The prepared nanoflowers have good selectivity to NH$_3$ with a detection limit of 0.5 ppm. This study attributes the excellent performance of the SnS$_2$ sensor for NH$_3$ to the unique thin-layer flower-like nanostructure, which is beneficial to the carrier transfer process and gas adsorption/desorption process [23].

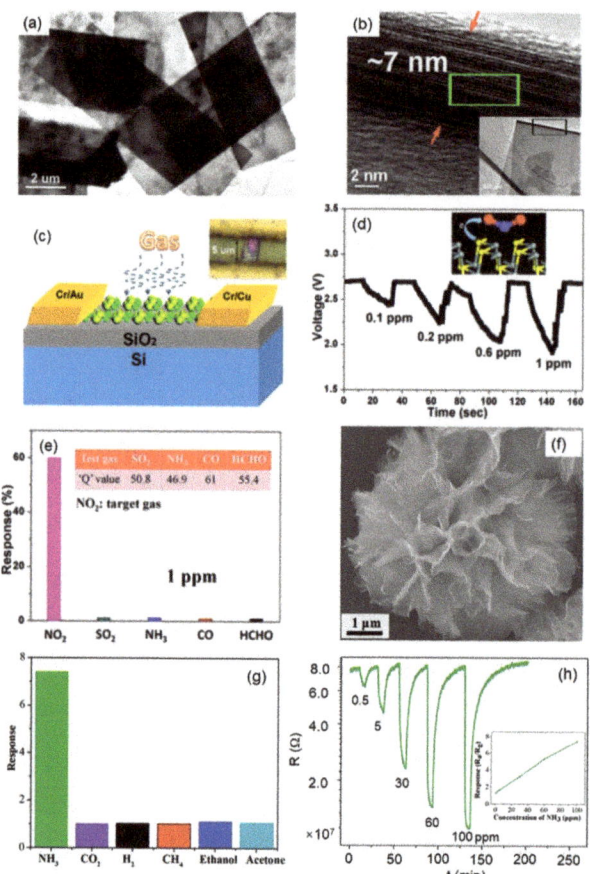

Figure 7. TEM (**a**) and HRTEM (**b**) images of 2D thin SnS crystals. Inset in (**b**) is the corresponding low-magnification TEM image. (**c**) Schematic structure of SnS thin-crystal-based gas-sensor device. The inset shows the optical image of the device. (**d**) Real-time voltage response after exposure of the device to NO_2 gas with increased concentration. The inset schematically illustrates the electron transfer process from SnS to NO_2. (**e**) Selectivity of the sensor to a series of gases of 1 ppm. Inset shows the Q values of the SnS thin crystal sensor for NO_2 as a target gas. Reprinted with permission from Ref. [79]. Copyright 2016 ACS. (**f**) Representative FESEM image of the flower-like SnS_2 synthesized by a facile solvothermal technique. (**g**) Sensor responses of the SnS_2 based sensor upon exposure to six kinds of gases at 200 °C. (**h**) Typical response-recovery characteristic of the SnS_2 based sensor to different concentrations of NH_3 gas at 200 °C (Inset shows the corresponding response curve). Reprinted with permission from Ref. [80]. Copyright 2018 Elsevier.

In order to further improve the gas sensing performance of TMD materials, people have improved the gas-sensing performance through external energy strategies (ultraviolet-assisted irradiation or applying a bias voltage, etc.) or composite strategies with other materials. For example, Late et al. found that 5-layer MoS_2 has a more sensitive response to both NH_3 and NO_2 than 2-layer MoS_2 with the assistance of a bias voltage (+15 V). The photoconverted radiation of 4 mW/cm² can increase the sensitivity of NO_2 gas sensing, while the light intensity of 15 mW/cm² can reduce the recovery time [78]. Wu et al. [81] prepared $MoTe_2$ nanosheets by mechanical exfoliation, and the sensitivity to NO_2 under 254 nm UV light was significantly improved by one order of magnitude compared with the dark condition, and the detection limit was significantly reduced to 252 ppt. Gu et al. [82] synthesized 2D SnS_2 nanosheets by a solvothermal method, and after irradiation with a 520 nm green LED lamp, realized NO_2 detection at room temperature, with good repeatability and selectivity for 8 ppm NO_2 in a dry environment and the response value was 10.8. Cheng et al. [83] combined the excellent sensing performance and gas adsorption capacity of 2D SnS_2 hexagonal nanosheets with the good electrical properties of graphene, and used the excellent electrical conductivity of graphene to make up for the shortcoming of the poor conductivity of SnS_2 at room temperature and prepared a high-performance RGO/SnS_2 heterojunction-based NO_2 sensor. Compared with the single SnS_2 gas sensor, the graphene-doped sensor exhibited better selectivity to NO_2, while effectively reducing the optimal operating temperature of the device, and its response to 5 ppm NO_2 gas increased by nearly one order of magnitude, and the response recovery time was reduced to less than a minute. Compared with the single SnS_2 gas sensor, the graphene-doped sensor exhibited good selectivity to NO_2, while effectively reducing the optimal operating temperature of the device, and its response to 5 ppm NO_2 gas increased by nearly one order of magnitude, and the response recovery time was shortened to less than one minute.

4.3. 2D Metal Oxide Based Gas Sensors

2D semiconductor oxide nanosheets are also commonly used 2D materials in the field of gas sensors. Among them, layered MoO_3, WO_3 and SnO_2 have attracted much attention due to their stability in high-temperature air [84–86]. For example, Cho et al. [87] reported the preparation of MoO_3 nanosheets by ultrasonic spray pyrolysis, and studied their gas-sensing properties, and found that there was still a response even when the gas concentration of trimethylamine was lower than 45 ppb. The super sensitivity to trimethylamine gas is inseparable from its larger specific surface area, as ultrathin nanosheets with the larger specific surface area can provide a larger electron depletion layer and a faster gas diffusion rate across the nanosheets. In addition, MoO_3 is an acidic oxide, which is more likely to react with basic gas preferentially, so it has super selective properties for basic gas trimethylamine. Wang et al. [88] prepared WO_3 porous nanosheet arrays by chemical bath deposition, and found that WO_3 arrays composed of 20 nm ultrathin porous nanosheets had better low-temperature NO_2 gas sensing properties. At an operating temperature of 100 °C, the response to 10 ppm NO_2 was as high as 460.

For 2D metal oxide nanomaterials, the difference in exposed crystal planes will affect their gas sensing properties. For example, Kaneti et al. used a simple and effective hydrothermal method to prepare ZnO nanosheets. By simulating the adsorption of gas molecules on the surfaces of different ZnO crystals, it was found that the enhanced gas-sensing performance of ZnO nanosheets is related to the exposed surface, the $(10\bar{1}0)$ face of ZnO possesses better adsorption capacity for n-butanol than the $(11\bar{2}0)$ face and (0001) face, showing higher responsiveness, better selectivity, and higher stability [89]. In addition, Wang et al. [90] used a two-step method to synthesize ultrathin porous In_2O_3 nanosheets with uniform mesopores and found that they exhibited an ultra-high response to 10 ppb NOx at a lower operating temperature (120 °C), its sensitivity response value was 213, and the response time was 4 s. The thickness of the ultrathin nanosheets is about 3.7 nm, with a large number of active reaction sites, which can enhance the response to NOx, and the porous structure can shorten the gas transmission path and enhance the gas diffusion efficiency, thereby improving the gas sensing performance. Wang et al. synthesized Co_3O_4 mesh nanosheet arrays for the detection of NH_3 [91]. The porous mesh structure promoted gas diffusion and provided a larger active reactive surface to react with the target gas, thereby improving the gas sensing performance. Therefore, even when the NH_3 concentration is 0.2 ppm, the sensor still has obvious response characteristics. The response/recovery time of Co_3O_4 nanosheet arrays to 0.2 ppm NH_3 is 9 s/134 s, showing good reproducibility and long-term room temperature stability.

2D nanostructures have shown great potential in the field of gas sensing due to their high specific surface area and highly efficient active sites on exposed surfaces. To further enhance the gas-sensing properties of 2D metal oxides, ion doping or surface modification on them is a valuable approach to enhance the response and recovery properties. For example, Chen et al. [92] prepared 2D Cd-doped porous Co_3O_4 nanosheets by microwave-assisted solvothermal method and in situ annealing process, and investigated their sensing performance for NO_2 at room temperature. It was found that 5% Cd-doped Co_3O_4 nanosheets significantly improved the response to NO_2 at room temperature (3.38), decreased the recovery time (620 s), and lowered the detection limit to 154 ppb. The reason for the performance improvement is that Cd doping mainly promotes the adsorption of NO_2 through a series of factors such as enhancing the electronic conductivity, increasing the concentration of oxygen vacancies, and forming Co^{2+} - O^{2-}, thus promoting its excellent room temperature sensing performance.

4.4. Other 2D Material-Based Gas Sensors

MXene is a new type of 2D material with layered structure discovered in recent years, which is generally transition metal carbide or carbon-nitrogen compound, and is a MAX ternary phase material. Its general structural formula is $M_{n+1}AX_n$ (n = 1, 2 or 3), where M is one of transition metal elements, A is one of the main group elements (mainly III, IV group elements), X It is carbon or nitrogen, and there are more than 70 kinds of MAX materials. MXene materials are 2D materials formed by extracting element A in MAX. The general formula is $M_{n+1}X_n$ (n = 1, 2 or 3) [93,94]. Due to the characteristics of conventional semiconductor materials and the fact that a large number of functional groups and other active sites remain on the surface after etching, which facilitates subsequent modification, such materials have great application potential in the field of sensing [94,95].

Xiao et al. applied MXene nanomaterials to the detection of NH_3 gas in 2015 [83]. Since the successful synthesis of 2D compound MXenes by Gogotsi et al. in 2011, the application of 2D MXene nanomaterials in gas sensing has been continuously developed [96–98]. For example, Lee et al. [99] reported a $Ti_3C_2T_x$-based gas sensor. After studying the room-temperature gas sensing performance of $Ti_3C_2T_x$ nanosheets on flexible polyimide, it was found that the $Ti_3C_2T_x$ sensor exhibited p-type sensing behavior for reduced gases, with a theoretical detection limit of 9.27 ppm for acetone. Based on the charge interaction between gas molecules and $Ti_3C_2T_x$ surface functional groups -O and -OH, the sensing mechanism of $Ti_3C_2T_x$ is proposed. Chae et al. [100] investigated the dominant factors affecting the oxidation rate of $Ti_3C_2T_x$ flakes and their corresponding sensing properties. In order to improve the sensing performance of MXenes, the gas sensing performance of MXene-based sensors has been realized by surface chemistry and composite structure. Yang et al. [101] prepared organic-like $Ti_3C_2T_x$ by HF acid etching, added the prepared powder to NaOH solution, and used alkali treatment to demonstrate the effect of surface groups on its sensing performance. It was found that the response of the alkali-treated $Ti_3C_2T_x$ sensor to 100 ppm NH_3 at room temperature was two times higher than that of the untreated one. This is due to the adsorption of N atoms in NH_3 molecules on top of Ti atoms in $Ti_3C_2T_x$ to form strong N-Ti bonds. Alkaline treatment increased the -O end, increased N-Ti bond, and promoted the increase of NH_3 adsorption. Furthermore, after oxygen functionalization, $Ti_3C_2T_x$ increased the resistance by transitioning to a semiconductor, thereby increasing the gas response signal. Besides $Ti_3C_2T_x$, 2D MXenes such as V_2CT_x and Mo_2C have also been investigated for gas sensors [102,103]. The 2D V_2CT_x sensor composed of monolayer or multilayer 2D V_2CT_x on polyimide film fabricated by Lee et al. [102] can measure polar gases (hydrogen sulfide, ammonia, acetone, and ethanol) and non-polar gases at room temperature (hydrogen and methane). The V_2CT_x sensor shows ultrahigh sensitivity for non-polar gases, with minimum detection limits of 2 ppm and 25 ppm for hydrogen and methane, respectively.

5. Metal Oxide Nanomaterials-Based Gas Sensors

Because of its large specific surface area, high surface activity, many active sites and sensitive to the surrounding environment, the gas sensor prepared by metal oxide nanomaterials has high response sensitivity and fast response-recovery speed. According to the semiconductor type, metal oxide semiconductors can be divided into n-type and p-type. In n-type semiconductors, including SnO_2, ZnO, TiO_2, $In2O3$, etc., the carriers are mainly free electrons. However, in p-type semiconductors, such as CuO, NiO, Co_3O_4, etc., the carriers are mainly holes. When n-type semiconductors are exposed to reducing gases (such as ethanol, NH_3, H_2, etc.), the resistance of the materials will decrease, while when exposed to oxidizing gases (such as NO_3), the resistance of the materials will increase. In contrast to n-type semiconductors, the resistance of p-type semiconductors is higher when exposed to reducing gas and decreases when exposed to oxidizing gas. At present, the most studied metal oxide materials for gas sensing are SnO_2, ZnO, TiO_2, CuO, WO_3 and so on [57,104–107].

5.1. SnO_2-Based Gas Sensors

SnO_2 is a kind of direct band gap wide band gap n-type semiconductor (band gap ~ 3.6 eV), whose carriers are free electrons. The interaction with the reducing gas will increase the electrical conductivity. However, the oxidized gas will consume the sensing layer of charged electrons, resulting in a decrease in electrical conductivity [108]. SnO_2 nanomaterials are widely used in the field of gas sensing because of their simple preparation, low cost, easy control of morphology, and microstructure, good thermal/chemical stability, shallow donor energy level (0.03–0.15 eV), potential barrier of oxygen adsorption on the surface is 0.3–0.6 eV, oxygen vacancy and excellent gas-sensing properties [109–111]. Thanks to its high sensitivity to different gases, SnO_2 sensor can detect low concentration gases, but its selectivity is low.

In order to improve the sensitivity, stability, and selectivity of SnO_2-based gas sensors and reduce the working temperature, researchers modified SnO_2 materials by a variety of methods. One method is to control the morphology and size of SnO_2 materials to prepare zero-dimensional (0D), one-dimensional (1D), 2D, three-dimensional (3D) and porous hollow SnO_2 nanomaterials for the detection of various gases [112–117]. For example, Zhang et al. successfully synthesized leaf-like SnO_2 hierarchical architectures by using a simple template-free hydrothermal synthesis method. The sensor based on this unique leaf-like SnO_2 hierarchical structure had a high response and good selectivity to NO_2 at low operating temperature [118]. Feng et al. synthesized mesoporous SnO_2 nanomaterials with different pore sizes (4.1, 6.1, 8.0 nm) by carbon-assisted synthesis. The gas sensing properties of the three materials showed high sensitivity and ideal response recovery time to ethanol gas, and the detection limit was as low as ppb [119].

Using doping modification technology, SnO_2 nanomaterials are used as the matrix materials of gas sensors, which are modified by doping precious metals (such as Pt, Pd and Au) or other metal ions (such as Ni, Fe and Cu). It is another important means to improve the gas sensing properties of SnO_2 to CO, CH_4, NO_2 and other gases. For example, Dong et al. prepared SnO_2 nanofibers and Pt-doped SnO_2 nanofibers by electrospinning, which were used to test the sensitivity to H_2S. It was found that the response of Pt-doped SnO_2 nanofibers to H_2S gas was significantly improved. The response of 0.08 wt% Pt-doped SnO_2 nanofibers to 4–20 ppm H_2S was 25.9–40.6 times higher than that of pure SnO_2 nanofibers [120]. Chen et al. prepared Pd-doped SnO_2 nanoparticles by the coprecipitation method. Compared with pure SnO_2 nanoparticles, the response characteristics of SnO_2 to CO were significantly improved [121]. Lee et al. used Pd nanoparticles to modify the surface of SnO_2 nanorod thin films, and studied their sensing properties for H_2 and ethanol gas [122]. It was found that compared with the undoped samples, the responsiveness of Pd-doped SnO_2 nanorod thin films to 1000 ppm H_2 and ethanol at 300 °C was increased by 6 and 2.5 times, respectively. They assumed that the improved gas sensing properties are due to the formation of the electron depletion layer and the enhanced catalytic dissociation of molecular adsorbates on the surface of Pd nanoparticles. Shen et al. also studied the gas-sensing properties of SnO_2 by Pd doping [123]. SnO_2 nanowires with a tetragonal structure were synthesized by thermal evaporation. The morphology, crystal structure, and H_2 gas-sensing properties of undoped and Pd-doped SnO_2 nanowires were studied. It was found that with the increase of Pd doping concentration, the working temperature decreased and the response of the sensor to H_2 increased. Similarly, doping Au into SnO_2 thin films can change the morphology of SnO_2 thin films, reduce the grain size of SnO_2 thin films, decrease the working temperature of the sensor, and improve the sensitivity and selectivity of SnO_2 to reducing gases such as CO [124]. Zhao et al. carried out Cu doping on SnO_2 nanowires. Compared with undoped SnO_2 nanoscale arrays, the sensitivity and selectivity of the sensor to SO_2 in a dry environment were improved significantly [125].

In addition, the researchers synthesized composite nanomaterials containing two different energy band structure materials to form heterostructures to improve the gas sensing performance of SnO_2-based gas sensors. For example, Chen et al. prepared Fe_2O_3@SnO_2 composite nanorods with multi-stage structure by a two-step hydrothermal method and found that the composite structure has good selectivity for ethanol [126]. Xue et al., using SnO_2 nanorods synthesized by hydrothermal method as carriers, obtained SnO_2 composite nanorods loaded with CuO nanoparticles by ultrasonic and subsequent calcination in $Cu(NO_3)_2$ solution. The gas sensing properties of the materials for the detection of H_2S were studied. It is found that the sensitivity of the sensor to 10 ppm H_2S can reach 9.4×10^6 at 60 °C [127]. The ultra-high sensitivity of the composite is attributed to the p-n junction formed between CuO and SnO_2. In the air, the formation of heterojunction increases the height of the energy barrier, hinders the flow of electrons, resulting in an increase in the resistance of the material. When the material is in contact with H_2S and reacts, it can form CuS, which is similar to metal conductivity, which greatly enhances the electrical conductivity of the material. Fu et al. prepared NiO-modified SnO_2 nanoparticles, which increased the thickness of the electron depletion layer on the surface of SnO_2 through the formation of p-n heterojunction in air. While in the SO_2 atmosphere, NiO reacted with SO_2 to form NiS, which promoted the release of electrons from the surface adsorbed O^- to SnO_2, thus enhancing the response of the device to SO_2 gas and improving the gas sensitivity of SnO_2 materials to SO_2 [128].

5.2. ZnO-Based Gas Sensors

ZnO is an n-type metal oxide semiconductor material with a wide band gap (3.3 eV). ZnO is widely used in the field of gas sensors because of its good chemical stability and low resistivity. Yuliarto et al. successfully synthesized ZnO nanorod thin films on Al_2O_3 substrates by chemical bath deposition (CBD) [129]. ZnO thin films with different thicknesses were prepared by different times of CBD processes. By optimizing the thickness of ZnO thin films, the response performance of ZnO-based gas sensors to SO_2 was improved. The gas sensing response of the ZnO film of two CBD to 70 ppm SO_2 at 300 °C is 93%, which is 15% higher than that of the ZnO film of one CBD. At different operating temperatures, the response of ZnO nanorods prepared by two CBD deposition is 20–40% higher than that of ZnO nanorods deposited by one CBD deposition. Wang et al. successfully prepared three kinds of ZnO nanostructures (nanorods, flowers, and spheres) with different morphologies by a simple hydrothermal and water-bath method, and studied their sensing properties of NO_2 at room temperature under UV (365 nm LED) excitation, as shown in Figure 8 [130]. It was found that ZnO nanospheres have the highest response (29.4) to 5 ppm NO_2 (Figure 8d,e), which was mainly due to the largest specific surface area and the largest number of oxygen ions adsorbed on the surface of ZnO nanospheres. However, due to the high crystallinity, few surface defects and unidirectional electron transfer path, the response speed and recovery speed of ZnO nanorods are the fastest (9 s and 18 s, respectively) (Figure 8f,g). For ZnO nanoflowers, the gas sensing response, response and recovery rate are between ZnO nanorods and ZnO nanospheres. All three kinds of ZnO have good selectivity and repeatability for NO_2 (Figure 8h,i). The good selectivity of ZnO to NO_2 is attributed to the following two points: (1) NO_2 molecule has an unpaired electron, which is beneficial to its chemisorption on ZnO surface; (2) NO_2 molecule has the smallest bond energy, which is about 312.7 kJ/mol. The smaller the bond energy is, the more favorable the sensing reaction is, especially for the sensors working at room temperature.

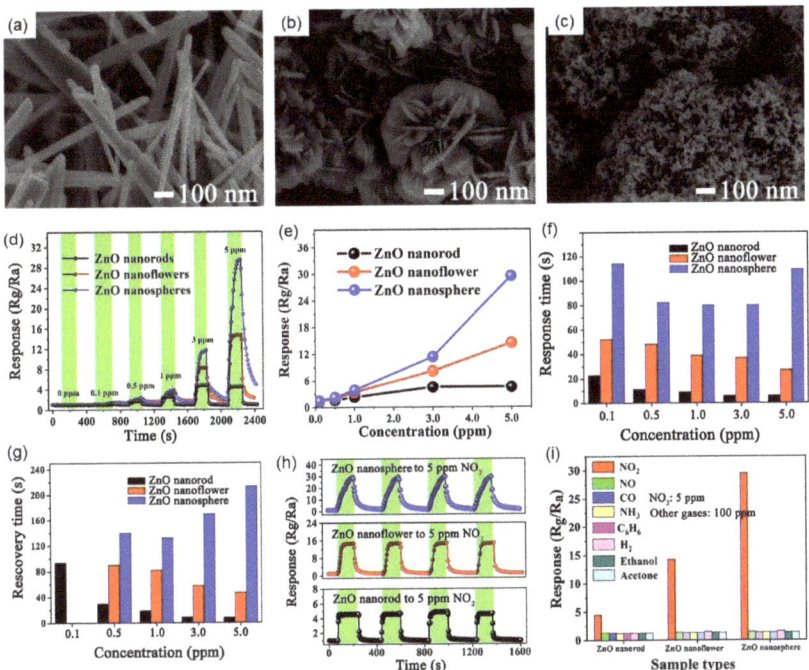

Figure 8. SEM images of (**a**) ZnO nanorods, (**b**) ZnO nanoflowers and (**c**) ZnO nanospheres. (**d**) Dynamic response curves with time of three different ZnO nanostructures; (**e**) the response curves with NO$_2$ concentration of three different ZnO nanostructures; (**f**,**g**) the response and recovery time of three different ZnO nanostructures; (**h**) the repeatability of three different ZnO nanostructures to 5 ppm NO$_2$; (**i**) the selectivity of three different ZnO nanostructures to other harmful gases. Reprinted with permission from Ref. [130]. Copyright 2021 Elsevier.

Similar to the SnO$_2$-based gas sensor, researchers changed the cell parameters of the original ZnO by element doping, making it produce lattice deformation, cause the surface defects of the gas sensing materials, and increase the surface active sites, so as to improve the gas sensing properties of the sensitive materials [131,132]. For example, Chaitra et al. prepared Al-doped ZnO thin films by the sol–gel method and spin-coating technique [133]. It was found that 2 at.% Al-doped ZnO thin films have the highest sensitivity to 3 ppm SO$_2$ gas at 300 °C, which was lower than the threshold limit. Kolhe et al. prepared Al-doped ZnO thin films by chemical spray pyrolysis [134]. It was found that the doping of Al in ZnO led to the fracture of thin nanofilms, resulting in more active sites. Al doping also leads to the increase of oxygen vacancy-related defects and the change of crystal size due to the difference of ion radius between Al^{3+} and Zn^{2+} ions. The doped sensor has enhanced sensing characteristics, which also leads to the decrease of the optimal operating temperature. Xiang et al. used the photochemical method to embed Ag nanoparticles into ZnO nanorods and studied their gas-sensing properties [135]. It was found that Ag nanoparticles embedded on the surface of ZnO nanorods could improve the performance of the sensor. The response of ZnO nanorods to 50 ppm ethanol was almost three times that of pure ZnO nanorods, and had long-term stability. After 100 days of exposure to ethanol in 30 ppm, the response of the sensor had no obvious degradation.

The heterostructure is an important means to improve the gas sensing properties of semiconductor oxides, which usually includes two kinds of semiconductor oxides with different Fermi levels. When two kinds of semiconductor oxides come into contact with each other, the free electrons will change from the oxidation stream with a higher Fermi level to the oxide with a lower Fermi level. Compared with the single semiconductor oxide, the electron transfer efficiency of the two semiconductor oxides is higher, and a thicker electron depletion layer and higher resistance can be formed at the contact interface. Therefore, the introduction of heterojunction can effectively improve the performance of the semiconductor gas sensor. For example, Kim et al. synthesized p-n CuO/ZnO core–shell nanowires by thermal oxidation and atomic layer deposition, and studied their sensing properties to reduce gas by controlling the thickness of the ZnO shell [136]. When the shell thickness is less than or equal to Debye wavelength (λ_D), a complete electron depletion layer will be formed. When exposed to the reducing gas (CO), the desorption of surface oxygen releases electrons back into the conduction band of the shell, returning the conduction band to its original state and significantly improving the conductivity (Figure 9a). When the thickness of the shell is higher than λ_D, only part of the electron loss will be caused. When reducing gases are introduced, they are adsorbed on the partially depleted shell, and the resistance changes only slightly, as shown in Figure 9b. Therefore, for p-n heterostructure nanowires, controlling the shell thickness plays an important role in improving the performance of gas sensors. Zhou et al. prepared NiO/ZnO nanowires by one-step hydrothermal method and tested their gas sensing properties to SO_2 [137]. It was found that at the optimum operating temperature of 240 °C, the response of NiO/ZnO nanowires to 50 ppm SO_2 was 28.57, and the gas detection range was 5–800 ppm. The response time, response time and recovery time of the prepared NiO-ZnO nanowires gas sensor to 20 ppm SO_2 gas were 16.25, 52, and 41 s, respectively.

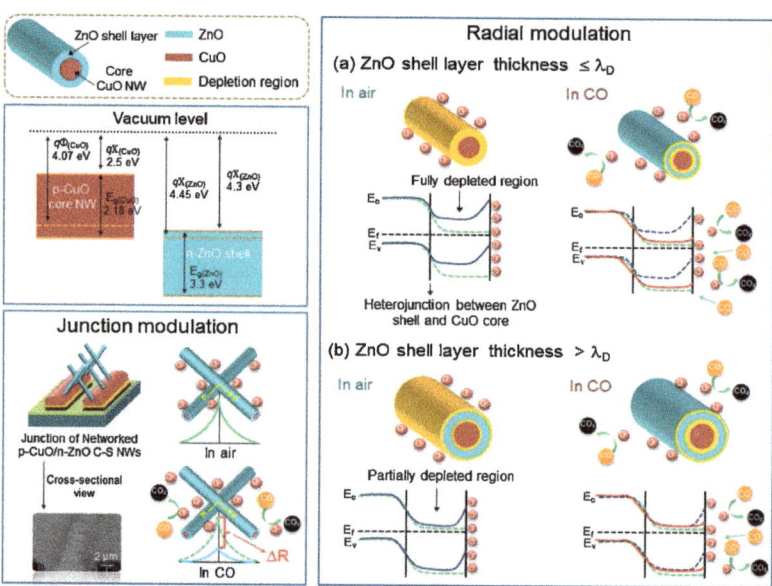

Figure 9. Schematic of the reducing gas sensing mechanism in the CuO–ZnO C–S NWs. Ec and EF indicate the conduction band energy and Fermi energy level, respectively, in cases of ZnO shell layers (**a**) thinner and (**b**) thicker than ZnO's Debye length. Reprinted with permission from Ref. [136]. Copyright 2016 Elsevier.

5.3. CuO-Based Gas Sensors

CuO is a typical p-type semiconductor oxide material with a band gap of 1.2–1.9 eV. Because of its good electrical properties, chemical stability, catalytic activity and other physical and chemical properties, CuO has been widely studied in the fields of catalysis, optoelectronic devices, gas sensors and so on. CuO can respond to reducing gases at lower operating temperatures, which attracts researchers to prepare different morphologies of CuO, doped CuO and heterostructure CuO for gas sensors to study their gas sensing properties. For example, Li et al. prepared porous CuO nanosheets on alumina tubes by hydrothermal method, which were used to make gas sensors to detect H_2S [138]. It was found that when the concentration of H_2S is as low as 10 ppb, the response sensitivity of the sensor was 1.25, and the response and recovery time were 234 and 76 s, respectively. Navale et al. synthesized CuO thin films that composed of CuO nanocubes on quartz substrates by simple and catalyst-free thermal evaporation technique, and studied their gas sensing properties [139]. It was found that CuO thin films have strong selectivity for NO_2 gas, and the response speed and recovery time are fast. At 150 °C, the maximum response value of CuO sensor film to NO_2 of 100 ppm was 76, the detection limit was 1 ppm, and the response time was only 6 s, but the recovery time was 1200 s. Huang et al. prepared CuO hollow microspheres by precipitation annealing at 270 °C using $CuSO_4$, Na_2CO_3, and cetyltrimethyl ammonium bromide (CTAB) as raw materials [140]. The CuO hollow microspheres showed good gas sensitivity to ethanol. The response sensitivity to ethanol at 250 °C was 5.6 and the response and recovery times were 17.0 and 11.9 s, respectively. Hu et al. fabricated CuO nanoneedle arrays directly on commercial ceramic tubes by magnetron sputtering, wet chemical etching and annealing, which have good selectivity, reproducibility and long-term stability for low concentration H_2S (10 ppm) [141]. For metal oxide semiconductors, high specific surface area and exposed crystal plane are two key factors that determine their gas sensing properties. In order to study the effect of surface structure on gas sensing properties, Huo et al. obtained CuO nanotubes on (111) exposed surfaces and CuO nanocubes on (110) exposed surfaces in Cu nanowires and Cu_2O nanocubes, respectively, which were used to detect the gas sensing properties of CO gas, as shown in Figure 10 [142]. The results indicated that compared with CuO nanocubes, CuO nanotubes have lower optimal operating temperatures and higher sensitivity for CO gas detection.

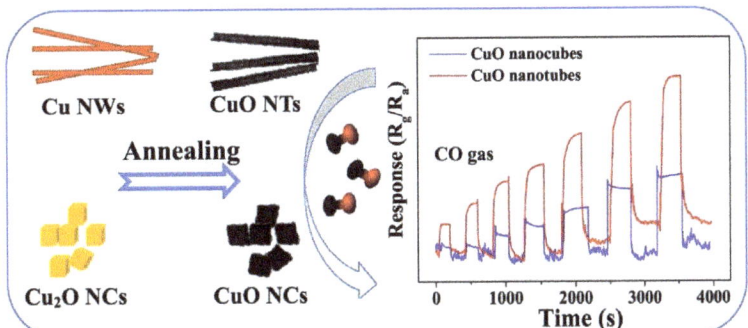

Figure 10. The preparation of CuO NTs and CuO NCs and CO gas-sensing behaviors of CuO NTs and CuO NCs at the operation temperature of 175 °C with different CO concentrations (50–1000 ppm). Reprinted with permission from Ref. [142]. Copyright 2018 Elsevier.

Although pure CuO as a sensitive material can be used to detect a variety of toxic and harmful gases. However, it still faces some problems in practical applications, such as low sensitivity, high working temperature, poor selectivity, long response/recovery time and so on. For this reason, researchers use doping, recombination and other methods to improve the gas sensing performance of CuO-based gas sensors, and achieved some remarkable results. The doping of precious metals or rare earth elements can greatly increase the active sites of CuO gas-sensing reaction, which is beneficial to the adsorption of gas molecules on the sensitive material surface, and most of the dopants have strong catalytic activity, which can further enhance the gas sensing reaction. For example, Hu et al. prepared CuO nanoflowers with different Pd-doping concentrations by simple water-bath heating method [143]. Compared with pure CuO, the specific surface area of CuO nanoflowers with a size of about 400 nm prepared when the mass fraction of Pd was 1.25% increased by 1.8 times, and the response (Rg/Ra) to 50 ppm H_2S at 80 °C was 123.4, which was 7.9 times that of pure CuO. In addition, the gas sensor has good stability and repeatability. Tang et al. prepared Pt-doped CuO nanoflowers by the same method, which significantly improved the gas sensing performance of the sensor to H_2S gas [144]. When the amount of Pt doping was 1.25 wt.%, the response of the sensor to 10 ppm H_2S at 40 °C was 135.1, which was 13.1 times that of pure CuO. The researchers also selected other metal elements to dope CuO, and achieved excellent results. For example, Mnethu et al. reported a highly sensitive and selective Zn-doped CuO nano-chip-based sensor [145]. At 150 °C, the response of 0.1 at.% Zn-doped CuO samples to 100 ppm xylene gas was 53. Bhuvaneshwari et al. reported a Cr-doped CuO nanoboat, which significantly improves the sensing performance of NH_3 in the concentration range of 100–600 ppm at room temperature [146]. The gas sensing test results show that the sensitivity of CuO nanospheres doped with atomic fraction 6% Cr to NH_3 at room temperature was 2.5 times higher than that of undoped nanospheres. The enhanced gas sensing performance is attributed to the increase of oxygen vacancy caused by chromium doping, which makes the nanospheres absorb more surface oxygen, and chromium doping also reduces the activation energy of the sensor at low temperatures. Al-doped CuO [147], In-doped CuO [148], and Ag-doped CuO [149] also showed excellent gas sensing properties for target gases.

The composite gas sensor can integrate the unique properties of the material and improve the performance of the sensor through complementary enhancement. Researchers have designed a variety of CuO-based composite gas sensors to improve the selectivity of target gases, enhance gas sensing properties, shorten response/recovery time and reduce the optimal operating temperature, especially semiconductor oxides with heterostructures. For example, Sui et al. used the template-free hydrothermal method to grow multi-layer heterogeneous CuO/NiO nanowires on ceramic tubes for the detection of H_2S gas [150]. The CuO/NiO-based sensor has a wide linear range in the 50~1000 ppb range and has good repeatability, selectivity and long-term stability. At 133 °C, the 2.84 at.% CuO modified NiO showed good sensing properties, and the response to 5 ppm H_2S was 36.9, which was 5.6 times higher than that of NiO. The detection limit of H_2S is further reduced from 1 ppb of pure NiO sensor to 0.5 ppb. Park et al. synthesized SnO_2-CuO hollow nanofibers by electrospinning and thermal processing, which can be used in the field of H_2S gas sensing [151]. The electrospun nanofiber materials have the advantages of large surface area, high porosity and permeability to air or moisture, which is conducive to ionic diffusion and suitable for applications in gas sensors, lithium-ion batteries and wound healing [152,153]. SnO_2-CuO nanotubes increase the specific surface area, decrease the working temperature and improve the sensing performance of H_2S. At the working temperature of 200 °C, the sensitivity of hollow SnO_2-CuO nanotubes to 5 ppm H_2S was 1395 and the response time was 5.27 s. Liang et al. also prepared the heterostructure of In_2O_3 nanofibers supported on CuO by electrospinning and studied the sensing properties of H_2S [154]. The gas sensor based on the heterostructure had a high sensitivity to 5 ppm H_2S gas at 150 °C, which was 225 times higher than that based on pure In_2O_3, even at room temperature. The above research results show that the construction of heterojunction composites can effectively improve the sensitivity and selectivity of the gas sensor, reduce the working temperature, accelerate the response/recovery speed and prolong the life of the sensor.

5.4. Other Metal Oxide-Based Gas Sensors

WO_3 is a kind of n-type wide band gap semiconductor oxide, which has the advantages of photoelectric conversion, electrochromism, photocatalysis, and gas sensitivity, so it is used in a variety of optoelectronic devices. WO_3 is more likely to form oxygen defects and unsaturated coordination bonds. When WO_3 is heated in the air, it is easy for O_2 to seize e^- to form O^-. The formed O^- is chemically adsorbed on the surface of WO_3 and forms an electron depletion layer. When operating at a low temperature, it is easy to form O^{2-}. when it is at a higher operating temperature, it is easy to form O^- and O^{2-} [155,156]. Therefore, WO_3 is an effective gas sensing material, especially more sensitive to reducing gas. For example, Hu et al. synthesized WO_3 nanorods with needle-shape via a hydrothermal mehod and subsequent calcination, which showed high performance for triethylamine gas sensing [157]. By comparison, it was found that WO_3 nanosheet devices showed the highest response and the shortest response time to 1–10 ppm SO_2. Li et al. proposed a method for the synthesis of WO_3 particles assisted by ionic liquids. The hollow sphere structure composed of WO_3 nanorods, nanoparticles and nanosheets was synthesized. Their gas sensing properties for various organic compounds (methanol, ethanol, isopropanol, ethyl acetate and toluene) were studied. It was found that it has remarkable sensitivity, low detection limit and fast response/recovery time [158]. Li et al. prepared SnO_2-WO_3 hollow nanospheres with a diameter of about 550 nm and a thickness of about 30 nm by hydrothermal method, and studied the temperature dependence of humidity sensors prepared at different relative humidity and temperature. It is found that compared with the original WO_3 nanoparticles and SnO_2 nanoparticles, SnO_2-WO_3 hollow nanospheres have excellent sensing properties [159].

α-Fe_2O_3 is also a typical n-type semiconductor with a narrow band gap (2.2 eV), low cost, high stability, high corrosion resistance, and non-toxicity, so it has attracted great attention as a gas sensing material [160,161]. Liang et al. successfully synthesized ultrafine and highly monodisperse α-Fe_2O_3 nanoparticles with an average particle size of 3 nm by a simple reverse microemulsion method, which showed high sensitivity, high selectivity and good stability to acetone [162]. Shoorangiz et al. synthesized α-Fe_2O_3 nanoparticles by a sol–gel method and evaluated their gas-sensing properties for ethanol and other gases [163]. At the optimal sensing temperature of 150 °C, it has good selectivity to ethanol gas, and the response to 100 ppm ethanol gas was 14.5%. Qu et al. reported high-performance gas sensors based on MoO_3 nanoribbons that were modified by Fe_2O_3 nanoparticles [164]. Compared with the original MoO_3 nanoribbons, the reaction of p-xylene in the Fe_2O_3 nanoribbons modified by Fe_2O_3 nanoparticles increased by 2–4 times due to the formation of heterojunction between Fe_2O_3 and MoO_3.

As a p-type single metal oxide semiconductor, Co_3O_4 is also used to test gas sensitivity. It has been found to have a good gas sensing response to H_2S in some studies [91,165]. Among different types of metal oxides, p-type Co_3O_4 is also considered to be the best candidate for ethanol gas sensing. For example, Li et al. reported that Co_3O_4 nanotubes are sensitive to ethanol gas at room temperature, and Co_3O_4 sensors show excellent repeatability after more than 50 tests [166]. Sun et al. obtained monodisperse porous Co_3O_4 microspheres by solvothermal method and thermal decomposition [167]. The gas sensing properties of these Co_3O_4 microspheres were compared with those of commercial Co_3O_4 nanoparticles. These Co_3O_4 microspheres showed higher ethanol sensitivity and selectivity at relatively low temperatures. In addition, Zhang et al. synthesized three kinds of Co_3O_4 with different morphologies (cube, rod, and sheet) by a hydrothermal method, and studied their sensing properties to toluene [168]. It was found that the sensor based on the Co_3O_4 flake structure had better sensing performance for toluene than the other two sensors. At the working temperature of 180 °C, the fabricated sensor showed higher sensitivity, faster response and recovery speed, as well as better selectivity.

6. 2D Materials/Metal Oxide-Based Gas Sensors

As a class of important materials for various sensors, semiconducting metal oxides have been widely used in various redox gas sensing due to their high sensitivity, simple preparation, and low price. However, disadvantages such as high temperature and easy agglomeration can also be found. Especially the disadvantage of high working temperature makes it extremely demanding on the working conditions of the environment, which greatly affects the service life. 2D materials such as graphene have unique physical and chemical properties, which can generate gas-sensitive responses at room temperature and have good selectivity. 2D materials can not only provide active sites for catalysts and sensors, but also serve as flat building blocks for forming complex nanostructures. Using 2D materials as a matrix to support semiconducting metal oxides can reduce the agglomeration of metal oxides and expose more adsorption and reaction sites. Due to the large specific surface area and high porosity of 2D materials, the response sensitivity and selectivity of metal oxides to specific gases can be further improved, and the working temperature can be reduced. Meanwhile, the synergistic effect and heterostructure of layered 2D materials and metal oxides can not only bring out the greatest advantages of both components but also overcome their respective defects, thereby improving the comprehensiveness of gas sensing performance. Therefore, the combination of 2D materials and metal oxides has become an important research direction in the field of gas sensors.

6.1. Synthesis of 2D Materials/Metal Oxide Composites

The properties of a material have a great relationship with its morphology, structure, and its composition. Single-component nanomaterials are far from meeting the development and application needs of modern nanotechnology, while multi-component composite materials combine the characteristics of different materials show better performance than their single components. It is of great significance in developing new materials, studying novel properties of materials, and constructing functional gas sensors. The preparation methods of 2D materials/metal oxide composites mainly include hydrothermal synthesis, microwave-assisted synthesis, self-assembly, chemical reduction method, and others. In this section, we would like to present a brief introduction to these potential synthesis methods for 2D materials/metal oxide composites.

Hydrothermal synthesis is a common method for the preparation of 2D materials/metal oxide nanocomposites, which has the advantages of simple operation, mild conditions, and low cost. Usually, the precursor solution is put into a high-pressure reactor, hydrothermally reacted under high temperature and high pressure, and then the composite material is prepared by post-processing methods such as separation, washing, and drying. Many 2D materials/metal oxide nanocomposites have been synthesized and used for the fabrication of gas sensors. For example, Chen et al. [169] prepared a core–shell structure composed of TiO_2 nanoribbons and Sn_3O_4 nanosheets by a two-step hydrothermal reaction. Sn_3O_4 nanosheets were uniformly immobilized onto the surface of porous TiO_2 nanoribbons. It was found that the structural morphology of the products in the hydrothermal process is affected by the reaction time. Chen et al. [170] investigated the effect of hydrothermal reaction temperature on the morphology of oxide materials. At a relatively high temperature, they obtained SnO_2-decorated TiO_2 nanoribbons. Wang et al. [171] obtained a composite structure based on SnO_2 nanoparticles and TiO_2 nanoribbons by controlling the precursor solution. Precise control of the hydrothermal synthesis conditions is a key factor for the preparation of high-quality and diversely shaped metal oxide nanostructures. In addition to metal oxide-based 2D composites, the composites of metal oxides and other 2D materials such as graphene have also been prepared by hydrothermal methods. For example, Chen et al. [172] prepared Co_3O_4/rGO composites by the hydrothermal method, and studied their gas-sensing properties to NO2 and methanol at room temperature. Chen and co-workers [173] synthesized SnO_2 nanorods/rGO composite nanostructures by hydrothermal method and investigated their NH_3 sensing properties. Liu et al. [174] fabricated a layered flower-like In_2O_3/rGO composites by a one-step hydrothermal method. The synthesized materials were further utilized for the fabrication of gas sensors, which exhibited improved sensing performance for 1 ppm NO_2 at room temperature compared with pure In_2O_3-based gas sensors. The hydrothermal synthesis usually reacts at high temperatures (>150 °C). When the metal oxide is coupled with 2D materials for gas sensing, the operation temperature will be decreased, which is lower than the materials preparation temperature. Thus, the thermal stability and applicability of the heterostructures for gas sensors can be improved.

The microwave-assisted synthesis of materials uses microwaves to provide energy for the reaction, and it is different from the traditional heating method. It uses microwaves to make the reactants generate heat by themselves to promote the reaction. It has the characteristics of uniform heating and high heating efficiency. For example, Pienutsa et al. synthesized SnO_2 decorated RGO and further used the created SnO_2-RGO composite for the real-time monitoring of ethanol vapor [175]. In another similar study, Kim et al. [176] obtained SnO_2/graphene heterostructured composites by microwave-treating graphene/SnO_2 nanocomposites, in which graphene-enhanced efficient transmission of microwave energy and facilitated the evaporation and redeposition of SnO_x nanoparticles.

Chemical reduction has been often used to synthesize composite nanostructures of RGO and metal oxides. Usually, a metal salt solution is used as a precursor to be mixed with a graphene oxide dispersion solution, and a chemical reducing agent is used to reduce it in one step to obtain RGO. This method usually relies on a microwave, hydrothermal reaction, and other sources to provide energy. For example, Russo et al. [177] synthesized SnO_2/rGO composites using $SnCl_4$ and GO as precursors. Under the irradiation of microwave, GO and $SnCl_4$ were reduced to form SnO_2/rGO composites, which were further reacted with H_2PtCl_6 to form Pt-SnO_2/rGO composites. The created composites exhibited high sensitivity to hydrogen at room temperature.

Besides the above-introduced methods, other techniques such as self-assembly can also be utilized for the synthesis of 2D materials/metal oxide composites. For instance, Zhang et al. fabricated rGO/TiO_2 multilayer composite films using a layer-by-layer self-assembly process, and the fabrication process is shown in Figure 11 [178]. The rGO/TiO_2 multilayer composite films were fabricated by alternately depositing TiO_2 nanospheres and GO via the layer-by-layer self-assembly technique to form nanostructures, and then thermally reducing GO to rGO. Since p-type rGO and n-type TiO_2 form a p-n heterojunction at the interface, the depletion layer generated by the built-in electric field will be beneficial to control the carrier transport process inside the material. It is clear that SO_2 acts as an electron donor, which increases the electron concentration of the composite material, resulting in a decrease in device resistance. The device could be used to detect SO_2 gas at as low as 1 ppb at room temperature with good selectivity and stability. The response of the fabricated gas sensor to 1 ppm SO_2 was 10.08%, and the response and recovery times were 95 and 128 s, respectively.

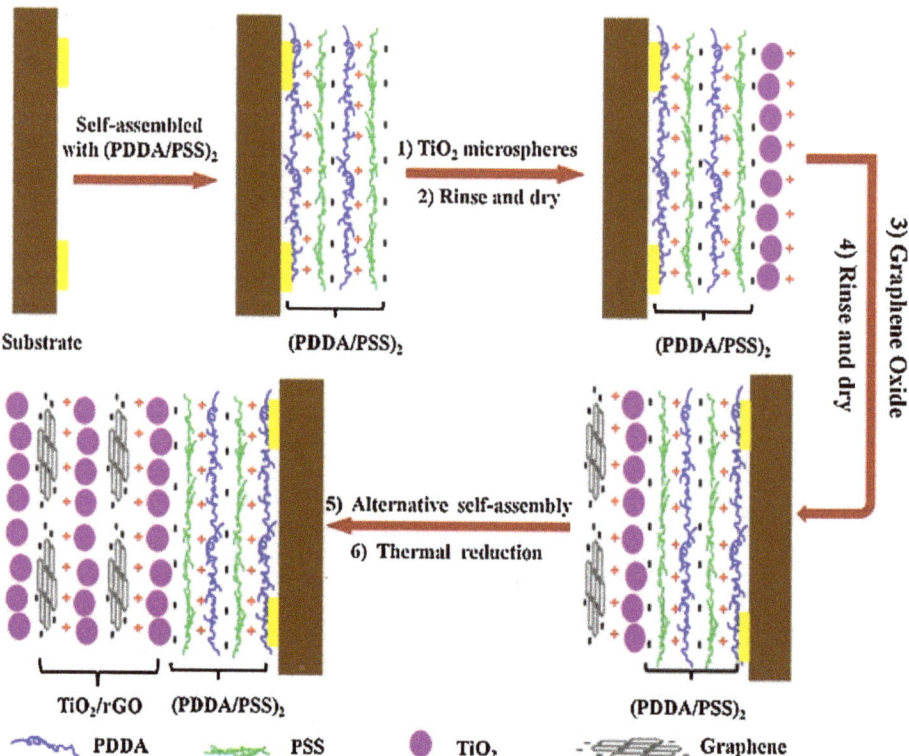

Figure 11. Fabrication illustration of TiO$_2$/rGO multilayer hybrid film by layer-by-layer self-assembly. Reprinted with permission from Ref. [178]. Copyright 2017 Elsevier.

6.2. Graphene/Metal Oxide Composite-Based Gas Sensors

Metal oxide-based gas sensors have the advantages of low production cost, good stability, wide application range, and easy integration with portable devices, and have been widely used in the measurement and monitoring of toxic and harmful gases [128,130,148,163]. However, since metal oxide-based sensors are limited by the required high operating temperature, they often bring additional energy consumption. In order to reduce the temperature for gas detection, composite materials are used as gas sensing materials. Taking advantage of the excellent gas sensing properties of metal oxides and the unique electrical, mechanical and thermodynamic properties of graphene, the formed graphene/metal oxide composites revealed high potential in the field of gas sensing.

Compared with traditional semiconductor gas sensing materials, graphene/metal oxide composite materials combine the advantages of the two materials and can produce synergistic effects for gas sensing, often with higher sensitivity, faster response/recovery speed, and lower noise signal. For example, Li et al. [179] prepared the rGO/ZnO hollow spheres by a one-step solvothermal method, in which the ZnO hollow spheres were uniformly immobilized on the surface of rGO nanosheets. When used as a material for the gas sensor to detect NO$_2$, the composite material exhibited fast response and high sensitivity to NO$_2$ at room temperature. In another case, Liu et al. synthesized rGO/ZnO composites by a redox method, and the prepared gas sensor had a response value of 25.6% to 5 ppm NO$_2$ with a response time of 165 s and a recovery time of 499 s [180]. Sun et al. used PVP to assist the synthesis of the composite material rGO/ZnO nanowires. The gas-sensing performance test indicated that the rGO/ZnO composite material can respond to 500 ppb NH$_3$ at room temperature, which is helpful for achieving the ultra-sensitive and high-accuracy detection of harmful gases [181]. Wang et al. used a one-step hydrothermal method to synthesize rGO/CuO/ZnO ternary composites to form nanoscale p-n junctions on rGO substrates. The gas sensors prepared by using the created materials showed excellent response characteristics and good selectivity to acetone, which was almost 1.5 and 2.0 times higher than those of CuO/ZnO and rGO/ZnO-based gas sensors, respectively [182]. It has become a research hotspot in the field of gas sensing to improve the performance of gas sensing materials by combining metal oxide semiconductor materials with 2D graphene materials.

Wang et al. successfully assembled SnO_2 onto the surface of GO, and studied the gas-sensing properties of formaldehyde. It was found that SnO_2 can be assembled on the surface of GO in a large area, and has good gas-sensing response properties to formaldehyde [183]. Wang et al. synthesized SnO_2 nanoparticles onto RGO through a hydrothermal reduction to form SnO_2-RGO composites, which exhibited promising application for room-temperature gas sensing of NO_2 [184]. Yin et al. [185] synthesized SnO_2/rGO nanocomposites with the SnO_2 particle sizes of 3–5 nm uniformly immobilized on rGO nanosheets through a heteronuclear growth process by a simple redox reaction under microwave irradiation. The SnO_2/rGO nanostructure on the surface has a sesame cake-like layered structure and an ultra-high specific surface area of 2110.9 $m^2 \cdot g^{-1}$. Compared with SnO_2 nanocrystals (5–10 nm), the designed SnO_2/rGO nanostructures have stronger gas-sensing behavior due to the unique hierarchical structure, high specific surface area, and synergistic effect of SnO_2 nanoparticles and rGO nanosheets. At the optimal operating temperature of 100 °C, the SnO_2/rGO-based gas sensor has a sensitivity as high as 78 and a response time as short as 7 s when exposed to 10 ppm H_2S. In a similar study, Kim et al. [176] also synthesized a graphene/SnO_2 composite material by the microwave-assisted method, and then sprayed the material onto SiO_2 substrate to fabricate a NO_2 gas sensor. At the optimal operating temperature of 150 °C, the response value of 1 ppm NO_2 was 24.7. Its excellent gas-sensing response may be related to the homojunction between SnO_2, the heterojunction between SnO_2 and graphene, and the interstitial defects of Sn atoms in the SnO_2 lattice.

Using the novel properties of SnO_2, multi-walled carbon nanotubes (MWCNTs), and rGO, Tyagi et al. developed a hybrid nanocomposite sensor for efficient detection of SO_2 gas [186]. The rGO-SnO_2 and MWCNT-SnO_2 composites were prepared by physical mixing and spin-coated onto the surface of Pt interdigital electrodes for SO_2 gas detection. The sensing response of the bare SnO_2 sensor to 500 ppm SO_2 gas at 220 °C was 1.2. However, for the same concentration of SO_2 gas, the enhanced sensing response of the MWCNT-SnO_2 sensor was 5 at 60 °C, while the maximum sensing response of the rGO-SnO_2 sensor was 22 at 60 °C. The enhanced SO_2 gas sensing performance of these composites is mainly attributed to the p-n heterojunction formed at the interface between n-type SnO_2 and p-type rGO or MWCNT.

Yu et al. [187] successfully synthesized α-Fe_2O_3@graphene nanocomposites using a simple low-temperature hydrolysis and calcination process, and fabricated the synthesized materials into gas sensors to detect different gases. Their prepared α-Fe_2O_3@graphite nanocomposites consist of porous α-Fe_2O_3 nanorods stably and orderly grown on graphitic nanosheets, as shown in Figure 12a,b. The length of α-Fe_2O_3 nanorods is related to the reaction time. When the reaction time was 12 h, the length reached a maximum value of about 200–300 nm, and the pore size was about 3.7 nm. Compared with pure α-Fe_2O_3, the α-Fe_2O_3@graphite nanocomposite-based sensor exhibited higher sensing performance for acetone. At the optimal temperature of 260 °C, the response of α-Fe_2O_3@graphite nanocomposites to 50 ppm acetone reaches a maximum value of 16.9, which was 2.2 times that of α-Fe_2O_3, as shown in Figure 12c. The high sensing performance was attributed to the porous structure, high specific surface area, and p-n heterostructure of α-Fe_2O_3@graphite nanocomposites and the high temperature stability of graphite. When α-Fe_2O_3 was recombined with graphite, due to the large gradient of the same carrier concentration, the electrons in α-Fe_2O_3 and the holes in graphite diffuse in opposite directions, so that a built-in electric field is formed between the interfaces, and the electrons in the depletion layer.

The energy bands bend until the system reaches equilibrium at the Fermi level (E$_F$), leading to the formation of a p-n heterojunction. Once the α-Fe$_2$O$_3$@graphite heterojunction sensor is exposed to acetone gas, the oxygen anions adsorbed on the sample surface undergo a redox reaction with acetone molecules and release electrons back into α-Fe$_2$O$_3$, resulting in a decrease in the resistance of the sensor. At the same time, acetone releases electrons to combine with holes in p-type graphite, resulting in a decrease in hole concentration. The reduction of holes in graphite leads to an increase in electrons and reduces the concentration gradient of the same carriers on both sides of the p-n heterojunction. Therefore, the diffusion of carriers was weakened and the barrier height of the depletion layer was reduced, which further reduced the resistance of the α-Fe$_2$O$_3$@graphite sensor, as shown in Figure 12e.

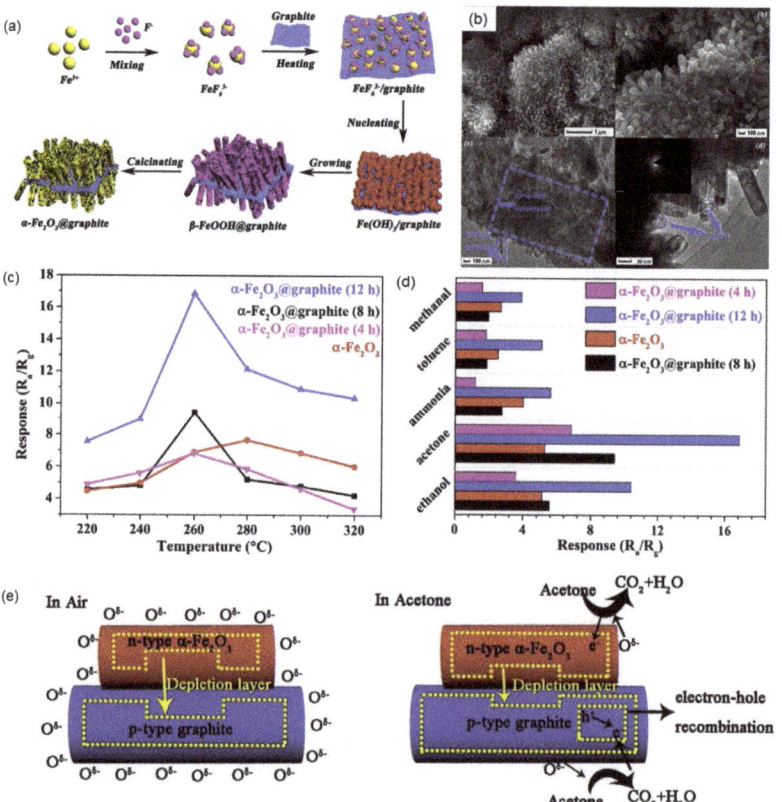

Figure 12. (**a**) Synthetic diagram of α-Fe$_2$O$_3$@graphite nanocomposites. (**b**) FE-SEM and TEM images of α-Fe$_2$O$_3$@graphite. (**c**) The responses to 50 ppm acetone of sensor based on α-Fe$_2$O$_3$ and α-Fe$_2$O$_3$@graphite (with different reaction times) operated at different operating temperatures. (**d**) The response of α-Fe$_2$O$_3$ and α-Fe$_2$O$_3$@graphite (with different reaction times) sensors to 50 ppm of different gases at 260 °C. (**e**) Reaction mechanism diagram of α-Fe$_2$O$_3$@graphite based gas sensor (O$^{\delta-}$ means O$^-$, O^{2-} and O$_2^-$). Reprinted with permission from Ref. [187]. Copyright 2019 Elsevier.

Co$_3$O$_4$, as a direct bandgap p-type metal oxide semiconductor material, has also received extensive attention in gas sensors and other fields due to its outstanding advantages such as strong corrosion resistance and non-toxicity. For instance, Zhou et al. [188] studied the performance of a Co$_3$O$_4$-based gas sensor and found that it can only work at temperatures over 200 °C. Zhang et al. [189] proposed that the rGO/Co$_3$O$_4$ nanocomposite-based sensor can realize the detection of NO_2 gas at room temperature. Srirattanapibul et al. [190] prepared a Co_3O_4-modified rGO (rGO/Co_3O_4) nanocomposite-based gas sensor by a solvothermal method. Co_3O_4 nanoparticles were distributed on and between the rGO flakes, and their dosage changed the bandgap and gas sensing properties of rGO. Decorating rGO with Co_3O_4 nanoparticles promoted the formation of the Co-C bridges, which enable the exchange of charge carriers between Co_3O_4 nanoparticles and rGO flakes, thereby increasing the number of sites for gas reactions to occur and improving gas sensing performance. The as-prepared 25% rGO/Co_3O_4-based gas sensor has a sensitivity of 1.78% and a response time of 351 s towards 20 ppm NH_3.

Many studies on the synthesis of the composites by combing graphene with other metal oxides have also been carried out. For example, Hao et al. [191] synthesized WO_3/rGO porous nanocomposites using a simple hydrothermal and annealing process. The material-based gas sensor showed good sensitivity to NO_2 and some volatile organic compounds. In another study, Ye et al. [192] fabricated uniform TiO_2/rGO membranes with enhanced NH_3 responsiveness by stepwise deposition of GO and TiO_2 layers followed by simple thermal treatment. To make it more clear, here we summarize the gas sensing properties of the above graphene/metal oxide nanocomposite-based gas sensors, as shown in Table 1.

Table 1. Sensors based on MOS modified with graphene/GO/rGO gas sensing performances.

Sensor Materials	Analyte	Response	Working Temperature	Refs.
ZnO/rGO	NO_2	17.4% (100 ppm)	RT	[179]
ZnO/rGO	NO_2	25.6% (5 ppm)	RT	[180]
ZnO/rGO	NH_3	7.2% (1 ppm)	RT	[181]
SnO_2/GO	HCHO	32 (100 ppm)	120 °C	[183]
SnO_2/rGO	H_2S	78 (10 ppm)	100 °C	[185]
Graphite/SnO_2	NO_2	24.7 (1 ppm)	150 °C	[176]
rGO/SnO_2	SO_2	22 (500 ppm)	60 °C	[186]
α-Fe_2O_3@graphite	C_3H_6O	16.9 (50 ppm)	260 °C	[187]
rGO/Co_3O_4	NO_2	26.8% (5 ppm)	RT	[189]
rGO/Co_3O_4	NH_3	1.78% (20 ppm)	RT	[190]
WO_3/rGO	NO_2	4.3 (10 ppm)	90 °C	[191]
TiO_2/rGO	NH_3	0.62 (10 ppm)	RT	[192]

6.3. 2D TMD/Metal Oxide Composite-Based Gas Sensors

Compared with graphene with a zero-band gap, the electronic structure of 2D transition metal dichalcogenides (TMDs) almost spans the whole range of the electronic structure, showing richer physical properties and a good application potential in gas sensing. However, TMD nanosheets are easy to form a dense stack structure in the process of forming a conductive network, which is not conducive to the full contact between the thin sheet and gas molecules in the conductive network, so its sensitivity and response recovery speed at room temperature need to be improved. TMDs and other metal oxide semiconductor materials form p-n heterojunction as a new type of electronic device, which can improve its gas sensing performance. The unique characteristics of TMDs make the composites an ideal choice for high-performance sensing materials at low temperatures.

In the research of gas sensing applications, SnO_2 has been the most widely used n-type semiconductor, so the combination of SnO_2 and MoS_2 is an effective way to improve the sensing performance of MoS_2. For instance, Qiao et al. loaded MoS_2 nanosheets onto SnO_2 nanofibers by hydrothermal synthesis. Through the optimal regulation of MoS_2/SnO_2 heterostructure, they achieved high-performance detection of trimethylamine at 230 °C, showing excellent sensing selectivity and long-term stability [193]. In the field of room temperature gas sensors, Cui et al. reported a new hybrid material by decorating MoS_2 nanosheets with SnO_2 nanocrystals, which achieved high-performance sensing of NO_2 at room temperature, and the loaded SnO_2 nanoparticles could significantly enhance the stability of MoS$_2$ nanosheets in the air (Figure 13) [194]. In the composite, SnO$_2$ nanocrystals acted as a strong p-type dopant at the top of MoS$_2$, resulting in the formation of p-type channels in MoS$_2$ nanosheets. In terms of sensing mechanism, they believed that SnO$_2$ nanocrystals may be the main gas adsorption center, while MoS$_2$ acted as a conductive channel at room temperature. The close electrical contact between two different semiconductor materials led to the formation of charge transfer and charge depletion layer. Because the work function of SnO$_2$ (5.7 eV) was larger than that of MoS$_2$ (5.2 eV), electrons are transferred from MoS$_2$ to SnO$_2$, resulting in a depletion layer and a Schottky barrier (Figure 13f). These electron depletion regions can be connected to each other on the surface of MoS$_2$ nanosheets and act as a passivation layer to prevent the interaction between oxygen and MoS$_2$, thus enhancing the stability of MoS$_2$ nanosheets in dry air. The fabricated gas sensor exhibited high sensitivity, excellent repeatability, and excellent selectivity to NO$_2$ in the actual dry air environment, and the detection limit could reach 0.5 ppm.

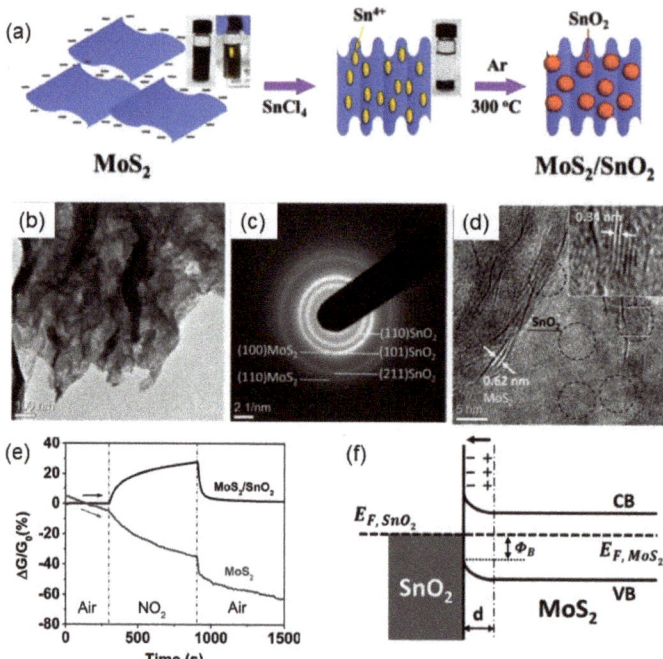

Figure 13. (a) Schematic illustration of the preparation process for MoS$_2$/SnO$_2$ nanohybrids. The inset photographs show the MoS$_2$ suspension in water before and after adding the SnCl$_4$ solution. (b) TEM images, (c) SAED pattern and (d) HRTEM images of the MoS$_2$/SnO$_2$ nanohybrids. The inset of (d) shows a typical SnO$_2$ nanocrystal on the MoS$_2$ surface. (e) The room temperature dynamic sensing response of MoS$_2$ nanosheets with and without SnO$_2$ NC decoration against 10 ppm NO$_2$ in a dry air environment, indicating the SnO$_2$ NCs significantly enhanced the stability of MoS$_2$ in the dry air. (f) Band diagram of the MoS$_2$/SnO$_2$ nanohybrid. The EF, SnO$_2$ and EF, MoS$_2$ are Fermi levels of SnO$_2$ and MoS$_2$, respectively. The CB and VB are the conductance and valance band edges of MoS$_2$, respectively. d is the thickness of the electron depletion zone, and ΦB is the Schottky barrier height. Reprinted with permission from Ref. [194]. Copyright 2015 John Wiley & Sons, Inc.

In addition to SnO$_2$, ZnO is another wide band gap n-type semiconductor for the fabrication of high-performance gas sensors. Yan et al. synthesized the ZnO/MoS$_2$ composite structure by coating ZnO nanoparticles onto MoS$_2$ nanosheets through a two-step hydrothermal method [195]. Among the composites, MoS_2 is a multi-stage structure composed of nanosheets with a thickness of 5~10 nm, and the size of ZnO particles was about 8 nm. The response value of the composite-based gas sensor to 50 ppm ethanol reached 42.8 at the operating temperature of 260 °C Han et al. designed a MoS_2/ZnO heterostructure on the MoS_2 nanosheets obtained by liquid phase exfoliation by a wet chemical method, and achieved efficient detection of NO_2 gas at room temperature [196]. After surface modification, ZnO nanoparticles had a good response to 5 ppm NO_2, and the response value reached 3050%, which was 11 times higher than that of pure phase MoS_2 nanoparticles. In addition, the recoverability of the heterostructure was improved to more than 90% without auxiliary means, and the sensor also had the characteristics of fast response speed (40 s), reliable long-term stability within 10 weeks, good selectivity, and a low detection concentration of 50 ppb. The enhanced sensing performance of MoS_2/ZnO heterostructure can be attributed to the unique 2D/0D heterostructure, synergistic effect and the p-n heterojunction between ZnO nanoparticles and MoS_2 nanosheets.

Besides, Zhao et al. reported the hydrothermal synthesis of MoS_2-modified TiO_2 nanotube composites and studied their gas sensing properties [76]. TiO_2 nanotubes are filled and covered by 1–3 layers of flake MoS_2 nanosheets. The formed MoS_2-TiO_2 composites revealed excellent sensing properties and high sensitivity to ethanol vapor at low operating temperatures, and their sensitivity was almost 11 times that of TiO_2 nanotubes. The response to 100 ppm ethanol gas was ~14.2 and the optimum working temperature was as low as 150 °C. Zhang et al. prepared CuO/MoS_2 heterostructure sensing films on the substrate by layer-by-layer self-assembly technique [197]. Compared with the pure phase CuO and MoS_2, the formed CuO/MoS_2 composite structure exhibited higher response, shorter response/recovery time, better repeatability, higher selectivity, and longer-term stable H_2S detection performance. The excellent H_2S sensing properties are mainly due to the existence of a large number of oxygen and sulfur vacancies in the composite structure of CuO nanorods and MoS_2 nanosheets, which brings a large number of active sites for gas adsorption. In addition, the synergistic effect of binary nanostructures and the modulation of electron transfer by the formation of p-n heterojunction at the material interface between p-type CuO semiconductors and n-type MoS_2 semiconductors promoted the performance of the composite structure. Ikram et al. prepared MoO_2/MoS_2 nanonetworks by controllable vulcanization and successfully applied them to the efficient detection of NO_2 gas at room temperature [198]. The response value of the composite structure to 100 ppm NO_2 gas was 19.4, and it had ultra-fast response speed and recovery speed, and the response and recovery time were 1.06 s and 22.9 s, respectively. The excellent gas-sensing performance of the sensor can be attributed to the synergistic effect between MoS_2 nanosheets and MoO_2 nanoparticles. The defects in the synthesis process provided more active sites for NO_2 gas molecules, and the formation of p-n heterojunction accelerated the charge transfer between NO_2 and gas molecules.

In addition to MoS_2, other transition metal dichalcogenides and metal oxide composites have been also often used as sensitive materials for gas detection. For example, Qin et al. prepared 2D WS_2 nanosheets/TiO_2 quantum dots composites by chemical stripping method, which have been successfully used in room temperature NH_3 sensing. The fabricated gas sensors exhibited a quicker sensing response to 250 ppm NH_3, which was almost 17 times that of the original WS_2 [199]. Gu et al. prepared SnO_2/SnS_2 heterojunction nanocomposites by the in situ high-temperature oxidizer SnS_2, which significantly improved the response to NO_2 and decreased the working temperature [200]. To make it more clear, the gas sensing properties of the transition metal dichalcogenides/metal oxide composite-based gas sensors are shown in Table 2.

Table 2. Sensors based on MOS modified with TMDs gas sensing performances.

Sensor Materials	Analyte	Response	Working Temperature	Refs.
SnO_2/MoS_2	C_3H_9N	106.3 (200 ppm)	230 °C	[193]
SnO_2/MoS_2	NO_2	28% (10 ppm)	RT	[194]
ZnO/MoS_2	C_2H_6O	42.8 (50 ppm)	260 °C	[195]
ZnO/MoS_2	NO_2	3050% (5 ppm)	RT	[196]
MoS_2/TiO_2	C_2H_6O	14.2 (100 ppm)	150 °C	[76]
CuO/MoS_2	H_2S	61 (30 ppm)	RT	[197]
MoO_2/MoS_2	NO_2	19.4 (100 ppm)	RT	[198]
TiO_2 QDs/WS_2	NH_3	43.7% (250 ppm)	RT	[199]
SnO_2/SnS_2	NO_2	5.3 (8 ppm)	80 °C	[200]

6.4. Other 2D Material/Metal Oxide Composites-Based Gas Sensors

As an important wide band gap and semiconductor gas sensing material with a special layered structure, MoO_3 is easy to form metal phase MoS_2 in the process of reacting with H_2S, which makes MoO_3 have unique response characteristics to H_2S gas. However, single-phase MoO_3 has some shortcomings, such as high working temperature and poor limit detection ability [201,202], so it needs to be compounded with other materials to improve its gas-sensing performance. For example, Gao et al. used graphene as a sacrificial template and prepared porous MoO_3/SnO_2 composite nanosheets with n-n heterostructure by hydrothermal method, and studied their gas sensing properties, as shown in Figure 14 [203]. The TEM characterizations indicated that the composite is a porous structure composed of MoO_3 and SnO_2 nanoparticles, and the lattice distortion at the interface indicates the existence of MoO_3-SnO_2 n-n heterojunction (Figure 14a,b). The gas-sensing performance test showed that compared with SnO_2 nanosheets, the introduction of MoO_3 and the formation of heterojunctions lead to the change of energy band structure and carrier separation transfer rate, and the optimum operating temperature of MoO_3/SnO_2 nanosheet sensor is lower and the sensitivity is higher. The detection limit of MoO_3/SnO_2 nanoparticles for H_2S was as low as 100 ppb at 115 °C, and the response and recovery times were 22 s and 10 s, respectively (Figure 14c–e). The excellent gas-sensing performance of the MoO_3/SnO_2 composite nanosheet gas sensor was attributed to the following aspects.

Firstly, in the composites, the n-n heterojunction formed at the interface between MoO_3 and SnO_2, and the charge accumulation layer and charge consumption layer are formed on both sides of the heterojunction, respectively, thus forming an internal electric field and hindering the free transport of free electrons in the semiconductor. The sensing mechanism of the composite material is shown in Figure 14f,g. When the material is exposed to air, the electrons in the semiconductor MoO_3 and SnO_2 conduction bands adsorb the oxygen in the air to the surface of the semiconductor gas sensing material to form the adsorbed oxygen. This leads to the loss of electrons in the conduction bands of semiconductors MoO_3 and SnO_2 (Figure 14f). On the energy band diagram, the energy bands of semiconductors MoO_3 and SnO_2 bend upward, and the barrier height increases. In the atmosphere of H_2S, when H_2S reacts with the adsorbed oxygen on the surface of semiconductor sensitive materials, the electrons captured by adsorbed oxygen are released back into the semiconductor MoO_3 and SnO_2 conduction bands, which leads to a significant decrease in the barrier height at the n-n heterojunction and a significant decrease in the resistance of the sensor (Figure 14g). The electrical conductivity of composite semiconductor materials is inversely proportional to the barrier height, and the barrier height can effectively control the gas sensing response characteristics of sensitive materials with heterojunctions. Therefore, a large change of barrier height has a significant contribution to the H_2S performance of the MoO_3/SnO_2 gas sensor. Secondly, the large specific surface area of MoO_3/SnO_2 composite nanosheets means that when exposed to H_2S gas, the MoO_3/SnO_2 sensor with a large specific surface area can contact more H_2S gas molecules and react with them, resulting in greater resistance change and higher sensitivity response. Thirdly, the composite has high porosity, and the larger pore volume can not only promote the adsorption of H_2S gas molecules to the sensitive material surface more quickly, but also provide a shorter gas conduction path and promote the gas molecules to break away from the sensitive material surface and then achieve the ideal state of rapid response and rapid recovery.

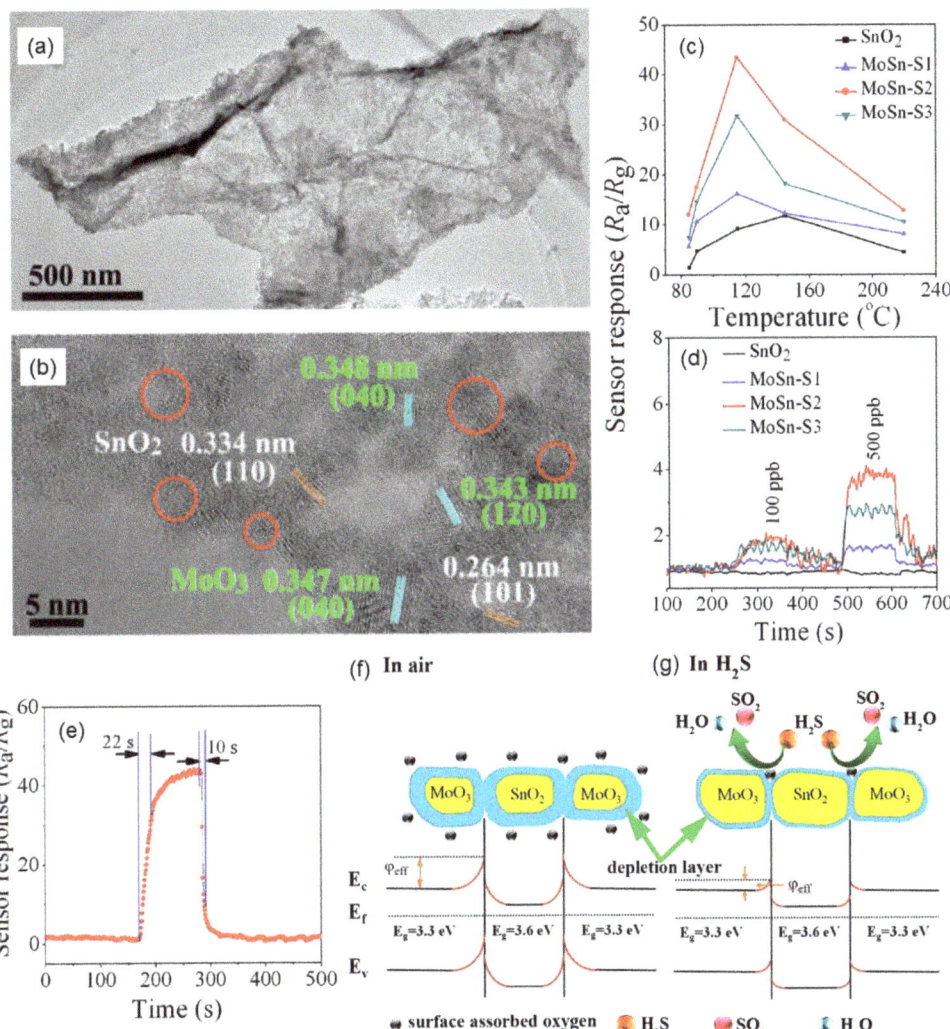

Figure 14. (a,b) Low-magnification TEM images and HRTEM images of MoSn-S_2 nanoflakes. Red circles in (b) shows the lattice distortions at the interfaces between MoO_3 and SnO_2. (c) Sensor responses of as-prepared samples as a function of different operating temperatures to 10 ppm of H_2S concentration. (d) Typical sensor responses of SnO_2, MoSn-S1, MoSn-S_2, and MoSn-S_3 toward 100 and 500 ppb of H_2S gas at optimal working temperature, and (e) response and recovery times curve of MoSn-S_2 NFs to 10 ppm of H_2S at 115 °C. (f,g) Diagram of energy band structure of MoSn−nanocomposites in (f) air and (g) H_2S. Reprinted with permission from Ref. [203]. Copyright 2019 ACS.

Yin et al. [204] synthesized hierarchical Fe_2O_3/WO_3 nanocomposites with ultra-high specific surface area composed of Fe_2O_3 nanoparticles and single-crystal WO_3 nanosheets through microwave heating and in situ growth. The BET specific surface area of the sample that prepared with 5 wt.% Fe_2O_3/WO_3 by this process was as high as 1207 $m^2 \cdot g^{-1}$, which was 5.9 times that of the corresponding WO_3 nanosheets (203 $m^2 \cdot g^{-1}$). The significant enhancement of the specific surface area of the $Fe_2O_3@WO_3$ samples was attributed to the hierarchical structure of the prepared composite materials, in which the monolayer and unconnected Fe_2O_3 nanoparticles are tightly anchored to the surface of the WO_3 nanosheets, so that the inner surface or interface of the aggregated polycrystal is entirely the outer surface. The gas-sensing performance tests indicated that $Fe_2O_3@WO_3$ nanocomposites exhibited high response and selectivity towards H_2S at low operating temperatures due to the synergistic effect of the components of $Fe_2O_3@WO_3$ nanocomposites and the hierarchical microstructure with ultra-high specific surface area. At 150 °C, the fabricated gas sensor showed a response to 10 ppm H_2S of as high as 192, which was four times that of the WO_3 nanosheet-based gas sensor.

7. Conclusions and Future Perspectives

In this paper, we summarize the research progress of gas sensors using 2D materials, metal oxides, and their composites as sensitive materials. The gas sensing mechanism, main factors affecting sensing performance, and the applications of various gas sensors are presented and discussed in detail. It can be concluded that the 2D material/metal oxide-based gas sensor can efficiently identify and detect toxic and harmful gases. Compared with pure metal oxide semiconductors, composite materials have higher carrier rates, larger high mechanical strength, and large specific surface area, and the synergistic effect of the two components can further enhance the gas sensing performance. The -OH, -O, and other functional groups on the surface of 2D materials not only provide chemical bonds to form composite metal oxide materials during the composite process of the material, but also give more active sites for the gas sensing process, thereby further improving the gas sensing performance. In addition, the composite materials can effectively reduce the working temperature.

Although nanomaterial-based gas sensors have made great progress in the past few decades, the operating temperature of metal oxides is too high, and the selectivity of 2D materials is still unsatisfactory. In the future, effective strategies such as building composite structures are highly needed to improve the selectivity, reduce the operating temperature, and improve the sensitivity and other properties. In addition, the research on the combination of metal oxides with 2D materials is still at an early stage, and its sensing mechanisms of the composite-based gas sensors should be further studied. Only when the mechanism and process are clear, the preparation and assembly of nanomaterial-based gas sensors can be achieved purposefully. In addition, it is necessary for researchers to develop new design strategies to further optimize metal oxide nanomaterials with 2D nanomaterials to make them more suitable for gas sensing. Finally, it is expected that facile assembly and fabrication processes will be developed to enable batch fabrication of gas sensors with high stability, selectivity, sensitivity, reproducibility, and quick response in the future.

Author Contributions: Conceptualization, T.L., W.Y. and G.W.; reference analysis, T.L., Y.S., S.G., P.X. and S.W.; resources, all authors; writing—original draft preparation, T.L.; writing—review and editing, T.L., H.K., G.Y. and G.W.; supervision, G.W.; project administration, T.L. and G.W.; funding acquisition, G.W. All authors have read and agreed to the published version of the manuscript.

Funding: Taishan Scholars Program of Shandong Province (No. tsqn201909104) and the High-Grade Talents Plan of Qingdao University.

Data Availability Statement: The data presented in this study are available in insert article.

Acknowledgments: The authors thank the financial support from Taishan Scholars Program of Shandong Province (No. tsqn201909104) and the High-Grade Talents Plan of Qingdao University.

Conflicts of Interest: The authors declare no conflict of interest.

References

1. Chatterjee, S.G.; Chatterjee, S.; Ray, A.K.; Chakraborty, A.K. Graphene-metal oxide nanohybrids for toxic gas sensor: A review. *Sens. Actuators B Chem.* **2015**, *221*, 1170–1181. [CrossRef]
2. Ab Kadir, R.; Li, Z.Y.; Sadek, A.; Rani, R.A.; Zoolfakar, A.S.; Field, M.R.; Ou, J.Z.; Chrimes, A.F.; Kalantar-zadeh, K. Electrospun Granular Hollow SnO_2 Nanofibers Hydrogen Gas Sensors Operating at Low Temperatures. *J. Phys. Chem. C* **2014**, *118*, 3129–3139. [CrossRef]
3. Zhang, D.Z.; Wu, J.F.; Li, P.; Cao, Y.H. Room-temperature SO_2 gas-sensing properties based on a metal-doped MoS_2 nanoflower: An experimental and density functional theory investigation. *J. Mater. Chem. A* **2017**, *5*, 20666–20677. [CrossRef]
4. Zhang, Y.F.; Thorburn, P.J. Handling missing data in near real-time environmental monitoring: A system and a review of selected methods. *Future Gener. Comp. Syst.* **2022**, *128*, 63–72. [CrossRef]
5. Di Natale, C.; Paolesse, R.; Martinelli, E.; Capuano, R. Solid-state gas sensors for breath analysis: A review. *Anal. Chim. Acta* **2014**, *824*, 1–17. [CrossRef] [PubMed]
6. Guntner, A.T.; Pineau, N.J.; Mochalski, P.; Wiesenhofer, H.; Agapiou, A.; Mayhew, C.A.; Pratsinis, S.E. Sniffing Entrapped Humans with Sensor Arrays. *Anal. Chem.* **2018**, *90*, 4940–4945.
7. Righettoni, M.; Amann, A.; Pratsinis, S.E. Breath analysis by nanostructured metal oxides as chemo-resistive gas sensors. *Mater. Today* **2015**, *18*, 163–171. [CrossRef]
8. Ponzoni, A.; Comini, E.; Concina, I.; Ferroni, M.; Falasconi, M.; Gobbi, E.; Sberveglieri, V.; Sberveglieri, G. Nanostructured Metal Oxide Gas Sensors, a Survey of Applications Carried out at SENSOR Lab, Brescia (Italy) in the Security and Food Quality Fields. *Sensors* **2012**, *12*, 17023–17045. [CrossRef]
9. Kim, W.S.; Kim, H.C.; Hong, S.H. Gas sensing properties of MoO_3 nanoparticles synthesized by solvothermal method. *J. Nanopart. Res.* **2010**, *12*, 1889–1896. [CrossRef]
10. Chen, Y.J.; Xiao, G.; Wang, T.S.; Zhang, F.; Ma, Y.; Gao, P.; Zhu, C.L.; Zhang, E.D.; Xu, Z.; Li, Q.H. α-MoO_3/TiO_2 core/shell nanorods: Controlled-synthesis and low-temperature gas sensing properties. *Sens. Actuators B Chem.* **2011**, *155*, 270–277. [CrossRef]
11. Lim, S.K.; Hwang, S.H.; Kim, S.; Park, H. Preparation of ZnO nanorods by microemulsion synthesis and their application as a CO gas sensor. *Sens. Actuators B Chem.* **2011**, *160*, 94–98. [CrossRef]
12. Yang, B.; Wang, C.; Xiao, R.; Yu, H.Y.; Wang, J.X.; Xu, J.L.; Liu, H.M.; Xia, F.; Xiao, J.Z. CO Response Characteristics of $NiFe_2O_4$ Sensing Material at Elevated Temperature. *J. Electrochem. Soc.* **2019**, *166*, B956–B960. [CrossRef]
13. Kim, H.J.; Lee, J.H. Highly sensitive and selective gas sensors using p-type oxide semiconductors: Overview. *Sens. Actuators B Chem.* **2014**, *192*, 607–627. [CrossRef]
14. Volanti, D.P.; Felix, A.A.; Orlandi, M.O.; Whitfield, G.; Yang, D.J.; Longo, E.; Tuller, H.L.; Varela, J.A. The Role of Hierarchical Morphologies in the Superior Gas Sensing Performance of CuO-Based Chemiresistors. *Adv. Funct. Mater.* **2013**, *23*, 1759–1766. [CrossRef]
15. Virji, S.; Huang, J.X.; Kaner, R.B.; Weiller, B.H. Polyaniline nanofiber gas sensors: Examination of response mechanisms. *Nano Lett.* **2004**, *4*, 491–496. [CrossRef]
16. Gilbertson, L.M.; Busnaina, A.A.; Isaacs, J.A.; Zimmerman, J.B.; Eckelman, M.J. Life Cycle Impacts and Benefits of a Carbon Nanotube-Enabled Chemical Gas Sensor. *Environ. Sci. Technol.* **2014**, *48*, 11360–11368. [CrossRef]
17. Ma, H.Y.; Li, Y.W.; Yang, S.X.; Cao, F.; Gong, J.; Deng, Y.L. Effects of Solvent and Doping Acid on the Morphology of Polyaniline Prepared with the Ice-Templating Method. *J. Phys. Chem. C* **2010**, *114*, 9264–9269. [CrossRef]
18. Patil, V.L.; Vanalakar, S.A.; Patil, P.S.; Kim, J.H. Fabrication of nanostructured ZnO thin films based NO_2 gas sensor via SILAR technique. *Sens. Actuators B Chem.* **2017**, *239*, 1185–1193. [CrossRef]
19. Meng, F.L.; Zheng, H.X.; Chang, Y.L.; Zhao, Y.; Li, M.Q.; Wang, C.; Sun, Y.F.; Liu, J.H. One-Step Synthesis of Au/SnO_2/RGO Nanocomposites and Their VOC Sensing Properties. *IEEE Trans. Nanotechnol.* **2018**, *17*, 212–219. [CrossRef]
20. Choi, S.J.; Lee, I.; Jang, B.H.; Youn, D.Y.; Ryu, W.H.; Park, C.O.; Kim, I.D. Selective Diagnosis of Diabetes Using Pt-Functionalized WO_3 Hemitube Networks As a Sensing Layer of Acetone in Exhaled Breath. *Anal. Chem.* **2013**, *85*, 1792–1796. [CrossRef]
21. Navale, S.T.; Liu, C.; Yang, Z.; Patil, V.B.; Cao, P.; Du, B.; Mane, R.S.; Stadler, F.J. Low-temperature wet chemical synthesis strategy of In_2O_3 for selective detection of NO_2 down to ppb levels. *J. Alloys Compd.* **2018**, *735*, 2102–2110. [CrossRef]
22. Fine, G.F.; Cavanagh, L.M.; Afonja, A.; Binions, R. Metal Oxide Semi-Conductor Gas Sensors in Environmental Monitoring. *Sensors* **2010**, *10*, 5469–5502. [CrossRef] [PubMed]
23. Suematsu, K.; Shin, Y.; Ma, N.; Oyama, T.; Sasaki, M.; Yuasa, M.; Kida, T.; Shimanoe, K. Pulse-Driven Micro Gas Sensor Fitted with Clustered Pd/SnO_2 Nanoparticles. *Anal. Chem.* **2015**, *87*, 8407–8415. [CrossRef] [PubMed]
24. Chen, Y.; Yang, G.Z.; Liu, B.; Kong, H.; Xiong, Z.; Guo, L.; Wei, G. Biomineralization of $ZrO2$ nanoparticles on graphene oxide-supported peptide/cellulose binary nanofibrous membranes for high-performance removal of fluoride ions. *Chem. Eng. J.* **2022**, *430*, 132721. [CrossRef]
25. Liu, B.; Jiang, M.; Zhu, D.Z.; Zhang, J.M.; Wei, G. Metal-organic frameworks functionalized with nucleic acids and amino acids for structure- and function-specific applications: A tutorial review. *Chem. Eng. J.* **2022**, *428*, 131118. [CrossRef]

26. Liu, X.H.; Ma, T.T.; Pinna, N.; Zhang, J. Two-Dimensional Nanostructured Materials for Gas Sensing. *Adv. Funct. Mater.* **2017**, *27*, 1702168. [CrossRef]
27. Neri, G. Thin 2D: The New Dimensionality in Gas Sensing. *Chemosensors* **2017**, *5*, 21. [CrossRef]
28. Anichini, C.; Czepa, W.; Pakulski, D.; Aliprandi, A.; Ciesielski, A.; Samorì, P. Chemical sensing with 2D materials. *Chem. Soc. Rev.* **2018**, *47*, 4860–4908. [CrossRef]
29. Mao, S.; Chang, J.B.; Pu, H.H.; Lu, G.H.; He, Q.Y.; Zhang, H.; Chen, J.H. Two-dimensional nanomaterial-based field-effect transistors for chemical and biological sensing. *Chem. Soc. Rev.* **2017**, *46*, 6872–6904. [CrossRef]
30. Kim, T.H.; Kim, Y.H.; Park, S.Y.; Kim, S.Y.; Jang, H.W. Two-Dimensional Transition Metal Disulfides for Chemoresistive Gas Sensing: Perspective and Challenges. *Chemosensors* **2017**, *5*, 15. [CrossRef]
31. Kakanakova-Georgieva, A.; Giannazzo, F.; Nicotra, G.; Cora, I.; Gueorguiev, G.K.; Persson, P.O.A.; Pecz, B. Material proposal for 2D indium oxide. *Appl. Surf. Sci.* **2021**, *548*, 149275. [CrossRef]
32. Dos Santos, R.B.; Rivelino, R.; Gueorguiev, G.K.; Kakanakova-Georgieva, A. Exploring 2D structures of indium oxide of different stoichiometry. *CrystEngComm* **2021**, *23*, 6661–6667. [CrossRef]
33. Wu, J.; Yang, Y.; Yu, H.; Dong, X.T.; Wang, T.T. Ultra-efficient room-temperature H_2S gas sensor based on $NiCo_2O_4$/r-GO nanocomposites. *New J. Chem.* **2019**, *43*, 10501–10508. [CrossRef]
34. Zhang, D.Z.; Jiang, C.X.; Sun, Y.E. Room-temperature high-performance ammonia gas sensor based on layer-by-layer self-assembled molybdenum disulfide/zinc oxide nanocomposite film. *J. Alloys Compd.* **2017**, *698*, 476–483. [CrossRef]
35. Nakate, U.T.; Ahmad, R.; Patil, P.; Bhat, K.S.; Wang, Y.S.; Mahmoudi, T.; Yu, Y.T.; Suh, E.K.; Hahn, Y.B. High response and low concentration hydrogen gas sensing properties using hollow ZnO particles transformed from polystyrene@ZnO core-shell structures. *Int. J. Hydrogen Energy* **2019**, *44*, 15677–15688. [CrossRef]
36. Nakate, U.T.; Lee, G.H.; Ahmad, R.; Patil, P.; Hahn, Y.B.; Yu, Y.T.; Suh, E.K. Nano-bitter gourd like structured CuO for enhanced hydrogen gas sensor application. *Int. J. Hydrogen Energy* **2018**, *43*, 22705–22714. [CrossRef]
37. Nakate, U.T.; Lee, G.H.; Ahmad, R.; Patil, P.; Bhopate, D.P.; Hahn, Y.B.; Yu, Y.T.; Suh, E.K. Hydrothermal synthesis of p-type nanocrystalline NiO nanoplates for high response and low concentration hydrogen gas sensor application. *Ceram. Int.* **2018**, *44*, 15721–15729. [CrossRef]
38. Yang, S.X.; Jiang, C.B.; Wei, S.H. Gas sensing in 2D materials. *Appl. Phys. Rev.* **2017**, *4*, 021304. [CrossRef]
39. Yue, Q.; Shao, Z.Z.; Chang, S.L.; Li, J.B. Adsorption of gas molecules on monolayer MoS_2 and effect of applied electric field. *Nanoscale Res. Lett.* **2013**, *8*, 425. [CrossRef]
40. Wang, D.; Tian, L.; Li, H.; Wan, K.; Yu, X.; Wang, P.; Chen, A.; Wang, X.; Yang, J. Mesoporous Ultrathin SnO_2 Nanosheets in Situ Modified by Graphene Oxide for Extraordinary Formaldehyde Detection at Low Temperatures. *ACS Appl. Mater. Interfaces* **2019**, *11*, 12808–12818. [CrossRef]
41. Rothschild, A.; Komem, Y. The effect of grain size on the sensitivity of nanocrystalline metal-oxide gas sensors. *J. Appl. Phys.* **2004**, *95*, 6374–6380. [CrossRef]
42. Xu, C.N.; Tamaki, J.; Miura, N.; Yamazoe, N. Grain-size effects on gas sensitivity of porous SnO_2-based elements. *Sens. Actuators B Chem.* **1991**, *3*, 147–155. [CrossRef]
43. Sun, Y.F.; Liu, S.B.; Meng, F.L.; Liu, J.Y.; Jin, Z.; Kong, L.T.; Liu, J.H. Metal oxide nanostructures and their gas sensing properties: A review. *Sensors* **2012**, *12*, 2610–2631. [CrossRef] [PubMed]
44. Han, M.A.; Kim, H.J.; Lee, H.C.; Park, J.S.; Lee, H.N. Effects of porosity and particle size on the gas sensing properties of SnO_2 films. *Appl. Surf. Sci.* **2019**, *481*, 133–137. [CrossRef]
45. Min, B.K.; Choi, S.D. SnO_2 thin film gas sensor fabricated by ion beam deposition. *Sens. Actuators B Chem.* **2004**, *98*, 239–246. [CrossRef]
46. Shoyama, M.; Hashimoto, N. Effect of poly ethylene glycol addition on the microstructure and sensor characteristics of SnO_2 thin films prepared by sol-gel method. *Sens. Actuators B Chem.* **2003**, *93*, 585–589. [CrossRef]
47. Boudiba, A.; Zhang, C.; Bittencourt, C.; Umek, P.; Olivier, M.G.; Snyders, R.; Debliquy, M. SO_2 gas sensors based on WO_3 nanostructures with different morphologies. *Procedia Eng.* **2012**, *47*, 1033–1036. [CrossRef]
48. Jia, Q.Q.; Ji, H.M.; Wang, D.H.; Bai, X.; Sun, X.H.; Jin, Z.G. Exposed facets induced enhanced acetone selective sensing property of nanostructured tungsten oxide. *J. Mater. Chem. A* **2014**, *2*, 13602–13611. [CrossRef]
49. Lu, Y.Y.; Zhan, W.W.; He, Y.; Wang, Y.T.; Kong, X.J.; Kuang, Q.; Xie, Z.X.; Zheng, L.S. MOF-Templated Synthesis of Porous Co_3O_4 Concave Nanocubes with High Specific Surface Area and Their Gas Sensing Properties. *ACS Appl. Mater. Interfaces* **2014**, *6*, 4186–4195. [CrossRef]
50. Wang, L.L.; Zhang, R.; Zhou, T.T.; Lou, Z.; Deng, J.N.; Zhang, T. Concave Cu_2O octahedral nanoparticles as an advanced sensing material for benzene (C_6H_6) and nitrogen dioxide (NO_2) detection. *Sens. Actuators B Chem.* **2016**, *223*, 311–317. [CrossRef]
51. Dos Santos, R.B.; Rivelino, R.; Mota, F.D.; Gueorguiev, G.K.; Kakanakova-Georgieva, A. Dopant species with Al-Si and N-Si bonding in the MOCVD of AlN implementing trimethylaluminum, ammonia and silane. *J. Phys. D Appl. Phys.* **2015**, *48*, 295104. [CrossRef]
52. Kakanakova-Georgieva, A.; Gueorguiev, G.K.; Yakimova, R.; Janzen, E. Effect of impurity incorporation on crystallization in AlN sublimation epitaxy. *J. Appl. Phys.* **2004**, *96*, 5293–5297. [CrossRef]
53. Muller, S.A.; Degler, D.; Feldmann, C.; Turk, M.; Moos, R.; Fink, K.; Studt, F.; Gerthsen, D.; Barsan, N.; Grunwaldt, J.D. Exploiting Synergies in Catalysis and Gas Sensing using Noble Metal-Loaded Oxide Composites. *ChemCatChem* **2018**, *10*, 864–880. [CrossRef]

54. Fedorenko, G.; Oleksenko, L.; Maksymovych, N.; Skolyar, G.; Ripko, O. Semiconductor Gas Sensors Based on Pd/SnO$_2$ Nanomaterials for Methane Detection in Air. *Nanoscale Res. Lett.* **2017**, *12*, 329. [CrossRef]
55. Barbosa, M.S.; Suman, P.H.; Kim, J.J.; Tuller, H.L.; Varela, J.A.; Orlandi, M.O. Gas sensor properties of Ag$^-$ and Pd-decorated SnO micro-disks to NO$_2$, H$_2$ and CO: Catalyst enhanced sensor response and selectivity. *Sens. Actuators B Chem.* **2017**, *239*, 253–261. [CrossRef]
56. Zhang, X.L.; Song, D.L.; Liu, Q.; Chen, R.R.; Liu, J.Y.; Zhang, H.S.; Yu, J.; Liu, P.L.; Wang, J. Designed synthesis of Co-doped sponge-like In$_2$O$_3$ for highly sensitive detection of acetone gas. *CrystEngComm* **2019**, *21*, 1876–1885. [CrossRef]
57. Ma, J.H.; Ren, Y.; Zhou, X.R.; Liu, L.L.; Zhu, Y.H.; Cheng, X.W.; Xu, P.C.; Li, X.X.; Deng, Y.H.; Zhao, D.Y. Pt Nanoparticles Sensitized Ordered Mesoporous WO$_3$ Semiconductor: Gas Sensing Performance and Mechanism Study. *Adv. Funct. Mater.* **2018**, *28*, 1705268. [CrossRef]
58. Ju, D.X.; Xu, H.Y.; Xu, Q.; Gong, H.B.; Qiu, Z.W.; Guo, J.; Zhang, J.; Cao, B.Q. High triethylamine-sensing properties of NiO/SnO$_2$ hollow sphere P–N heterojunction sensors. *Sens. Actuators B Chem.* **2015**, *215*, 39–44. [CrossRef]
59. Kida, T.; Nishiyama, A.; Hua, Z.Q.; Suematsu, K.; Yuasa, M.; Shimanoe, K. WO$_3$ Nano lamella Gas Sensor: Porosity Control Using SnO$_2$ Nanoparticles for Enhanced NO$_2$ Sensing. *Langmuir* **2014**, *30*, 2571–2579. [CrossRef]
60. Mishra, R.K.; Murali, G.; Kim, T.H.; Kim, J.H.; Lim, Y.J.; Kim, B.S.; Sahay, P.P.; Lee, S.H. Nanocube In$_2$O$_3$@RGO heterostructure based gas sensor for acetone and formaldehyde detection. *RSC Adv.* **2017**, *7*, 38714–38724. [CrossRef]
61. Schedin, F.; Geim, A.K.; Morozov, S.V.; Hill, E.W.; Blake, P.; Katsnelson, M.I.; Novoselov, K.S. Detection of individual gas molecules adsorbed on graphene. *Nat. Mater.* **2007**, *6*, 652–655. [CrossRef] [PubMed]
62. Singh, E.; Meyyappan, M.; Nalwa, H.S. Flexible Graphene-Based Wearable Gas and Chemical Sensors. *ACS Appl. Mater. Interfaces* **2017**, *9*, 34544–34586. [CrossRef] [PubMed]
63. Basu, S.; Bhattacharyya, P. Recent developments on graphene and graphene oxide based solid state gas sensors. *Sens. Actuators B Chem.* **2012**, *173*, 1–21. [CrossRef]
64. Varghese, S.; Lonkar, S.; Singh, K.; Swaminathan, S.; Abdala, A. Recent advances in graphene based gas sensors. *Sens. Actuators B Chem.* **2015**, *218*, 160–183. [CrossRef]
65. Ricciardella, F.; Massera, E.; Polichetti, T.; Miglietta, M.L.; Di Francia, G. Calibrated Graphene-Based Chemi-Sensor for Sub Parts-Per-Million NO$_2$ Detection Operating at Room Temperature. *Appl. Phys. Lett.* **2014**, *104*, 183502. [CrossRef]
66. Rumyantsev, S.; Liu, G.X.; Shur, M.S.; Potyrailo, R.A.; Balandin, A.A. Selective Gas Sensing with a Single Pristine Graphene Transistor. *Nano Lett.* **2012**, *12*, 2294–2298. [CrossRef]
67. Choi, H.; Choi, J.S.; Kim, J.S.; Choe, J.H.; Chung, K.H.; Shin, J.W.; Kim, J.T.; Youn, D.H.; Kim, K.C.; Lee, J.I.; et al. Flexible and Transparent Gas Molecule Sensor Integrated with Sensing and Heating Graphene Layers. *Small* **2014**, *10*, 3685–3691. [CrossRef]
68. Nomani, M.W.K.; Shishir, R.; Qazi, M.; Diwan, D.; Shields, V.B.; Spencer, M.G.; Tompa, G.S.; Sbrockey, N.M.; Koley, G. Highly sensitive and selective detection of NO$_2$ using epitaxial graphene on 6H-SiC. *Sens. Actuators B Chem.* **2010**, *150*, 301–307. [CrossRef]
69. Yang, G.; Kim, H.Y.; Jang, S.; Kim, J. Transfer-Free Growth of Multilayer Graphene Using Self-Assembled Monolayers. *ACS Appl. Mater. Interfaces* **2016**, *8*, 27115–27121. [CrossRef]
70. Shen, F.P.; Wang, D.; Liu, R.; Pei, X.F.; Zhang, T.; Jin, J. Edge-tailored graphene oxide nanosheet-based field effect transistors for fast and reversible electronic detection of sulfur dioxide. *Nanoscale* **2013**, *5*, 537–540. [CrossRef]
71. Ghosh, R.; Midya, A.; Santra, S.; Ray, S.K.; Guha, P.K. Chemically Reduced Graphene Oxide for Ammonia Detection at Room Temperature. *ACS Appl. Mater. Interfaces* **2013**, *5*, 7599–7603. [CrossRef] [PubMed]
72. Hu, N.T.; Wang, Y.Y.; Chai, J.; Gao, R.G.; Yang, Z.; Kong, E.S.W.; Zhang, Y.F. Gas sensor based on p-phenylenediamine reduced graphene oxide. *Sens. Actuators B Chem.* **2012**, *163*, 107–114. [CrossRef]
73. Dua, V.; Surwade, S.P.; Ammu, S.; Agnihotra, S.R.; Jain, S.; Roberts, K.E.; Park, S.; Ruoff, R.S.; Manohar, S.K. All-organic vapor sensor using inkjet-printed reduced graphene oxide. *Angew. Chem. Int. Ed.* **2010**, *49*, 2154–2157. [CrossRef]
74. Baek, D.H.; Kim, J. MoS$_2$ gas sensor functionalized by Pd for the detection of hydrogen. *Sens. Actuators B Chem.* **2017**, *250*, 686–691. [CrossRef]
75. Donarelli, M.; Prezioso, S.; Perrozzi, F.; Bisti, F.; Nardone, M.; Giancaterini, L.; Cantalini, C.; Ottaviano, L. Response to NO$_2$ and other gases of resistive chemically exfoliated MoS$_2$-based gas sensors. *Sens. Actuators B Chem.* **2015**, *207*, 602–613. [CrossRef]
76. Zhao, P.X.; Tang, Y.; Mao, J.; Chen, Y.X.; Song, H.; Wang, J.W.; Song, Y.; Liang, Y.Q.; Zhang, X.M. One-Dimensional MoS$_2$-Decorated TiO$_2$ nanotube gas sensors for efficient alcohol sensing. *J. Alloys Compd.* **2016**, *674*, 252–258. [CrossRef]
77. Li, H.; Yin, Z.Y.; He, Q.Y.; Li, H.; Huang, X.; Lu, G.; Fam, D.W.H.; Tok, A.I.Y.; Zhang, Q.; Zhang, H. Fabrication of Single- and Multilayer MoS$_2$ Film-Based Field-Effect Transistors for Sensing NO at Room Temperature. *Small* **2012**, *8*, 63–67. [CrossRef]
78. Late, D.J.; Huang, Y.K.; Liu, B.; Acharya, J.; Shirodkar, S.; Luo, J.; Yan, A.; Carles, J.; Waghmare, U.V.; Dravid, V.P.; et al. Sensing Behavior of Atomically Thin-Layered MoS$_2$ Transistors. *ACS Nano* **2013**, *7*, 4879–4891. [CrossRef]
79. Wang, J.; Lian, G.; Xu, Z.; Fu, C.; Lin, Z.; Li, L.; Wang, Q.; Cui, D.; Wong, C.P. Growth of Large-Size SnS Thin Crystals Driven by Oriented Attachment and Applications to Gas Sensors and Photodetectors. *ACS Appl. Mater. Interfaces* **2016**, *8*, 9545–9551. [CrossRef]
80. Xiong, Y.; Xu, W.; Ding, D.; Lu, W.; Zhu, L.; Zhu, Z.; Wang, Y.; Xue, Q. Ultra-sensitive NH$_3$ sensor based on flower-shaped SnS$_2$ nanostructures with sub-ppm detection ability. *J. Hazard. Mater.* **2018**, *341*, 159–167. [CrossRef]

81. Wu, E.X.; Xie, Y.; Yuan, B.; Zhang, H.; Hu, X.D.; Liu, J.; Zhang, D.H. Ultrasensitive and Fully Reversible NO_2 Gas Sensing Based on p-Type $MoTe_2$ under Ultraviolet Illumination. *ACS Sens.* **2018**, *3*, 1719–1726. [CrossRef] [PubMed]
82. Gu, D.; Wang, X.Y.; Liu, W.; Li, X.G.; Lin, S.W.; Wang, J.; Rumyantseva, M.N.; Gaskov, A.M.; Akbar, S.A. Visible-light activated room temperature NO_2 sensing of SnS_2 nanosheets based chemiresistive sensors. *Sens. Actuators B Chem.* **2020**, *305*, 127455. [CrossRef]
83. Cheng, M.; Wu, Z.P.; Liu, G.N.; Zhao, L.J.; Gao, Y.; Zhang, B.; Liu, F.M.; Yan, X.; Liang, X.S.; Sun, P.; et al. Highly sensitive sensors based on quasi-2D rGO/SnS_2 hybrid for rapid detection of NO_2 gas. *Sens. Actuators B* **2019**, *291*, 216–225. [CrossRef]
84. Xu, S.P.; Sun, F.Q.; Gu, F.L.; Zuo, Y.B.; Zhang, L.H.; Fan, C.F.; Yang, S.M.; Li, W.S. Photochemistry-Based Method for the Fabrication of SnO_2 Monolayer Ordered Porous Films with Size-Tunable Surface Pores for Direct Application in Resistive-Type Gas Sensor. *ACS Appl. Mater. Interfaces* **2014**, *6*, 1251–1257. [CrossRef]
85. Ji, F.X.; Ren, X.P.; Zheng, X.Y.; Liu, Y.C.; Pang, L.Q.; Jiang, J.X.; Liu, S.Z. 2D-MoO_3 nanosheets for superior gas sensors. *Nanoscale* **2016**, *8*, 8696–8703. [CrossRef]
86. Boudiba, A.; Zhang, C.; Bittencourt, C.; Umek, P.; Olivier, M.G.; Snyders, R.; Debliquy, M. Hydrothermal Synthesis of Two Dimensional WO_3 Nanostructures for NO_2 Detection in the ppb-level. *Proc. Eng.* **2012**, *47*, 228–231. [CrossRef]
87. Cho, Y.H.; Ko, Y.N.; Kang, Y.C.; Kim, I.D.; Lee, J.H. Ultraselective and ultrasensitive detection of trimethylamine using MoO_3 nanoplates prepared by ultrasonic spray pyrolysis. *Sens. Actuators B Chem.* **2014**, *195*, 189–196. [CrossRef]
88. Wang, M.S.; Wang, Y.W.; Li, X.J.; Ge, C.X.; Hussain, S.; Liu, G.W.; Qiao, G.J. WO_3 porous nanosheet arrays with enhanced low temperature NO_2 gas sensing performance. *Sens. Actuators B Chem.* **2020**, *316*, 128050. [CrossRef]
89. Kaneti, Y.V.; Yue, J.; Jiang, X.C.; Yu, A.B. Controllable Synthesis of ZnO Nanoflakes with Exposed (10(1)over-bar0) for Enhanced Gas Sensing Performance. *J. Phys. Chem. C* **2013**, *117*, 13153–13162. [CrossRef]
90. Wang, X.; Su, J.; Chen, H.; Li, G.D.; Shi, Z.F.; Zou, H.F.; Zou, X.X. Ultrathin In_2O_3 Nanosheets with Uniform Mesopores for Highly Sensitive Nitric Oxide Detection. *ACS Appl. Mater. Interfaces* **2017**, *9*, 16335–16342. [CrossRef]
91. Li, Z.J.; Lin, Z.J.; Wang, N.N.; Wang, J.Q.; Liu, W.; Sun, K.; Fu, Y.Q.; Wang, Z.G. High precision NH_3 sensing using network nano-sheet Co_3O_4 arrays based sensor at room temperature. *Sens. Actuators B Chem.* **2016**, *235*, 222–231. [CrossRef]
92. Chen, X.W.; Wang, S.; Su, C.; Han, Y.T.; Zou, C.; Zeng, M.; Hu, N.T.; Su, Y.J.; Zhou, Z.H.; Yang, Z. Two-dimensional Cd-doped porous Co_3O_4 nanosheets for enhanced roomtemperature NO_2 sensing performance. *Sens. Actuators B. Chem.* **2020**, *305*, 127393. [CrossRef]
93. Lin, H.; Gao, S.S.; Dai, C.; Chen, Y.; Shi, J.L. A Two-Dimensional Biodegradable Niobium Carbide (MXene) for Photothermal Tumor Eradication in NIR-I and NIR-II Biowindows. *J. Am. Chem. Soc.* **2017**, *139*, 16235–16247. [CrossRef] [PubMed]
94. He, T.T.; Liu, W.; Lv, T.; Ma, M.S.; Liu, Z.F.; Vasiliev, A.; Li, X.G. $MXene/SnO_2$ heterojunction based chemical gas sensors. *Sens. Actuators B Chem.* **2021**, *329*, 129275. [CrossRef]
95. Hermawan, A.; Zhang, B.; Taufik, A.; Asakura, Y.; Hasegawa, T.; Zhu, J.F.; Shi, P.; Yin, S. CuO Nanoparticles/$Ti_3C_2T_x$ MXene Hybrid Nanocomposites for Detection of Toluene Gas. *ACS Appl. Nano Mater.* **2020**, *3*, 4755–4766. [CrossRef]
96. Lee, E.; Kim, D.J. Review-Recent Exploration of Two-Dimensional MXenes for Gas Sensing: From a Theoretical to an Experimental View. *J. Electrochem. Soc.* **2020**, *167*, 037515. [CrossRef]
97. Aghaei, S.M.; Aasi, A.; Panchapakesan, B. Experimental and Theoretical Advances in MXene-Based Gas Sensors. *ACS Omega* **2021**, *6*, 2450–2461. [CrossRef]
98. Naguib, M.; Kurtoglu, M.; Presser, V.; Lu, J.; Niu, J.J.; Heon, M.; Hultman, L.; Gogotsi, Y.; Barsoum, M.W. Two-dimensional nanocrystals produced by exfoliation of Ti_3AlC_2. *Adv. Mater.* **2011**, *23*, 4248–4253. [CrossRef]
99. Lee, E.; Mohammadi, A.V.; Prorok, B.C.; Yoon, Y.S.; Beidaghi, M.; Kim, D.J. Room Temperature Gas Sensing of Two-Dimensional Titanium Carbide (MXene). *ACS Appl. Mater. Interfaces* **2017**, *9*, 37184–37190. [CrossRef]
100. Chae, Y.; Kim, S.J.; Cho, S.Y.; Choi, J.; Maleski, K.; Lee, B.J.; Jung, H.T.; Gogotsi, Y.; Lee, Y.; Ahn, C.W. An investigation into the factors governing the oxidation of two-dimensional Ti_3C_2 MXene. *Nanoscale* **2019**, *11*, 8387–8393. [CrossRef]
101. Yang, Z.J.; Liu, A.; Wang, C.L.; Liu, F.M.; He, J.M.; Li, S.Q.; Wang, J.; You, R.; Yan, X.; Sun, P.; et al. Improvement of Gas and Humidity Sensing Properties of Organ-like MXene by Alkaline Treatment. *ACS Sens.* **2019**, *4*, 1261–1269. [CrossRef] [PubMed]
102. Lee, E.; VahidMohammadi, A.; Yoon, Y.S.; Beidaghi, M.; Kim, D.J. Two-Dimensional Vanadium Carbide MXene for Gas Sensors with Ultrahigh Sensitivity Toward Nonpolar Gases. *ACS Sens.* **2019**, *4*, 1603–1611. [CrossRef] [PubMed]
103. Cho, S.Y.; Kim, J.Y.; Kwon, O.; Kim, J.; Jung, H.T. Molybdenum carbide chemical sensors with ultrahigh signal-to-noise ratios and ambient stability. *J. Mater. Chem. A* **2018**, *6*, 23408–23416. [CrossRef]
104. Wu, T.; Wang, Z.B.; Tian, M.H.; Miao, J.Y.; Zhang, H.X.; Sun, J.B. UV Excitation NO_2 Gas Sensor Sensitized by ZnO Quantum Dots at Room Temperature. *Sens. Actuators B Chem.* **2018**, *259*, 526–531. [CrossRef]
105. Lee, J.; Jung, Y.; Sung, S.H.; Lee, G.; Kim, J.; Seong, J.; Shim, Y.S.; Jun, S.C.; Jeon, S. High-performance gas sensor array for indoor air quality monitoring: The role of Au nanoparticles on WO_3, SnO_2, and NiO-based gas sensors. *J. Mater. Chem. A* **2021**, *9*, 1159–1167. [CrossRef]
106. Pan, F.J.; Lin, H.; Zhai, H.Z.; Miao, Z.; Zhang, Y.; Xu, K.L.; Guan, B.; Huang, H.; Zhang, H. Pd-doped TiO_2 Film Sensors Prepared by Premixed Stagnation Flames for CO and NH_3 Gas Sensing. *Sens. Actuators B Chem.* **2018**, *261*, 451–459. [CrossRef]
107. Zhao, Y.; Zhang, J.; Wang, Y.; Chen, Z. A highly sensitive and room temperature $CNTs/SnO_2/CuO$ sensor for H_2S gas sensing applications. *Nanoscale Res. Lett.* **2020**, *15*, 40. [CrossRef]

108. Schipani, F.; Aldao, C.M.; Ponce, M.A. Schottky barriers measurements through Arrhenius plots in gas sensors based on semiconductor films. *AIP Adv.* **2012**, *2*, 032138. [CrossRef]
109. Zhou, Q.; Chen, W.G.; Xu, L.N.; Kumar, R.; Gui, Y.G.; Zhao, Z.Y.; Tang, C.; Zhu, S.P. Highly sensitive carbon monoxide (CO) gas sensors based on Ni and Zn doped SnO_2 nanomaterials. *Ceram. Int.* **2018**, *44*, 4392–4399. [CrossRef]
110. Li, J.; Xian, J.B.; Wang, W.J.; Cheng, K.; Zeng, M.; Zhang, A.H.; Wu, S.J.; Gao, X.S.; Lu, X.B.; Liu, J.M. Ultrafast response and high-sensitivity acetone gas sensor based on porous hollow Ru-doped SnO_2 nanotubes. *Sens. Actuators B Chem.* **2022**, *352*, 131061. [CrossRef]
111. Wang, T.T.; Wang, Y.; Sun, Q.; Zheng, S.L.; Liu, L.Z.; Li, J.L.; Hao, J.Y. Boosted interfacial charge transfer in SnO_2/$SnSe_2$ heterostructures: Toward ultrasensitive room-temperature H_2S detection. *Inorg. Chem. Front.* **2021**, *8*, 2068–2077. [CrossRef]
112. Zhong, X.; Shen, Y.; Zhao, S.; Li, T.; Lu, R.; Yin, Y.; Han, C.; Wei, D.; Zhang, Y.; Wei, K. Effect of pore structure of the metakaolin-based porous substrate on the growth of SnO_2 nanowires and their H_2S sensing properties. *Vacuum* **2019**, *167*, 118–128. [CrossRef]
113. Hu, D.; Han, B.Q.; Han, R.; Deng, S.J.; Wang, Y.; Li, Q.; Wang, Y.D. SnO_2 nanorods based sensing material as an isopropanol vapor sensor. *New J. Chem.* **2014**, *38*, 2443–2450. [CrossRef]
114. Xu, R.; Zhang, L.X.; Li, M.W.; Yin, Y.Y.; Yin, J.; Zhu, M.Y.; Chen, J.J.; Wang, Y.; Bie, L.J. Ultrathin SnO_2 nanosheets with dominant high-energy {001} facets for low temperature formaldehyde gas sensor. *Sens. Actuators B Chem.* **2019**, *289*, 186–194. [CrossRef]
115. Ren, H.B.; Zhao, W.; Wang, L.Y.; Ryu, S.O.; Gu, C.P. Preparation of porous flower-like SnO_2 micro/nano structures and their enhanced gas sensing property. *J. Alloy Compd.* **2015**, *653*, 611–618. [CrossRef]
116. Sun, H.M.; Zhang, C.; Peng, Y.J.; Gao, W. Synthesis of double-shelled SnO_2 hollow cubes for superior isopropanol sensing performance. *New J. Chem.* **2019**, *43*, 4721–4726. [CrossRef]
117. Lee, J.H. Gas sensors using hierarchical and hollow oxide nanostructures: Overview. *Sens. Actuators B Chem.* **2009**, *140*, 319–336. [CrossRef]
118. Zhang, Y.Q.; Li, D.; Qin, L.G.; Zhao, P.L.; Liu, F.M.; Chuai, X.H.; Sun, P.; Liang, X.S.; Gao, Y.; Sun, Y.F.; et al. Preparation and gas sensing properties of hierarchical leaf-like SnO_2 materials. *Sens. Actuators B Chem.* **2018**, *255*, 2944–2951. [CrossRef]
119. Feng, X.Y.; Jiang, J.; Ding, H.; Ding, R.M.; Luo, D.; Zhu, J.H.; Feng, Y.M.; Huang, X.T. Carbon-assisted synthesis of mesoporous SnO_2 nanomaterial as highly sensitive ethanol gas sensor. *Sens. Actuators B Chem.* **2013**, *183*, 526–534. [CrossRef]
120. Dong, K.Y.; Choi, J.K.; Hwang, I.S.; Lee, J.W.; Kang, B.H.; Ham, D.J.; Lee, J.H.; Ju, B.K. Enhanced H_2S sensing characteristics of Pt doped SnO_2 nanofibers sensors with micro heater. *Sens. Actuators B Chem.* **2011**, *157*, 154–161. [CrossRef]
121. Chen, Y.P.; Qin, H.W.; Hu, J.F. CO sensing properties and mechanism of Pd doped SnO_2 thick-films. *Appl. Surf. Sci.* **2018**, *428*, 207–217. [CrossRef]
122. Lee, Y.C.; Huang, H.; Tan, O.K.; Tse, M.S. Semiconductor gas sensor based on Pd-doped SnO_2 nanorod thin films. *Sens. Actuators B Chem.* **2008**, *135*, 239–242. [CrossRef]
123. Shen, Y.B.; Yamazaki, T.; Liu, Z.F.; Meng, D.; Kikuta, T.; Nakatani, N.; Saito, M.; Mori, M. Microstructure and H-2 gas sensing properties of undoped and Pd-doped $SnO2$ nanowires. *Sens. Actuators B Chem.* **2009**, *135*, 524–529. [CrossRef]
124. Ramgir, N.S.; Hwang, Y.K.; Jhung, S.H.; Kim, H.K.; Hwang, J.S.; Mulla, I.S.; Chang, J.S. CO sensor derived from mesostructured Au-doped SnO_2 thin film. *Appl. Surf. Sci.* **2006**, *252*, 4298–4305. [CrossRef]
125. Zhao, C.H.; Gong, H.M.; Niu, G.Q.; Wang, F. Ultrasensitive SO_2 sensor for sub-ppm detection using Cu-doped SnO_2 nanosheet arrays directly grown on chip. *Sens. Actuators B Chem.* **2020**, *324*, 128745. [CrossRef]
126. Chen, Y.; Zhu, C.L.; Shi, X.L.; Cao, M.S.; Jin, H.B. The synthesis and selective gas sensing characteristics of SnO_2/alpha-Fe_2O_3 hierarchical nanostructures. *Nanotechnology* **2008**, *19*, 205603. [CrossRef]
127. Xue, X.Y.; Xing, L.L.; Chen, Y.J.; Shi, S.L.; Wang, Y.G.; Wang, T.H. Synthesis and H_2S sensing properties of CuO-SnO_2 core/shell PN-junction nanorods. *J. Phys. Chem. C* **2008**, *112*, 12157–12160. [CrossRef]
128. Haoyuan, F.Y.L.J.X. SnO_2 recycled fromtin slime for enhanced SO_2 sensing properties by NiO surface decoration. *Mater. Sci. Semicond. Process.* **2020**, *114*, 105073.
129. Yuliarto, B.; Ramadhani, M.F.; Nugraha; Septiani, N.L.W.; Hamam, K.A. Enhancement of SO_2 gas sensing performance using ZnO nanorod thin films: The role of deposition time. *J. Mater. Sci.* **2017**, *52*, 4543–4554. [CrossRef]
130. Wang, H.T.; Dai, M.; Li, Y.Y.; Bai, J.H.; Liu, Y.Y.; Li, Y.; Wang, C.C.; Liu, F.M.; Lu, G.Y. The influence of different ZnO nanostructures on NO_2 sensing performance. *Sens. Actuators B Chem.* **2021**, *329*, 129145. [CrossRef]
131. Liu, Y.X.; Hang, T.; Xie, Y.Z.; Bao, Z.; Song, J.; Zhang, H.L.; Xie, E.Q. Effect of Mg doping on the hydrogen-sensing characteristics of ZnO thin films. *Sens. Actuators B Chem.* **2011**, *160*, 266–270. [CrossRef]
132. Hassan, H.S.; Kashyout, A.B.; Soliman HM, A.; Uosif, M.A.; Afify, N. Effect of reaction time and Sb doping ratios on the architecturing of ZnO nanomaterials for gas sensor applications. *Appl. Surf. Sci.* **2013**, *277*, 73–82. [CrossRef]
133. Chaitra, U.; Ali, A.V.M.; Viegas, A.E.; Kekud, D.; Rao, K.M. Growth and characterization of undoped and aluminium doped zinc oxide thin films for SO_2 gas sensing below threshold value limit. *Appl. Surf. Sci.* **2019**, *496*, 143724. [CrossRef]
134. Kolhe, P.S.; Shinde, A.B.; Kulkarni, S.G.; Maiti, N.; Koinkar, P.M.; Sonawane, K.M. Gas sensing performance of Al doped ZnO thin film for H_2S detection. *J. Alloys Compd.* **2018**, *748*, 6–11. [CrossRef]
135. Xiang, Q.; Meng, G.F.; Zhang, Y.; Xu, J.Q.; Xu, P.C.; Pan, Q.Y.; Yu, W.J. Ag nanoparticle embedded-ZnO nanorods synthesized via a photochemical method and its gas-sensing properties. *Sens. Actuators B Chem.* **2010**, *143*, 635–640. [CrossRef]
136. Kim, J.H.; Katoch, A.; Kim, S.S. Optimum shell thickness and underlying sensing mechanism in p-n CuO-ZnO core-shell nanowires. *Sens. Actuators B Chem.* **2016**, *222*, 249–256. [CrossRef]

137. Zhou, Q.; Zeng, W.; Chen, W.G.; Xu, L.N.; Kumar, R.; Umar, A. High sensitive and low concentration sulfur dioxide (SO_2) gas sensor application of heterostructure NiO-ZnO nanodisks. *Sens. Actuators B Chem.* **2019**, *298*, 126870. [CrossRef]
138. Li, Z.J.; Wang, N.N.; Lin, Z.J.; Wang, J.Q.; Liu, W.; Sun, K.; Fu, Y.Q.; Wang, Z.G. Room-Temperature High-Performance H_2S Sensor Based on Porous CuO Nanosheets Prepared by Hydrothermal Method. *ACS Appl. Mater. Interfaces* **2016**, *8*, 20962–20968. [CrossRef]
139. Navale, Y.H.; Navale, S.T.; Galluzzi, M.; Stadler, F.J.; Debnath, A.K.; Ramgir, N.S.; Gadkari, S.C.; Gupta, S.K.; Aswal, D.K.; Patil, V.B. Rapid synthesis strategy of CuO nanocubes for sensitive and selective detection of NO_2. *J. Alloys Compd.* **2017**, *708*, 456–463. [CrossRef]
140. Huang, X.; Ren, Z.B.; Zheng, X.H.; Tang, D.P.; Wu, X.; Lin, C. A facile route to batch synthesis CuO hollow microspheres with excellent gas sensing properties. *J. Mater. Sci. Mater. Electron.* **2018**, *29*, 5969–5974. [CrossRef]
141. Hu, Q.; Zhang, W.J.; Wang, X.Y.; Wang, Q.; Huang, B.Y.; Li, Y.; Hua, X.H.; Liu, G.; Li, B.S.; Zhou, J.Y.; et al. Binder-free CuO nanoneedle arrays based tube-type sensor for H_2S gas sensing. *Sens. Actuators B Chem.* **2021**, *326*, 128993. [CrossRef]
142. Hou, L.; Zhang, C.M.; Li, L.; Du, C.; Li, X.K.; Kang, X.F.; Chen, W. CO gas sensors based on p-type CuO nanotubes and CuO nanocubes: Morphology and surface structure effects on the sensing performance. *Talanta* **2018**, *188*, 41–49. [CrossRef] [PubMed]
143. Hu, X.B.; Zhu, Z.G.; Chen, C.; Wen, T.Y.; Zhao, X.L.; Xie, L.L. Highly sensitive H_2S gas sensors based on Pd-doped CuO nanoflowers with low operating temperature. *Sens. Actuators B Chem.* **2017**, *253*, 809–817. [CrossRef]
144. Tang, Q.; Hu, X.-B.; He, J.; Xie, L.-L.; Zhu, Z.-G.; Wu, J.-Q. Effect of Platinum Doping on the Morphology and Sensing Performance for CuO-Based Gas Sensor. *Appl. Sci.* **2018**, *8*, 1091. [CrossRef]
145. Mnethu, O.; Nkosi, S.S.; Kortidis, I.; Motaung, D.E.; Kroon, R.E.; Swart, H.C.; Ntsasa, N.G.; Tshilongo, J.; Moyo, T. Ultra-sensitive and selective p-xylene gas sensor at low operating temperature utilizing Zn doped CuO nanoplatelets: Insignificant vestiges of oxygen vacancies. *J. Colloid Interface Sci.* **2020**, *576*, 364–375. [CrossRef]
146. Bhuvaneshwari, S.; Gopalakrishnan, N. Enhanced ammonia sensing characteristics of Cr doped CuO nanoboats. *J. Alloys Compd.* **2016**, *654*, 202–208. [CrossRef]
147. Molavi, R.; Sheikhi, M.H. Facile wet chemical synthesis of Al doped CuO nanoleaves for carbon monoxide gas sensor applications. *Mater. Sci. Semicond. Process.* **2020**, *106*, 104767. [CrossRef]
148. Zhang, H.; Li, H.R.; Cai, L.N.; Lei, Q.; Wang, J.N.; Fan, W.H.; Shi, K.; Han, G.L. Performances of In-doped CuO-based heterojunction gas sensor. *J. Mater. Sci. Mater. Electron.* **2020**, *31*, 910–919. [CrossRef]
149. Wang, Z.F.; Li, F.; Wang, H.T.; Wang, A.; Wu, S.M. An enhanced ultra-fast responding ethanol gas sensor based on Ag functionalized CuO nanoribbons at room-temperature. *J. Mater. Sci. Mater. Electron.* **2018**, *29*, 16654–16659. [CrossRef]
150. Sui, L.L.; Yu, T.T.; Zhao, D.; Cheng, X.L.; Zhang, X.F.; Wang, P.; Xu, Y.M.; Gao, S.; Zhao, H.; Gao, Y.; et al. In situ deposited hierarchical CuO/NiO nanowall arrays film sensor with enhanced gas sensing performance to H_2S. *J. Hazard. Mater.* **2020**, *385*, 121570. [CrossRef]
151. Park, K.R.; Cho, H.B.; Lee, J.; Song, Y.; Kim, W.B.; Choa, Y.H. Design of highly porous SnO_2-CuO nanotubes for enhancing H_2S gas sensor performance. *Sens. Actuators B Chem.* **2020**, *302*, 127179. [CrossRef]
152. Li, T.; Sun, M.; Wu, S. State-of-the-Art Review of Electrospun Gelatin-Based Nanofiber Dressings for Wound Healing Applications. *Nanomaterials* **2022**, *12*, 784. [CrossRef] [PubMed]
153. Sheng, X.; Li, T.; Sun, M.; Liu, G.; Zhang, Q.; Ling, Z.; Gao, S.; Diao, F.; Zhang, J.; Rosei, F.; et al. Flexible electrospun iron compounds/carbon fibers: Phase transformation and electrochemical properties. *Electrochim. Acta* **2022**, *407*, 139892. [CrossRef]
154. Liang, X.; Kim, T.H.; Yoon, J.W.; Kwak, C.H.; Lee, J.H. Ultrasensitive and ultraselective detection of H_2S using electrospun CuO-loaded In_2O_3 nanofiber sensors assisted by pulse heating. *Sens. Actuators B Chem.* **2015**, *209*, 934–942. [CrossRef]
155. Barsan, N.; Koziej, D.; Weimar, U. Metal oxide-based gas sensor research: How to? *Sens. Actuators B Chem.* **2007**, *121*, 18–35. [CrossRef]
156. Upadhyay, S.B.; Mishra, R.K.; Sahay, P.P. Enhanced acetone response in co-precipitated WO_3 nanostructures upon indium doping. *Sens. Actuators B Chem.* **2015**, *209*, 368–376. [CrossRef]
157. Hu, Q.; Chang, J.; Gao, J.; Huang, J.; Feng, L. Needle-shaped WO3 nanorods for triethylamine gas sensing. *ACS Appl. Nano Mater.* **2020**, *3*, 9046–9054. [CrossRef]
158. Li, Z.H.; Li, J.C.; Song, L.L.; Gong, H.Q.; Niu, Q. Ionic liquid-assisted synthesis of WO_3 particles with enhanced gas sensing properties. *J. Mater. Chem. A* **2013**, *1*, 15377–15382. [CrossRef]
159. Li, H.; Liu, B.; Cai, D.P.; Wang, Y.R.; Liu, Y.; Mei, L.; Wang, L.L.; Wang, D.D.; Li, Q.H.; Wang, T.H. High-temperature humidity sensors based on WO_3-SnO_2 composite hollow nanospheres. *J. Mater. Chem. A* **2014**, *2*, 6854–6862. [CrossRef]
160. Thu, N.T.A.; Cuong, N.D.; Nguyen, L.C.; Khieu, D.Q.; Nam, P.C.; Van Toan, N.; Hung, C.M.; Van Hieu, N. Fe_2O_3 nanoporous network fabricated from Fe_3O_4/reduced graphene oxide for high-performance ethanol gas sensor. *Sens. Actuators B Chem.* **2018**, *255*, 3275–3283. [CrossRef]
161. Li, Z.J.; Huang, Y.W.; Zhang, S.C.; Chen, W.M.; Kuang, Z.; Ao, D.Y.; Liu, W.; Fu, Y.Q. A fast response & recovery H_2S gas sensor based on α-Fe_2O_3 nanoparticles with ppb level detection limit. *J. Hazard. Mater.* **2015**, *300*, 167–174. [PubMed]
162. Liang, S.; Li, J.P.; Wang, F.; Qin, J.L.; Lai, X.Y.; Jiang, X.M. Highly sensitive acetone gas sensor based on ultrafine α-Fe_2O_3 nanoparticles. *Sens. Actuators B Chem.* **2017**, *238*, 923–927. [CrossRef]
163. Shoorangiz, M.; Shariatifard, L.; Roshan, H.; Mirzaei, A. Selective ethanol sensor based on alpha-Fe2O3 nanoparticles. *Inorg. Chem. Commun.* **2021**, *133*, 108961. [CrossRef]

164. Qu, F.D.; Zhou, X.X.; Zhang, B.X.; Zhang, S.D.; Jiang, C.J.; Ruan, S.P.; Yang, M.H. Fe_2O_3 nanoparticles-decorated MoO_3 nanobelts for enhanced chemiresistive gas sensing. *J. Alloys Compd.* **2019**, *782*, 672–678. [CrossRef]
165. Navale, S.T.; Liu, C.S.T.; Gaikar, P.S.; Patil, V.B.; Sagar, R.U.R.; Du, B.; Mane, R.S.; Stadler, F.J. Solution-processed rapid synthesis strategy of Co_3O_4 for the sensitive and selective detection of H_2S. *Sens. Actuators B Chem.* **2017**, *245*, 524–532. [CrossRef]
166. Li, W.Y.; Xu, L.N.; Chen, J. Co_3O_4 nanomaterials in lithium-ion batteries and gassensors. *Adv. Funct. Mater.* **2005**, *15*, 851–857. [CrossRef]
167. Sun, C.W.; Rajasekhara, S.; Chen, Y.J.; Goodenough, J.B. Facile synthesis of monodisperse porous Co_3O_4 microspheres with superior ethanol sensing properties. *Chem. Commun.* **2011**, *47*, 12852–12854. [CrossRef]
168. Zhang, R.; Gao, S.; Zhou, T.T.; Tu, J.C.; Zhang, T. Facile preparation of hierarchical structure based on p-type Co_3O_4 as toluene detecting sensor. *Appl. Surf. Sci.* **2020**, *503*, 144167. [CrossRef]
169. Chen, X.F.; Huang, Y.; Zhang, K.C.; Feng, X.S.; Wang, M.Y. Porous TiO_2 nanobelts coated with mixed transition-metal oxides Sn_3O_4 nanosheets core-shell composites as high-performance anode materials of lithium ion batteries. *Electrochim. Acta Mater.* **2018**, *259*, 131–142. [CrossRef]
170. Chen, G.; Ji, S.; Li, H.; Kang, X.; Chang, S.; Wang, Y.; Yu, G.; Lu, J.; Claverie, J.; Sang, Y.; et al. High-energy faceted SnO_2-coated TiO_2 nanobelt heterostructure for near-ambient temperature-responsive ethanol sensor. *ACS Appl. Mater. Interfaces* **2015**, *7*, 24950–24956. [CrossRef]
171. Wang, X.H.; Sang, Y.H.; Wang, D.Z.; Ji, S.Z.; Liu, H. Enhanced gas sensing property of SnO_2 nanoparticles by constructing the SnO_2-TiO_2 nanobelt heterostructure. *J. Alloys Compd.* **2015**, *639*, 571–576. [CrossRef]
172. Chen, N.; Li, X.G.; Wang, X.Y.; Yu, J.; Wang, J.; Tang, Z.A.; Akbar, S.A. Enhanced room temperature sensing of Co_3O_4-intercalated reduced graphene oxide based gas sensors. *Sens. Actuators B Chem.* **2013**, *188*, 902–908. [CrossRef]
173. Chen, Y.; Zhang, W.; Wu, Q.S. A highly sensitive room-temperature sensing material for NH_3: SnO_2-nanorods coupled by rGO. *Sens. Actuators B Chem.* **2017**, *242*, 1216–1226. [CrossRef]
174. Liu, J.; Li, S.; Zhang, B.; Wang, Y.L.; Gao, Y.; Liang, X.S.; Wang, Y.; Lu, G.Y. Flower-like In_2O_3 modified by reduced graphene oxide sheets serving as a highly sensitive gas sensor for trace NO_2 detection. *J. Colloid Interface Sci.* **2017**, *504*, 206–213. [CrossRef] [PubMed]
175. Pienutsa, N.; Roongruangsree, P.; Seedokbuab, V.; Yannawibut, K.; Phatoomvijitwong, C.; Srinives, S. SnO_2-graphene composite gas sensor for a room temperature detection of ethanol. *Nanotechnology* **2021**, *32*, 115502. [CrossRef]
176. Kim, H.W.; Na, H.G.; Kwon, Y.J.; Kang, S.Y.; Choi, M.S.; Bang, J.H.; Wu, P.; Kim, S.S. Microwave-Assisted Synthesis of Graphene-SnO_2 Nanocomposites and Their Applications in Gas Sensors. *ACS Appl. Mater. Interfaces* **2017**, *9*, 31667–31682. [CrossRef]
177. Russo, P.A.; Donato, N.; Leonardi, S.G.; Baek, S.; Conte, D.E.; Neri, G.; Pinna, N. Room-Temperature Hydrogen Sensing with Heteronanostructures Based on Reduced Graphene Oxide and Tin Oxide. *Angew. Chem. Int. Ed.* **2012**, *51*, 11053–11057. [CrossRef]
178. Zhang, D.Z.; Liu, J.J.; Jiang, C.X.; Li, P.; Sun, Y.E. High-performance sulfur dioxide sensing properties of layer-by-layer self-assembled titania-modified graphene hybrid nanocomposite. *Sens. Actuators B Chem.* **2017**, *245*, 560–567. [CrossRef]
179. Li, J.; Liu, X.; Sun, J.B. One step solvothermal synthesis of urchin-like ZnO nanorods/graphene hollow spheres and their NO_2, gas sensing properties. *Ceram. Int.* **2016**, *42*, 2085–2090. [CrossRef]
180. Liu, S.; Yu, B.; Zhang, H.; Fei, T.; Zhang, T. Enhancing NO_2 gas sensing performances at room temperature based on reduced graphene oxide-ZnO nanoparticles hybrids. *Sens. Actuators B Chem.* **2014**, *202*, 272–278. [CrossRef]
181. Sun, Z.; Huang, D.; Yang, Z.; Li, X.L.; Hu, N.T.; Yang, C.; Wei, H.; Yin, G.L.; He, D.N.; Zhang, Y.F. ZnO Nanowire-Reduced Graphene Oxide Hybrid Based Portable NH_3 Gas Sensing Electron Device. *IEEE Electron Dev. Lett.* **2015**, *36*, 1376–1379. [CrossRef]
182. Wang, C. Reduced graphene oxide decorated with CuO-ZnO hetero-junctions: Towards high selective gas-sensing property to acetone. *J. Mater. Chem. A* **2014**, *2*, 18635–18643. [CrossRef]
183. Wang, D.; Zhang, M.L.; Chen, Z.L.; Li, H.J.; Chen, A.Y.; Wang, X.Y.; Yang, J.H. Enhanced formaldehyde sensing properties of hollow SnO2 nanofibers by graphene oxide. *Sens. Actuators B Chem.* **2017**, *250*, 533–542. [CrossRef]
184. Wang, Z.; Hang, T.; Fei, T.; Liu, S.; Zhang, T. Investigation of microstructure effect on NO_2 sensors based on SnO_2 nanoparticles/reduced graphene oxide hybrids. *ACS Appl. Mater. Interfaces* **2018**, *10*, 41773–41783. [CrossRef] [PubMed]
185. Yin, L.; Chen, D.L.; Cui, X.; Ge, L.F.; Yang, J.; Yu, L.T.; Zhang, B.; Zhang, R.; Shao, G.S. Normal-pressure microwave rapid synthesis of hierarchical SnO_2@rGO nanostructures with superhigh surface areas as high-quality gas-sensing and electrochemical active materials. *Nanoscale* **2014**, *6*, 13690–13700. [CrossRef]
186. Tyagi, P.; Sharma, A.; Tomar, M.; Gupta, V. A comparative study of RGO-SnO_2 and MWCNT-SnO_2 nanocomposites based SO_2 gas sensors. *Sens. Actuators B Chem.* **2017**, *248*, 980–986. [CrossRef]
187. Yu, X.J.; Cheng, C.D.; Feng, S.P.; Jia, X.H.; Song, H.J. Porous α-Fe_2O_3 nanorods@graphite nanocomposites with improved high temperature gas sensitive properties. *J. Alloys Compd.* **2019**, *784*, 1261–1269. [CrossRef]
188. Zhou, T.T.; Zhang, T.; Deng, J.A.; Zhang, R.; Lou, Z.; Wang, L.L. P-type Co_3O_4 nanomaterials-based gas sensor: Preparation and acetone sensing performance. *Sens. Actuators B Chem.* **2017**, *242*, 369–377. [CrossRef]
189. Zhang, B.; Cheng, M.; Liu, G.N.; Gao, Y.; Zhao, L.J.; Li, S.; Wang, Y.P.; Liu, F.M.; Liang, X.S.; Zhang, T.; et al. Room temperature NO_2 gas sensor based on porous Co_3O_4 slices/reduced graphene oxide hybrid. *Sens. Actuators B Chem.* **2018**, *263*, 387–399. [CrossRef]
190. Srirattanapibul, S.; Nakarungsee, P.; Issro, C.; Tang, I.M.; Thongmee, S. Enhanced room temperature NH_3 sensing of rGO/Co_3O_4 nanocomposites. *Mater. Chem. Phys.* **2021**, *272*, 125033. [CrossRef]

191. Hao, Q.; Liu, T.; Liu, J.Y.; Liu, Q.; Jing, X.Y.; Zhang, H.Q.; Huang, G.Q.; Wang, J. Controllable synthesis and enhanced gas sensing properties of a single-crystalline WO_3-rGO porous nanocomposite. *RSC Adv.* **2017**, *7*, 14192. [CrossRef]
192. Ye, Z.B.; Tai, H.L.; Xie, T.; Su, Y.J.; Yuan, Z.; Liu, C.H.; Jiang, Y.D. A facile method to develop novel TiO_2/rGO layered film sensor for detecting ammonia at room temperature. *Mater. Lett.* **2016**, *165*, 127. [CrossRef]
193. Qao, X.Q.; Zhang, Z.W.; Hou, D.F.; Li, D.S.; Liu, Y.L.; Lan, Y.Q.; Zhang, J.; Feng, P.Y.; Bu, X.H. Tunable MoS_2/SnO_2 P–N Heterojunctions for an Efficient Trimethylamine Gas Sensor and 4-Nitrophenol Reduction Catalyst. *ACS Sustainable Chem. Eng.* **2018**, *6*, 12375–12384. [CrossRef]
194. Cui, S.M.; Wen, Z.H.; Huang, X.K.; Chang, J.B.; Chen, J.H. Stabilizing MoS_2 Nanosheets through SnO_2 Nanocrystal Decoration for High-Performance Gas Sensing in Air. *Small* **2015**, *11*, 2305–2313. [CrossRef]
195. Yan, H.H.; Song, P.; Zhang, S.; Yang, Z.X.; Wang, Q. Facile synthesis, characterization and gas sensing performance of ZnO nanoparticles-coated MoS_2 nanosheets. *J. Alloys Compd.* **2016**, *662*, 118–125. [CrossRef]
196. Han, Y.T.; Huang, D.; Ma, Y.J.; He, G.L.; Hu, J.; Zhang, J.; Hu, N.T.; Su, Y.J.; Zhou, Z.H.; Zhang, Y.F.; et al. Design of Hetero-Nanostructures on MoS_2 Nanosheets To Boost NO_2 Room-Temperature Sensing. *ACS Appl. Mater. Interfaces* **2018**, *10*, 22640–22649. [CrossRef]
197. Zhang, D.Z.; Wu, J.F.; Cao, Y.H. Ultrasensitive H_2S gas detection at room temperature based on copper oxide/molybdenum disulfide nanocomposite with synergistic effect. *Sens. Actuators B Chem.* **2019**, *287*, 346–355. [CrossRef]
198. Ikram, M.; Liu, L.J.; Liu, Y.; Ullah, M.; Ma, L.F.; Bakhtiar, S.U.; Wu, H.Y.; Yu, H.T.; Wang, R.H.; Shi, K.Y. Controllable synthesis of MoS_2@MoO_2 nanonetworks for enhanced NO_2 room temperature sensing in air. *Nanoscale* **2019**, *11*, 8554–8564. [CrossRef]
199. Qin, Z.Y.; Ouyang, C.; Zhang, J.; Wan, L.; Wang, S.M.; Xie, C.S.; Zeng, D.W. 2D WS_2 nanosheets with TiO_2 quantum dots decoration for high-performance ammonia gas sensing at room temperature. *Sens. Actuators B Chem.* **2017**, *253*, 1034. [CrossRef]
200. Gu, D.; Li, X.; Zhao, Y.; Wang, J. Enhanced NO_2 sensing of SnO_2/SnS_2 heterojunction based sensor. *Sens. Actuators B Chem.* **2017**, *244*, 67. [CrossRef]
201. Zhang, L.; Liu, Z.L.; Jin, L.; Zhang, B.B.; Zhang, H.T.; Zhu, M.H.; Yang, W.Q. Self-assembly gridding alpha-MoO_3 nanobelts for highly toxic H2S gas sensors. *Sens. Actuators B Chem.* **2016**, *237*, 350–357. [CrossRef]
202. Bai, S.L.; Chen, C.; Zhang, D.F.; Luo, R.X.; Li, D.Q.; Chen, A.F.; Liu, C.C. Intrinsic characteristic and mechanism in enhancing H_2S sensing of Cd-doped α-MoO_3 nanobelts. *Sens. Actuators B Chem.* **2014**, *204*, 754–762. [CrossRef]
203. Gao, X.M.; Ouyang, Q.Y.; Zhu, C.L.; Zhang, X.T.; Chen, Y.J. Porous MoO_3/SnO_2 Nanoflakes with n–n Junctions for Sensing H_2S. *ACS Appl. Nano Mater.* **2019**, *2*, 2418–2425. [CrossRef]
204. Yin, L.; Chen, D.L.; Feng, M.J.; Ge, L.F.; Yang, D.W.; Song, Z.H.; Fan, B.B.; Zhang, R.; Shao, G.S. Hierarchical Fe_2O_3@WO_3 nanostructures with ultrahigh specific surface areas: Microwave-assisted synthesis and enhanced H_2S-sensing performance. *RSC Adv.* **2015**, *5*, 328–337. [CrossRef]

Review

Recent Advances in Fabrication of Flexible, Thermochromic Vanadium Dioxide Films for Smart Windows

Jongbae Kim and Taejong Paik *

School of Integrative Engineering, Chung-Ang University, Seoul 06974, Korea; jbkim0406@gmail.com
* Correspondence: paiktae@cau.ac.kr

Abstract: Monoclinic-phase VO_2 ($VO_2(M)$) has been extensively studied for use in energy-saving smart windows owing to its reversible insulator–metal transition property. At the critical temperature ($T_c = 68\ °C$), the insulating $VO_2(M)$ (space group P21/c) is transformed into metallic rutile VO_2 ($VO_2(R)$ space group P42/mnm). $VO_2(M)$ exhibits high transmittance in the near-infrared (NIR) wavelength; however, the NIR transmittance decreases significantly after phase transition into $VO_2(R)$ at a higher T_c, which obstructs the infrared radiation in the solar spectrum and aids in managing the indoor temperature without requiring an external power supply. Recently, the fabrication of flexible thermochromic $VO_2(M)$ thin films has also attracted considerable attention. These flexible films exhibit considerable potential for practical applications because they can be promptly applied to windows in existing buildings and easily integrated into curved surfaces, such as windshields and other automotive windows. Furthermore, flexible $VO_2(M)$ thin films fabricated on microscales are potentially applicable in optical actuators and switches. However, most of the existing fabrication methods of phase-pure $VO_2(M)$ thin films involve chamber-based deposition, which typically require a high-temperature deposition or calcination process. In this case, flexible polymer substrates cannot be used owing to the low-thermal-resistance condition in the process, which limits the utilization of flexible smart windows in several emerging applications. In this review, we focus on recent advances in the fabrication methods of flexible thermochromic $VO_2(M)$ thin films using vacuum deposition methods and solution-based processes and discuss the optical properties of these flexible $VO_2(M)$ thin films for potential applications in energy-saving smart windows and several other emerging technologies.

Keywords: VO_2; phase change material; flexible thin film; thermochromics; energy efficient materials

Citation: Kim, J.; Paik, T. Recent Advances in Fabrication of Flexible, Thermochromic Vanadium Dioxide Films for Smart Windows. *Nanomaterials* 2021, 11, 2674. https://doi.org/10.3390/nano11102674

Academic Editor: Yuanbing Mao

Received: 30 August 2021
Accepted: 5 October 2021
Published: 11 October 2021

Publisher's Note: MDPI stays neutral with regard to jurisdictional claims in published maps and institutional affiliations.

Copyright: © 2021 by the authors. Licensee MDPI, Basel, Switzerland. This article is an open access article distributed under the terms and conditions of the Creative Commons Attribution (CC BY) license (https://creativecommons.org/licenses/by/4.0/).

1. Introduction

To address the rapidly increasing energy demand and growing environmental concerns, the development of renewable resources and smart-energy materials is receiving widespread attention [1]. Building energy consumption is estimated to account for 30–40% of the total global energy consumption, and this proportion is expected to continue increasing [2,3]. Windows are the most energy-inefficient component of a building; in this regard, smart windows offer the potential to reduce energy consumption by reducing the air-conditioning load via modulation of solar radiation [4]. Researchers have extensively studied the development of energy-efficient materials for smart windows to address the increasing energy needs. Monoclinic-phase VO_2 ($VO_2(M)$) was first reported by Morin in 1959 and is the most widely studied inorganic material owing to its switchable thermochromic properties [5]. VO_2 exhibits a first-order insulator–metal phase transition at the critical temperature ($T_c = 68\ °C$), accompanied by reversible phase-change properties in the transition from the insulating monoclinic (P21/c) phase to the metallic rutile (P42/mmm) phase [6,7]. Figure 1 shows the crystal structure and band diagram of monoclinic and rutile phase VO_2. The vanadium ions in the monoclinic phase dimerize to form zigzag atomic chains with two V-V distances of ≈3.12 and ≈2.65 Å. Conversely, in the rutile phase, straight and evenly distanced vanadium chains are formed along the c-axis with ≈2.85 Å of distance and V^{4+} ions surrounded by O^{2-} are located at the center and corner positions [8,9]. Dimerization of the vanadium ion causes the $d_{\parallel}$ band to split into a filled bonding ($d_{\parallel}$) and an empty antibonding ($d_{\parallel}^*$). Furthermore, the π^* orbitals shift to higher energies and make a forbidden band of approximately 0.7 eV between the $d_{\parallel}$ and π^* [10,11]. The Fermi level is located within the forbidden band, thereby forming the insulating $VO_2(M)$. When the temperature is higher than T_c, the density of the Fermi energy states in $VO_2(R)$ is formed by a mixture of π^* and $d_{\parallel}$ orbitals [9,12]. The electrons at the $d_{\parallel}$ state exhibit a behavior similar to that of free electrons, accomplishing a half-filled metallic state. The Fermi level form between the π^* and $d_{\parallel}$ bands, indicating an enhanced electrical conductivity of the $VO_2(R)$ [13]. Therefore, electrical and optical properties are considerably modulated during the phase transition. The phase transition of VO_2 can be induced by different types of stimuli, such as heat [5], electric fields [14], and mechanical strain [15]. The phase change in $VO_2(M)$ has also been utilized in various emerging technologies, including optical switches [16], thermoelectrics [17], hydrogen storage [18,19], sensors [20–22], transistors [23–25], active metamaterials [26–28], and photoelectric devices [29]. The application of $VO_2(M)$ in smart windows was investigated in the 1980s by Jorgenson et al. [7] and Babulanam et al. [30]. When the external temperature is lower than the phase-transition temperature, which is approximately 68 °C, $VO_2(M)$ exists in the insulating phase, exhibiting high transmittance of near-infrared (NIR) wavelengths in the solar spectrum. Conversely, when the temperature is higher than the phase-transition temperature, the crystal and band structures change because of the transition from the insulator phase ($VO_2(M)$) to the metallic phase (rutile VO_2 ($VO_2(R)$)), which significantly reduces the optical transmittance of NIR wavelengths. Therefore, thermochromic smart windows can reversibly modulate their solar transmittance at different temperatures and can reduce the room temperature during hot weather conditions; this will reduce the total energy consumption of the building. VO_2-based thermochromic smart windows offer characteristic advantages over other types of energy-saving windows, such as low-emissivity (low-e) glass [31,32] and electrochromic (EC) windows, owing to their ability to self-regulate solar transmission/reflection according to the external environment without utilizing an external energy supply [33–35]. Moreover, thermochromic windows have a relatively simple structure when compared with low-e or EC glass, thereby exhibiting potential for large-area installation and mass production for commercialization [36].

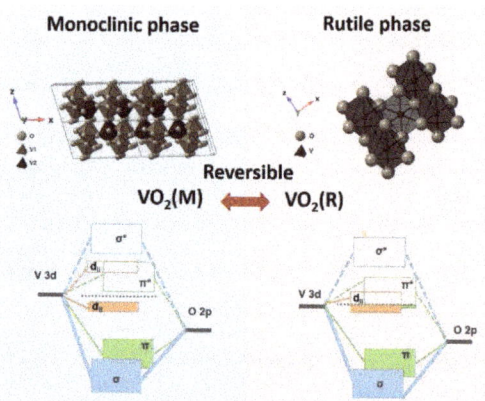

Figure 1. Schematic of the crystal structure and electronic band structure of the insulating $VO_2(M)$ and the metallic $VO_2(R)$. Adapted with permission from [13]. Copyright 2011, American Chemical Society.

The performance of VO_2 for smart windows is evaluated in terms of the luminous transmittance (T_{lum}) and solar modulation ability. Luminous transmittance refers to the integrated optical amount of visible-light transmittance, which is determined from the following equation:

$$T_{lum} = \int \Phi_{lum}(\lambda) T d\lambda / \int \Phi_{lum}(\lambda) d\lambda, \ (380 \text{ to } 780 \text{ nm})$$

$T(\lambda)$ and Φ_{lum} characterize the transmittance of the wavelength λ and photopic luminous efficiency function in the visible region, respectively [30,37]. Solar-energy modulation ability (ΔT_{sol}) is also a critical feature for determining the energy-saving capability of material. ΔT_{sol} is defined as the difference in the solar-energy transmittance (T_{sol}) values before and after phase transition in the 240 to 2500 nm spectrum, which is estimated using the follow equations [38]:

$$T_{sol} = \int \Phi_{sol}(\lambda) T d\lambda / \int \Phi_{sol}(\lambda) d\lambda, \ (250 \text{ to } 2600 \text{ nm})$$

$$\Delta T_{sol} = T_{sol,\text{low temperature}} - T_{sol,\text{high temperature}}$$

where Φ_{sol} denotes the solar irradiance spectrum for an air mass of 1.5, which is equivalent to the presence of the sun at an angle of 37° from the horizon [37]; moreover, $T_{sol,\text{low temperature}}$ and $T_{sol,\text{high temperature}}$ represent the solar transmittance of VO_2 films at a low temperature in the monoclinic phase and at a high temperature in the rutile phase, respectively. T_{lum} should be greater than 40% to indicate the requirement for daylight across windows, and ΔT_{sol} should be sufficiently high, at least 10%, for energy saving [39]. Furthermore, the phase-transition temperature of VO_2 (T_c = 68 °C) should be reduced from 68 °C for efficient regulation of solar energy during daytime [40]. Therefore, a reduced phase-transition temperature (T_c), high luminous transmittance (T_{lum}), and strong solar-energy modulation ability (ΔT_{sol}) are important characteristics for energy-efficient smart windows. To fulfill the demand for practical applications of energy-saving smart windows, VO_2-based thermochromic thin films should possess the following features: the phase-transition temperature (T_c) should be reduced to near-ambient temperature, and a high luminous transmittance (T_{lum} > 40%) accompanied by a strong solar-energy modulation ability (ΔT_{sol} > 10%) should be available [41,42].

Several studies have been conducted to improve the energy-saving performance of VO_2-based smart windows. For example, reductions in T_c have been achieved by doping with metal ions [43–45], or by utilizing nonstoichiometric compounds [46], strains [47], and nano-size effects [48]. Among the aforementioned methods, doping with metal ions, such as W^{6+} [49], Al^{3+} [50], Mg^{2+} [51], Sn^{4+} [52], and Mo^{6+} [53,54], is considered the most efficient. However, an increase in the dopant content results in the deterioration of phase-transition behaviors, such as a reduction in ΔT_{sol} and a broadened phase-transition temperature range [55,56]. High values of T_{lum} and ΔT_{sol} are also required to accomplish high-energy modulation efficiency for smart windows; however, these parameters involve a tradeoff, and thus, it is difficult to enhance them simultaneously [57]. Various strategies have been suggested to improve T_{lum} and ΔT_{sol} simultaneously, such as doping with Mg^{2+} [56] and F^- [55], or utilizing nano-size thermochromic materials [58], photonic crystals [59], antireflective overcoating [60], porous films [60], and multilayered structures [60,61]. However, the fabrication of $VO_2(M)$ films with high T_{lum} (> 40%) and ΔT_{sol} (>10%) values as well as a sufficiently reduced T_c remains challenging, which limits the utilization of $VO_2(M)$ in practical applications [56,57,62]

Recently, the fabrication of flexible $VO_2(M)$ films has attracted widespread attention [39,56]. Flexible thermochromic films demonstrate significant potential for large-scale fabrication and commercialization [63–66]. For example, flexible $VO_2(M)$ films can be instantly applied to the windows of existing buildings and easily integrated onto curved surfaces, such as automobile windows. Moreover, flexible $VO_2(M)$ thin films show the potential for application in actuators and optical switches for future optical and electronic devices [63,67]. Thus far, high-quality $VO_2(M)$ thin films have been fabricated using vacuum-chamber-based techniques, such as chemical vapor deposition (CVD) [68–70], physical vapor deposition [56], radiofrequency (RF) magnetron sputtering [71], and pulsed laser deposition [72]. These deposition methods provide high-quality and highly crystalline $VO_2(M)$ films; however, they often require high-temperature deposition conditions or an additional thermal annealing process to yield phase-pure crystalline $VO_2(M)$ films [63]. The deposition temperature is typically higher than 400 °C, which exceeds the thermal resistance of most flexible polymeric substrates [51,73–75]. Therefore, chamber-based deposition processes are predominantly performed on rigid inorganic substrates with high thermal resistance, which limits the fabrication of crystalline $VO_2(M)$ films on flexible substrates. Flexible $VO_2(M)$ films can also be obtained via colloidal deposition using $VO_2(M)$ nanoparticles (NPs) [56,65,76]. Colloidal dispersion of $VO_2(M)$ enables solution-based deposition onto polymeric substrates or the formation of flexible composite films through mixing in a polymer matrix. Hydrothermal synthesis of colloidal $VO_2(M)$ NPs was reported in a recent study, which demonstrated the feasibility of producing flexible $VO_2(M)$ films through the solution-based deposition of NPs at room temperature [77,78]. However, for colloidal $VO_2(M)$ NPs synthesized hydrothermally, lowering T_c while maintaining favorable optical properties, such as a high T_{lum} and ΔT_{sol}, remains difficult [79,80]. Therefore, the fabrication of flexible VO_2 thin films using colloidal $VO_2(M)$ NPs with a reduced T_c, high T_{lum}, and high ΔT_{sol} is still significantly challenging. In this review, we focus on the recent advances in the fabrication methods for flexible thermochromic $VO_2(M)$ thin films. We systematically review the fabrication process, including chamber-based vacuum deposition on flexible substrates that possess high thermal resistance. In addition, we introduce film-transfer techniques used to transfer $VO_2(M)$ layers deposited on rigid substrates onto flexible polymer substrates. Finally, we introduce the solution-based deposition process using colloidal $VO_2(M)$ NPs. The optical properties and phase transition behaviors are discussed to investigate the potential of flexible $VO_2(M)$ films for application in energy-saving smart windows and other emerging technologies.

2. Fabrication Methods

2.1. Fabrication of Flexible Monoclinic-Phase VO_2 ($VO_2(M)$) Films via Chamber-Based Deposition

As discussed, the fabrication of stoichiometric and highly crystalline VO_2 films using vacuum deposition requires high-temperature conditions or an additional calcination process [81]. Therefore, rigid inorganic substrates, which have high thermal stability, such as SiO_2 [82], MgF_2 [83], and Al_2O_3 [84], are generally used for growing $VO_2(M)$ films. The fabrication of flexible $VO_2(M)$ films through chamber-based deposition of $VO_2(M)$ films has also demonstrated using flexible substrates with high thermal stability. For example, muscovite sheets were first used as substrates for the fabrication of $VO_2(M)$ films because such sheets possess a high thermal stability of over 500 °C and superior chemical resistance, which enable the formation of highly crystalline $VO_2(M)$ films through high-temperature sintering. High-quality, single-phase $VO_2(M)$ films can be grown epitaxially on (001) muscovite substrates with high crystallinity, leading to superior phase transition behaviors in terms of resistivity and infrared (IR) transmittance [85]. Li et al. also developed a process for depositing a VO_2 film directly on a flexible muscovite substrate [86]. First, V_2O_5 films were deposited on a native muscovite substrate through pulsed-laser deposition for 20 min; then, the films were annealed at 650 °C under a 5 mTorr oxygen atmosphere to obtain highly crystalline $VO_2(M)$ films (Figure 2a). The electrical resistance of the $VO_2(M)$ thin films was measured under various bending radii. During the phase transition, the electrical resistance of the films varied by an order of 10^3 or more ($\Delta R/R > 10^3$), and the change in luminous transmittance was higher than 50% ($\Delta T_r > 50\%$) (Figure 2b). Owing to the intrinsic transparency and flexibility of muscovite sheets, the VO_2/muscovite heterogeneous structures also exhibited superior flexibility and visible-light transparency. The electrical resistance of the VO_2/muscovite films remained the same even after the films were bent 1000 times; this confirmed the high mechanical stability of the films (Figure 2c). Thus, considering their enhanced electrical, thermal, optical, and mechanical properties, VO$_2$/muscovite films demonstrate considerable potential for application in flexible electronic devices, especially optical switches.

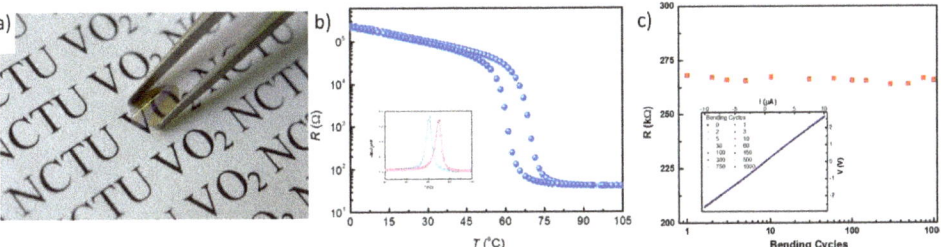

Figure 2. (a) Photograph of VO$_2$/muscovite thin film; (b) Temperature-dependent electrical resistance of VO$_2$/muscovite films; (c) Cyclability of VO$_2$/muscovite films over 1000 iterations in a bending test. Reproduced with permission from [86]. Copyright 2016, American Chemical Society.

VO$_2$(M) thin films grown on substrates, such as TiO$_2$, Al$_2$O$_3$, diamond, and SiO$_2$, have strong chemical bonds (ionic or covalent) between the VO$_2$(M) layers and the substrates. Thus, the VO$_2$ lattice is constrained, which is known as the substrate-clamping effect; this complicates the lattice rearrangement during phase transition [82,87,88]. Therefore, VO$_2$ films deposited on inorganic substrates typically require a higher energy to drive the metal–insulator transition (MIT). Conversely, VO$_2$(M) films deposited on mica sheets typically have weak van der Waals (vdW) bonds (0.1–10 kJ mol^{-1}) between VO$_2$(M) and the mica layer, which is 2–3 times weaker than the aforementioned ionic or covalent bonds (100–1000 kJ mol^{-1}) [89]. This weak vdW bonding between the VO$_2$ film and the mica sheet does not induce any significant lattice strain in the VO$_2$ layer. Therefore, the VO$_2$ film behaves as a nearly freestanding film on the mica sheet, which enables MIT with exceedingly low energy stimuli [90]. Moreover, owing to the weak vdW bonding between adjacent mica sheets, the thin mica sheet can be peeled off from the substrate, creating transparent and flexible VO$_2$(M)/mica sheets. Wang et al. also employed a mica sheet as a support for VO$_2$(M) to fabricate a mechanically flexible and electrically tunable flexible phase-change material for IR absorption [91]. First, 100-nm-thick Au thin films were deposited on a mica sheet through magnetron sputtering. Then, a 100-nm-thick vanadium film was deposited on the Au film via electron-beam evaporation and was thermally annealed in an oxygen atmosphere at temperatures of 430–470 °C. Au and mica sheet can withstand high-temperature annealing conditions. Finally, graphene thin films were transferred onto the VO$_2$ thin film to deposit the conductive electrode that induces the phase transition of the VO$_2$(M) thin films through Joule heating (Figure 3a). The IR absorption of this device can be continuously adjusted from 20% to 90% by changing the current applied to the graphene film. Moreover, this structure exhibited superior bending durability when it was bent up to 1500 times, without any noticeable deterioration in the optical properties (Figure 3b). Such tunable and flexible VO$_2$ devices have various application prospects in flexible photodetectors and active wearable devices.

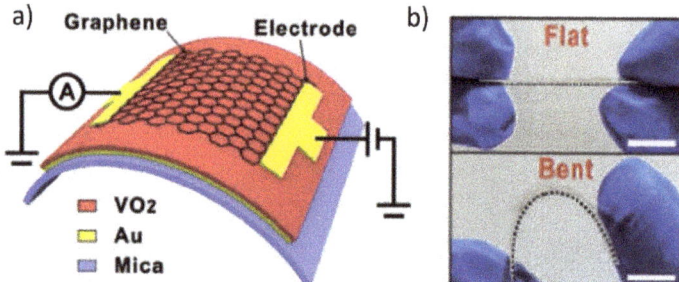

Figure 3. (a) Schematic representation of a flexible and electrically tuned flexible phase change material (FPCM) structure; (b) Mechanical flexibility of FPCM with 10 mm scale bars. Reproduced with permission from [91]. Copyright 2021, Wiley.

Chen et al. fabricated a flexible VO_2(M) thin film on a muscovite (mica) sheet directly through RF-plasma-assisted oxide molecular beam epitaxy (rf-OMBE) [92]. First, the VO_2 layer was grown using rf-OMBE on the (001) plane of mica sheets at 550 °C. Then, a layered single-walled carbon nanotube (SWNT) films was deposited using CVD on the high-quality VO_2/mica thin film. The SWNT layer exhibited superior conductivity and flexibility and can be employed as an efficient heater when a current/bias voltage is applied. The almost freestanding SWNTs/VO_2/mica (SVM) film was fabricated by peeling off the thin-layered SVM film from the substrates. Two Au electrodes were deposited on the flexible SVM thin film to provide a two-terminal electrode. The MIT process of the flexible VO_2(M) thin film can be easily controlled by heating SVM films with a bias current on Au electrodes, thereby enabling reversible modulation of IR transmission. When a bias current was applied, the transmittance decreased sharply from 70% and maintained an almost constant value of approximately 30% thereafter. When the input current was turned off, the transmittance quickly returned to its highest value of 70%; this confirms that direct modulation of the transmittance by applying a current is possible (Figure 4a,b). The MIT temperatures were 71 and 62 °C during the heating and cooling cycles, respectively (Figure 4c). Such ultrathin flexible SVM films with superior flexibility and transparency can be used for various applications involving future electrical devices.

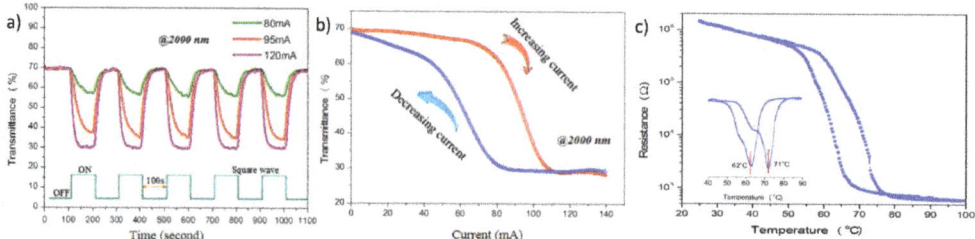

Figure 4. (a) Infrared (IR) response of flexible single-walled carbon nanotubes/VO_2/mica thin film with square-wave current; (b) IR performance as a function of applied current (2000 nm); (c) Resistance-dependent temperature curve for VO_2/mica thin film (the inset shows the differential curves during phase transition). Reproduced with permission from [92]. Copyright 2017, Elsevier.

In addition to mica sheets, carbon-based substrates, such as graphene sheets and networks of carbon nanotubes (CNTs), have also been utilized as flexible substrates for VO_2 deposition owing to their high thermal resistance. Xiao et al. reported the fabrication of VO_2/graphene/CNT (VGC) flexible thin films [93]. First, the graphene/CNT thin film was prepared by depositing graphene on a Cu substrate via low-pressure CVD. Then, the aligned CNT thin films were stacked on graphene substrates, followed by etching of the

Cu substrate to form flexible graphene/CNT flexible thin films. The VO$_x$ thin film was deposited on the graphene/CNT film through DC magnetron sputtering and was then thermally annealed at 450 °C in a low-pressure oxygen environment to obtain crystalline VO$_2$(M) thin films (Figure 5a). The phase transition of the VGC freestanding thin film can be induced by applying a current. The VGC films exhibited fast switching with low power consumption and highly reliable phase transition (Figure 5b,c). The drastic change in IR transmittance during the phase transition can potentially enable the application of VGC films in IR thermal camouflage, cloaking, and thermal optical modulators.

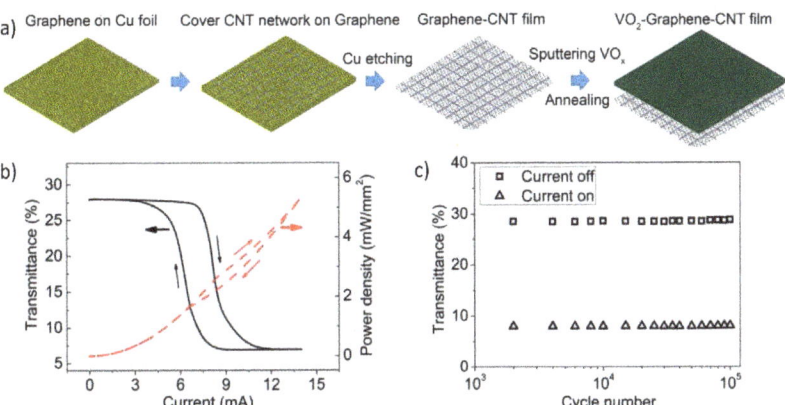

Figure 5. (a) Schematic of fabrication of VO$_2$/graphene/carbon nanotube (VGC) film; (b) Characterization of VGC film with current-dependent transmittance (1500 nm) (black line) and the correlated power consumption (red line); (c) Reliability measurement of the VGC films over 100,000 cycles with regard to current pulses. Reproduced with permission from [93]. Copyright 2015, American Chemical Society.

Chan et al. reported the fabrication of flexible VO$_2$(M)/Cr$_2$O$_3$/polyimide (PI) films using Cr$_2$O$_3$ as a buffer layer [94]. The Cr$_2$O$_3$ layer allows an epitaxial growth of the VO$_2$(M) layer, typically at approximately 300 °C, which enables the deposition of VO$_2$(M) on the PI polymer substrate at a relatively lower temperature (Figure 6a). The lattice constants for Cr$_2$O$_3$ are a = 0.496 nm, b = 0.496 nm, and c = 1.359 nm, and those for VO$_2$(R) are a = 0.455 nm, b = 0.455 nm, and c = 0.286 nm [95]. Therefore, Cr$_2$O$_3$ can act as a buffer layer owing to the similarity of its lattice constants with those of VO$_2$(R). Therefore, highly crystalline VO$_2$/Cr$_2$O$_3$ films can be successfully fabricated even under relatively low deposition conditions from 250 to 350 °C. Moreover, the refractive index of Cr$_2$O$_3$ is 2.2–2.3; hence, Cr$_2$O$_3$ behaves as an antireflective coating on top of the VO$_2$(M) layers, leading to a higher optical performance with T$_{lum}$ and ΔT$_{sol}$. The VO$_2$ film fabricated at 275 °C showed 42.4% of T$_{lum}$ and 0.4% of ΔT$_{sol}$; in contrast, the VO$_2$ film deposited with a 60-nm Cr$_2$O$_3$ buffer layer exhibits a high ΔT$_{sol}$ value of 6.7% at a similar T$_{lum}$ (43.7%). To fabricate flexible VO$_2$/Cr$_2$O$_3$/PI films, thin Cr$_2$O$_3$ layers were deposited on colorless PI films through magnetron sputtering; then, the VO$_2$ layers were directly deposited on Cr$_2$O$_3$/PI films using magnetron sputtering (Figure 6b). VO$_2$/Cr$_2$O$_3$/PI films exhibit minimal strain owing to the similar lattice parameters of the two layers. Therefore, flexible VO$_2$(M) films have a narrow and sharp hysteresis loop. The VO$_2$/Cr$_2$O$_3$/PI films exhibited superior IR modulation properties, i.e., approximately 60% variation at 2500 nm, when the VO$_2$ film thickness was approximately 80 nm (Figure 6c). The T$_c$ values of the films calculated during the heating and cooling cycles were 71.8 and 71.3 °C, respectively, and the transition width of the hysteresis loop was approximately 0.5 °C, which is significantly low for a phase transition (Figure 6d,e). Furthermore, the resistivity decreased by more than two orders of magnitude during the phase transition, indicating the high crystallinity of VO$_2$(M) films. However, the deposition temperature of >250 °C is still higher than the temperature that typical polymer films can withstand, which limits the utilization of various flexible polymeric substrates other than PI.

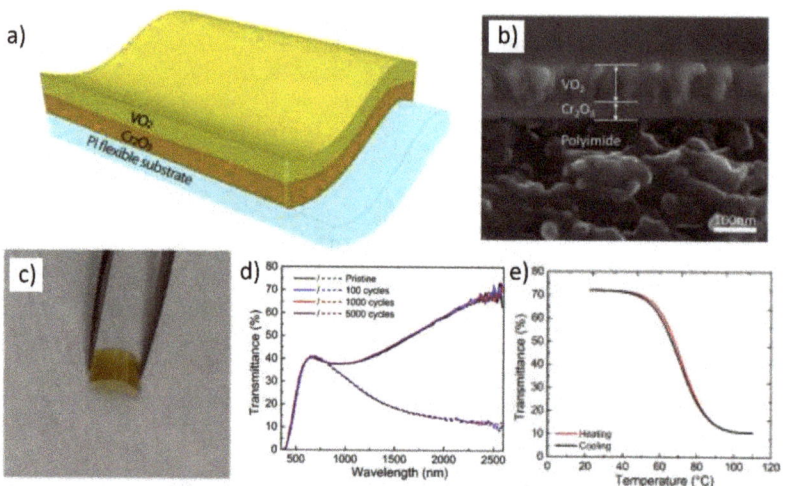

Figure 6. (**a**) Schematic representation and (**b**) cross-sectional scanning electron microscopy image of VO$_2$/Cr$_2$O$_3$/polyimide (PI) film; (**c**) Photograph of flexible VO$_2$/Cr$_2$O$_3$/PI film, (**d**) Ultraviolet–visible–near-IR (NIR) transmittance spectra of flexible VO$_2$/Cr$_2$O$_3$/PI film after multiple bending cycles; (**e**) Temperature-dependent transmittance hysteresis loop (2500 nm) of flexible VO$_2$/Cr$_2$O$_3$/PI film. Reproduced with permission from [94]. Copyright 2021, Elsevier.

Although direct deposition of VO$_2$(M) on substrates with high thermal resistance is a simple single-step process, only a limited number of substrates can be used under high-temperature deposition conditions. In contrast, the film-transfer process offers opportunities to utilize various types of polymeric substrates for the fabrication of flexible films [96]. In this process, VO$_2$(M) films are deposited on rigid substrates via high-temperature vacuum deposition and thermal annealing; then, the VO$_2$(M) thin films are transferred onto flexible polymeric substrates using the film-transfer process. As the VO$_2$(M) films are deposited under high-temperature conditions, they become highly crystalline, achieving enhanced optical properties (high T_{lum} and ΔT_{sol}) and improved stability under ambient conditions that persists for several months [97]. Moreover, polymer supports can impart enhanced mechanical stability and flexibility to films. The fabrication of flexible VO$_2$(M) films using the film-transfer process was first performed by Kim et al. [98]. In this process, an atomically thin, flexible graphene film was used to deposit a VO$_2$(M) layer for the transfer process. An amorphous VO$_x$ layer was first deposited on a graphene/Cu substrate through RF magnetron sputtering. Then, the VO$_x$ film on the graphene/Cu substrate was thermally annealed at 500 °C to transform VO$_x$ into crystalline VO$_2$ films. The Cu substrate was selectively etched, and the remaining VO$_2$(M)/graphene film was transferred to a polyethylene terephthalate (PET) film to fabricate flexible VO$_2$(M)/graphene/PET films. Because of the deposition on polymer films, the VO$_2$(M)/graphene/PET films exhibited high mechanical stability and flexibility while maintaining their reversible phase-transition property. These flexible VO$_2$/graphene/PET films exhibited a transmittance of 65.4% at a 550-nm wavelength; moreover, the variation in the transmittance during phase transition reached 53% at a wavelength of 2500 nm, with the transition band width being 9.8 °C (Figure 7a,b). The VO$_2$(M)/graphene/PET was integrated onto glass in a model house to investigate its ability to regulate the indoor temperature when functioning as a smart window. The VO$_2$/graphene films reduced the indoor room temperature by 5.8 °C compared with bare glass, thereby exhibiting the potential to function as an energy-efficient smart window (Figure 7c,d).

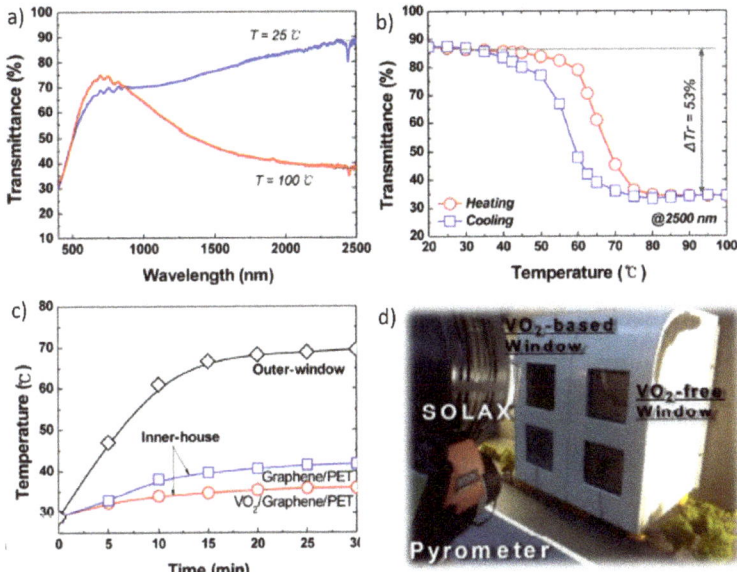

Figure 7. (**a**) Transmission spectra of VO$_2$/graphene/polyethylene terephthalate (PET) film at 25 and 100 °C; (**b**) Temperature-dependent transmittance of VO$_2$/graphene/PET film at 2500 nm; (**c**) Indoor temperature of a model house with VO$_2$/graphene/PET films; (**d**) Photograph of a model house coated with VO$_2$/graphene/PET films (VO$_2$-based windows) and graphene/PET films (VO$_2$-free windows). Reproduced with permission from [98]. Copyright 2013, American Chemical Society.

The fabrication of flexible VO$_2$(M) thin films with a reduced T$_c$ is still challenging owing to the difficulty in doping during the deposition process. Chae et al. reported a solution-based process to deposit VO$_2$(M) films using W-doped colloidal NPs, followed by film transfer, to fabricate flexible W-doped VO$_2$(M) films [74]. Colloidal VO$_x$ NPs were synthesized via high-temperature thermal decomposition with vanadium precursors, which were used for the deposition of VO$_2$(M) layers [99]. During the synthesis, W precursors were added into the reaction mixture for efficient doping of W during the formation of VO$_x$ NPs. Then, VO$_x$ NPs were deposited on mica substrates using the solution-based process and thermally annealed to form highly crystalline VO$_2$(M) films. Subsequently, the VO$_2$(M)/mica films were transferred onto polymeric substrates using adhesive-coated PET films. During the transfer process, a thin layer of mica sheet was peeled off and transferred to the polymer film to form transparent and flexible mica/VO$_2$(M)/PET films. As mica sheets are brittle, the polymer substrate can provide high mechanical strength and ensure reliable bending (Figure 8a). The W dopants were effectively doped into the VO$_2$(M) thin films, which resulted in the systematic reduction in T$_c$ depending on the different possible doping concentrations. The T$_c$ of flexible mica/VO$_2$(M)/PET films can be easily controlled at 25.6 °C when 1 at% W doping is used (Figure 8b). These flexible films exhibit superior optical properties—a T$_{lum}$ of 53% and a ΔT$_{sol}$ of 10%—at a T$_c$ of 29 °C when 1.3 at% Tungsten (W) doping is used. Thus, such films can be viable for use in energy-saving smart windows (Figure 8c,d).

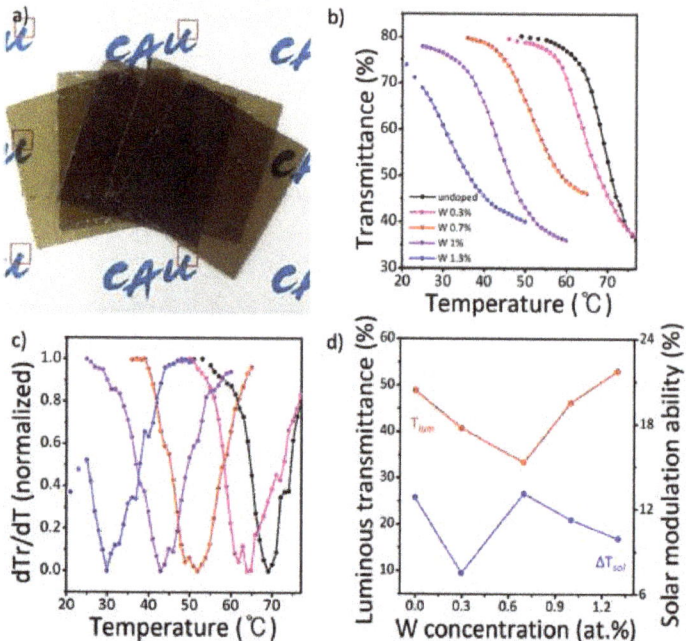

Figure 8. (a) Photograph of VO$_2$(M) thin films on PET substrates for various W doping concentrations; (b) Temperature dependence of transmittance; (c) First derivatives of transmittance; (d) Luminous transmittance (T$_{lum}$) and solar modulation ability (ΔT$_{sol}$) of VO$_2$(M)/mica thin films under various W doping concentrations (1900 nm). Reproduced with permission from [74]. Copyright 2021, Elsevier.

2.2. Fabrication of Flexible VO$_2$(M) Films through Solution-Based Deposition Process

Although the vacuum chamber-based deposition and film-transfer processes are highly effective for the fabrication of crystalline VO$_2$(M) films on flexible substrates, these processes are significantly complex, involving multiple deposition steps and often requiring an etching process, which can potentially limit large-scale fabrication and commercialization [100]. In contrast, the solution-based process enables simple, low-cost, and large-area fabrication of flexible VO$_2$(M) films [101]. Early examples of solution-processed VO$_2$(M) films were demonstrated via a sol–gel process [102]. Speck et al. were the first to demonstrate the sol–gel deposition of VO$_2$(M) films using molecular vanadium precursors [103]. In general, the sol–gel process of VO$_2$(M) thin films have been performed on thermally stable substrates, such as quartz, mica, or silicon wafers, owing to the high temperature thermal annealing process, typically above 400 °C [104]. Recent literature demonstrates that low temperature sol–gel deposition of VO$_2$(M) film processes can be achieved using deep ultraviolet photoactivation chemistry, which enable the fabrication of flexible VO$_2$(M)/Al$_2$O$_3$/PI films at 250 °C [105]. Not only has the sol–gel process been widely studied for the fabrication of flexible smart windows, but solution-based deposition using colloidal VO$_2$(M) NPs has too. Among a variety of synthetic methods based on colloidal VO$_2$(M) NPs, hydrothermal synthesis has attracted considerable attention owing to the high phase purity of the as-synthesized VO$_2$(M) NPs [38]. Hydrothermal synthesis involves a chemical reaction that yields high-quality crystals in a sealed pressurized reactor under high pressure and temperature. Hydrothermal growth of VO$_2$(M) films on the substrates has also been reported in the literature [106,107]. For example, VO$_2$(M) films have been fabricated via hydrothermal reactions by placing r-Al$_2$O$_3$ substrates in a hydrothermal reactor containing a solution mixture of ammonium metavanadate and oxalic acid [108]. The self-organized VO$_2$(M) films were formed with T$_{lum}$ of 65% and ΔT$_{sol}$

of ~11.82% [109]. However, for the direct hydrothermal deposition of VO$_2$(M) films on flexible substrates, the substrates should have high thermal and chemical resistance to ensure that they can withstand hydrothermal reaction conditions and calcination temperature [110–112]. Therefore, the use of VO$_2$(M) NPs for film depositions could have potential for large-area fabrication by mass-production processes using various substrate types. The single-step hydrothermal synthesis of VO$_2$(M) NPs was first demonstrated by Théobald et al. using a V$_2$O$_3$–V$_2$O$_5$–H$_2$O system, and the reaction was performed at a temperature of 20–400 °C under supercritical pressure [113]. There exist several stable vanadium oxide structures, such as VO$_2$, V$_2$O$_5$, V$_2$O$_3$, V$_5$O$_9$, V$_6$O$_{13}$, and V$_6$O$_{11}$, with various nonstoichiometric compounds [114]. Even in the stoichiometric compound, i.e., VO$_2$, several polymorphs exist, such as VO$_2$(A) [115], VO$_2$(B) [59], VO$_2$(D) [116], VO$_2$(P) [117], and VO$_2$(M) [118]. Therefore, hydrothermal synthesis of phase-pure and highly crystalline VO$_2$(M) is significantly challenging. Strong phase transition behaviors and favorable optical properties, including high values of T_{lum} and ΔT_{sol}, can be obtained using high-purity VO$_2$(M) NPs, in the absence of nonstoichiometry and impurities of metastable polymorphs. Therefore, careful control of synthetic procedures, including the hydrothermal reaction conditions, types of metal precursors, solvents, and additives, is a prerequisite for obtaining phase-pure VO$_2$(M) NPs.

To enhance phase purity and crystallinity, a two-step hydrothermal synthesis process to synthesize VO$_2$(M) NPs has widely studied. In this process, metastable VO$_2$ NPs are first synthesized hydrothermally and then thermally annealed for the conversion into the VO$_2$(M) phase. Phase-pure VO$_2$(M) NPs are obtained from various types of metastable VO$_2$ NPs and under different annealing conditions. Xie et al. first reported the hydrothermal synthesis of VO$_2$(D) with a size of 1–2 µm, using NH$_4$VO$_3$ and H$_2$C$_2$O$_4$. Hydrothermal synthesis was performed at 210 °C for 24 h, followed by a calcination process to transform the VO$_2$(D) into VO$_2$(M) [116]. Calcination of VO$_2$(D) was performed at temperatures as low as 300 °C for 2 h under a flow of high-purity nitrogen to obtain VO$_2$(R) NPs. These NPs also exhibit MIT near 68 °C. A two-step hydrothermal synthesis using VO$_2$(B) NPs has also been reported; however, the phase transformation from VO$_2$(B) to VO$_2$(M) occurs at a significantly higher annealing temperature, typically higher than 500 °C [119]. Corr et al. also studied the hydrothermal synthesis of VO$_2$(B) nanorods using V$_2$O$_5$ and formaldehyde solution at 180 °C for two days [120]. Then, thermal annealing was performed to convert VO$_2$(B) to VO$_2$(R) at 700 °C for 1 h in an argon atmosphere. Sun et al. reported the hydrothermal synthesis of VO$_2$(P) using VO(OC$_3$H$_7$)$_3$ and oleylamine at 220 °C for 48 h; then, they obtained VO$_2$(M) after thermal annealing at 400 °C for 40 or 60 s in a nitrogen or air atmosphere [121]. The size-dependent MIT property of VO$_2$(M) NPs was demonstrated through in situ variable-temperature IR spectroscopy. The authors observed that the variation in the transmittance of single-domain VO$_2$(M) NPs during phase transition systematically increased with a reduction in the size of the VO$_2$(M) NPs. Zhong et al. reported star-shaped VO$_2$(M) NPs that were hydrothermally synthesized using NH$_4$VO$_3$ and formic acid for two days at 200 °C. Then, the as-synthesized NPs were thermally annealed at 300–450 °C for 1 h to obtain VO$_2$(M) NPs. The VO$_2$(M) NP thin films were 325 nm thick and exhibited a T_{lum} and ΔT_{sol} of 44.18% and 7.32%, respectively [122]. Song et al. reported the hydrothermal synthesis of VO$_2$(D) using NH$_4$VO$_3$ and H$_2$C$_2$O$_4$·2H$_2$O at ~220 °C for ~18 h, followed by thermal annealing of VO$_2$(D) at 250–600 °C for 3 h, to obtain VO$_2$(M) nanoaggregates [123]. The as-synthesized VO$_2$(M) exhibited a low T$_c$ of approximately 41.0 °C and a thermal hysteresis width of approximately 6.6 °C. Li et al. demonstrated the electrothermochromicity of VO$_2$(M) NPs/Ag nanowire (NW) thin films deposited on glass and flexible PET substrates [124]. VO$_2$(M) NPs were hydrothermally synthesized using V$_2$O$_5$ and an oxalic acid dehydrate via at 220 °C for 36 h, followed by additional thermal annealing at 400 °C for 1 h in a vacuum chamber. The VO$_2$(M) NPs were deposited on top of Ag NW heaters. The optical response of the VO$_2$(M) NP films was then dynamically modulated by applying voltage on Ag NW. The infrared (IR) transmittance variation of the films from 0 V to 8 V of applied voltage is approximately 50%

at 1500 nm. Li et al. demonstrated the two-step hydrothermal synthesis of $VO_2(M)$ NPs using V_2O_5, $H_2C_2O_4$, and polyvinyl alcohol precursors [125]. The hydrothermal synthesis was performed at 220 °C for over 36 h, and the calcination was performed at 300–450 °C under vacuum. The $VO_2(M)$ film had a thickness of 463 nm and exhibited a high T_{lum} of over 70% at 700 nm; moreover, its IR transmittances at 1500 nm were approximately 89.5% and 53.8% before and after phase transition, respectively. The IR modulation exceeded 35%, which represents favorable optical properties for application in smart windows.

A single-step hydrothermal synthesis without a calcination process has also been reported. This method is a potentially simple, convenient, and low-cost process because it involves no additional post-annealing to obtain phase-pure $VO_2(M)$ NPs [126]. An additional thermal annealing processes induces grain growth in $VO_2(M)$ films. Size dependence of $VO_2(M)$ NPs on thermochromic properties have also been reported. Notably, a decrease in the size of $VO_2(M)$ NPs improves T_{lum} and ΔT_{sol} values [80]. Narayan et al. reported a phase-transition model in which the hysteresis width is directly proportional to the grain boundary area per unit volume [127]. Therefore, the hysteresis width is inversely correlated to the particle radius, and as the particle size increases, the phase transition temperature reduces, and the hysteresis width decreases. The smaller the nanoparticle size, the wider the hysteresis, and the VO_2 thermochromic performance is improved [128,129]. Therefore, single-step hydrothermal synthesis is more preferable to prevent particle coarsening by an additional thermal annealing process, hence sustaining a high thermochromic performance [130,131]. Gao et al. first demonstrated single-step hydrothermal synthesis of W-doped snowflake-shaped $VO_2(R)$ using V_2O_5 and $H_2C_2O_4$. The reaction was performed for seven days at 240 °C, and $VO_2(M)$ NPs were synthesized without a thermal annealing step [78]. The width of the $VO_2(R)$ nanocomposites was 200–300 nm, and the thickness was approximately 200–800 nm. Alie et al. also demonstrated single-step hydrothermal synthesis of star-shaped and spherical $VO_2(M)$ particles using $H_2C_2O_4$ and V_2O_5 in a molar ratio of 3:1 at 260 °C for 24 h [132]. The highly crystalline star-shaped $VO_2(M)$ particles exhibited a high thermal stability of up to ~300 °C and a >10% transmittance variation in the IR region during phase transition. Li et al. reported one-step hydrothermal synthesis of $VO_2(M)$ NPs using V_2O_5, TiO_2, and $H_2C_2O_4 \cdot 2H_2O$ at 240 °C for 24 h [133]. The $VO_2(M)$ NPs, with a size of approximately 50–100 nm, were further modified using $Zn(CH_3COO)_2$ to obtain a VO_2–ZnO structure. The $VO_2(M)$–ZnO films exhibited a low T_c of approximately 62.6 °C and a T_{lum} and ΔT_{sol} of approximately 52.2% and 9.3%, respectively. Ji et al. demonstrated the synthesis of $VO_2(M)$ using V_2O_5, N_2H_4, and H_2O_2 through a one-step hydrothermal process performed at 260 °C for 24 h (Figure 9a,b). The as-prepared $VO_2(M)$ NPs exhibited a transmittance change of approximately 50% at a wavelength of 2000 nm [134]. Moreover, as the concentration of the W dopant increased from 0% to 1%, the T_c of the $VO_2(M)$ NPs decreased from 55.5 to 37.1 °C (Figure 9c). Chen et al. reported the synthesis of phase-pure $V_{1-x}W_xO_2$ nanorods using $H_2C_2O_4 \cdot 2H_2O$ and V_2O_5 precursors. For W doping, $(NH_4)_5H_5[H_2(WO_4)_6]H_2O$ was added, and the reaction was performed at 260–280 °C for 6–72 h [135]. The T_{lum} of the 0.5 at% W-doped $VO_2(M)$ films was 60.6% at 20 °C, and ΔT_{sol} was 8.1%. Whittaker et al. reported the synthesis of W-doped $VO_2(M)$ nanobelts using V_2O_5 and $H_2C_2O_4$ precursors with H_2WO_4 for W doping [43]. The reaction was performed at 250 °C for 12 h to 7 days. W doping (0.90%) led to remarkable modulation of the T_c of $VO_2(M)$ films, from 68.0 to 33.8 °C. Shen et al. demonstrated that Zr doping significantly enhances optical properties while reducing T_c [118]. Moreover, Zr doping of $VO_2(M)$ reduces T_c while improving T_{lum} and ΔT_{sol}. However, T_c is only reduced from 68.6 to 64.3 °C with 9.8% Zr doping; conversely, Zr-doped VO_2 flexible films exhibit high values of T_{lum} (60.4%) and T_{sol} (14.1%). The optical bandgap, which is 1.59 eV for undoped $VO_2(M)$, increases to 1.89 eV after 9.8% Zr doping, resulting in a change in the apparent color of the $VO_2(M)$ films. Accordingly, the color of the Zr-doped $VO_2(M)$ flexible films is affected; the brown-yellow color of flexible $VO_2(M)$ film is brightened, along with an increase in T_{lum}. In addition, T_c is further reduced to 28.6 °C, and T_{lum} and ΔT_{sol} values of 48.6% and 4.9%, respectively, are achieved through W-Zr-co-doping.

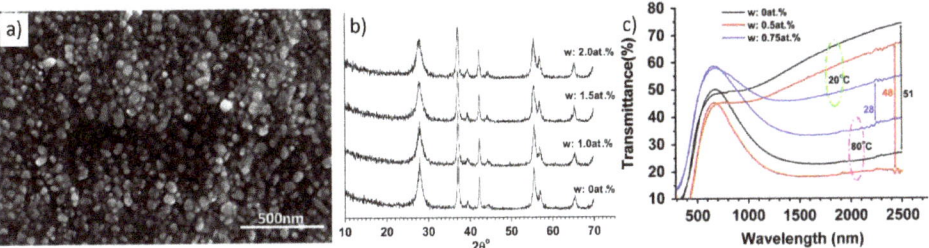

Figure 9. (a) SEM image and (b) XRD patterns of VO$_2$(M) NPs; (c) Temperature-dependent transmittance spectra during phase transition of VO$_2$(M) film. Reproduced with permission from [134]. Copyright 2011, Elsevier.

Several studies have been conducted to optimize the conditions for single-step hydrothermal synthesis to enhance the phase purity of VO$_2$(M) NPs and their optical properties, including T$_{lum}$ and ΔT$_{sol}$. Guo et al. performed a one-step hydrothermal synthesis process using VOSO$_4$ and N$_2$H$_4$·H$_2$O in the presence of H$_2$O$_2$ [77]. H$_2$O$_2$, a strong oxidizing agent, is separated after the reaction with the vanadium solution in a hydrothermal autoclave reactor. Then, H$_2$O$_2$ decomposes and evaporates at 150 °C to provide a moderately oxidizing environment. This facilitates the synthesis of stoichiometric and highly crystalline VO$_2$(M) NPs. The as-synthesized VO$_2$(M) NPs exhibited an average size of ~30 nm, with significant size uniformity (Figure 10a,b). For the preparation of flexible VO$_2$(M) films, the VO$_2$(M) NPs were dispersed in N,N-dimethylformamide with polyacrylonitrile polymers. Then, the solution was deposited on a flexible PET substrate. The flexible VO$_2$(M) films attained favorable optical properties, with a T$_{lum}$ of 54.26% and a ΔT$_{sol}$ of 12.34% (Figure 10c,d). In addition to optimizing the hydrothermal reaction conditions, the enhancement in the purity of vanadium precursors also produces VO$_2$(M) NPs with improved optical properties.

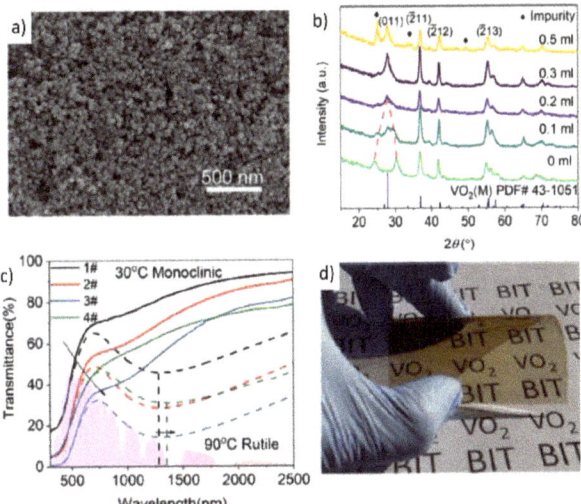

Figure 10. (a) SEM image of the VO$_2$ NPs with 0.2 mL of H$_2$O$_2$, (b) XRD patterns of as-synthesis VO$_2$ NPs with different amounts of H$_2$O$_2$, (c) temperature-dependent transmittance spectra of samples at 30 °C (bold line) and 90 °C (dashed line), and (d) the VO$_2$(M) NPs-based flexible films. Reproduced with permission from [77]. Copyright 2018, American Chemical Society.

Kim et al. demonstrated single-step hydrothermal synthesis of VO$_2$(M) NPs using phase-pure vanadium precursors [136]. After mixing the vanadium precursors, size-selective purification was performed to enhance the phase purity of the precursors, resulting in the formation of $VO_2(M)$ NPs with enhanced optical properties. The obtained phase-pure $VO_2(M)$ NPs exhibited an enhanced T_{lum} (55%) and ΔT_{sol} (18%), and the ΔT_{sol} value is one of the highest reported for hydrothermally synthesized $VO_2(M)$. Furthermore, W-doped $VO_2(M)$ NPs have been reported to exhibit superior phase-transition behaviors, while T_c is systematically reduced depending on the W doping concentration (Figure 11a,b). Flexible $VO_2(M)$ films were fabricated and deposited on PET polymer substrates over a large area using a spray coater (15 cm × 15 cm) (Figure 11c,d). In model house experiments under daytime solar irradiation, the W-doped $VO_2(M)$ films applied onto glass provided a significant reduction in the indoor temperature; thus, these films are potentially viable for practical applications.

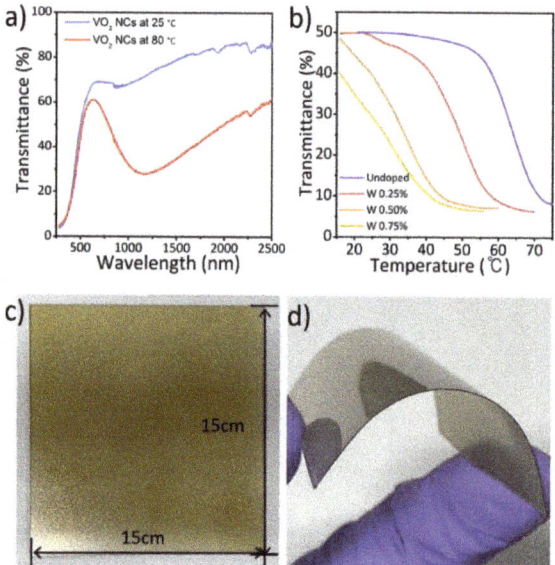

Figure 11. (a) Transmittance spectra of $VO_2(M)$ NP films before and after phase transition; (b) Temperature-dependent transmittance (1350 nm) of W-doped $VO_2(M)$ NP films during heating; Photographs of (c) 15 cm × 15 cm $VO_2(M)$ NP films on glass substrate and (d) flexible substrate obtained via spray-coating. Reproduced with permission from [136]. Copyright 2021, Elsevier.

Colloidal NPs enable convenient, large-scale fabrication of flexible $VO_2(M)$ film through the solution process, which is beneficial considering the expected requirement for large-scale fabrication techniques [63,65,76,137]. For the fabrication of flexible $VO_2(M)$ films, $VO_2(M)$ NPs have been coated on flexible polymer films or embedded into a polymer matrix [138]. Shen et al. reported a process for blade coating of $VO_2(M)$ NPs on indium tin oxide (ITO)-coated PET substrates to form flexible $VO_2(M)$ films [139]. Applying a current along the ITO layer induced ohmic heating, which resulted in the phase transition of $VO_2(M)$ layers and a change in IR transmittance. The obtained film showed well-controlled IR switching properties upon changing the input voltage, as well as superior thermochromic properties (T_{lum} of 57.3% and ΔT_{sol} of 13.8%). Under ohmic heating, the IR conversion properties did not show any evident deterioration, even after 10,000 bending cycles, which indicates superior stability and flexibility. Chen et al. demonstrated the preparation of $VO_2(M)$/polymer composite films by embedding $VO_2(M)$ NPs into a polymeric matrix. $VO_2(M)$ NPs were synthesized hydrothermally using V_2O_5 and N_2H_4 at approximately 180–400 °C for 15 h [80]. The size of the $VO_2(M)$ NPs ranged from approximately 25 to 45 nm. The synthesized $VO_2(M)$ NPs were dispersed in polyurethane

(PU) and coated onto PET to form flexible VO$_2$(M) films. These films achieved high optical performance, with a ΔT$_{sol}$ of 22.3% and a T$_{lum}$ of 45.6%. Similarly, Zhou et al. reported the roll-coating of Mg-doped VO$_2$(M) NPs that were hydrothermally synthesized using V$_2$O$_5$ and H$_2$C$_2$O$_4$ [140]. Mg-doped VO$_2$(M) composite foils were prepared by mixing NPs with PU solutions and were deposited on a PET substrate using a roll-coater. The flexible composite foils exhibited a high T$_{lum}$ and ΔT$_{sol}$ of 54.2% and 10.6%, respectively. Liang et al. reported the bar-coating of W-doped VO$_2$(M) nanorods; the nanorods were prepared via one-step hydrothermal synthesis for 48 h at 240 °C using V$_2$O$_5$, C$_4$H$_6$O$_6$, and ammonium tungstate (Figure 12a,b) [49]. The nanorods were then mixed with the tetraethyl orthosilicate and poly(ethyl methacrylate) solution. The solution mixture was cast on PET substrates using a stainless-steel coating bar to fabricate large-area, flexible VO$_2$(M) films (Figure 12c). The T$_c$ of the flexible VO$_2$(M) films could be systematically modulated by approximately 24.52 °C for 1 at% of W doping, and the mid-infrared transmission could be modulated by 31% at a T$_c$ of 37.3 °C. Inkjet printing has also been widely utilized as a useful direct-write technology to fabricate high-resolution, low-cost, large-area, and uniform-surface films on flexible substrates [141,142]. Haining et al. reported the fabrication of VO$_2$(M) smart windows via inkjet printing using hydrothermally synthesized VO$_2$(M) NPs [143,144]. Large-area VO$_2$(M) films were fabricated on polyethylene substrates with a T$_{lum}$ of 56.96% and a ΔT$_{sol}$ of 5.21%.

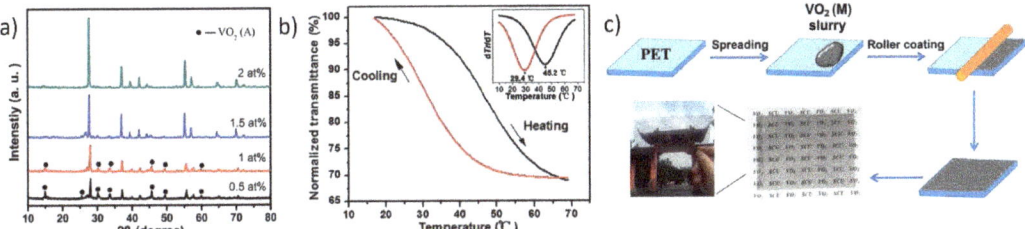

Figure 12. (a) XRD patterns of W doped VO$_2$(M) films, (b) Transmittance hysteresis loops and first derivatives of transmittance for W doped VO$_2$ (M) films recorded at a wavelength of 9 µm, (c) Schematic diagram of film deposition with W doped VO$_2$(M) NPs on PET substrates. Reproduced with permission from [49]. Copyright 2016, American Chemical Society.

The chemical instability of VO$_2$(M) NPs can potentially limit their long-term usage as smart windows in real-world environments [145]. To enhance the chemical stability of VO$_2$(M) NPs, core–shell structures, in which VO$_2$(M) NPs are overcoated with chemically inert shells, have been developed. Gao et al. reported a core–shell structure with VO$_2$@SiO$_2$ NPs. VO$_2$(M) was synthesized through a hydrothermal reaction, and SiO$_2$ shells were overcoated using the Stöber method [56]. SiO$_2$ is chemically inert and optically transparent, which is ideal for protecting VO$_2$(M) NPs. VO$_2$@SiO$_2$ NPs exhibit improved chemical resistance to oxidation. The SiO$_2$ shell of VO$_2$ NPs serves as an oxygen diffusion barrier layer, which can prevent the VO$_2$ from changing to V$_2$O$_5$. This phenomenon was confirmed through experiments conducted with VO$_2$ NPs and VO$_2$@SiO$_2$ NPs after annealing in an air atmosphere for 2 h at 300 °C. Flexible films were fabricated by embedding VO$_2$@SiO$_2$ NPs into a PU matrix; then, the VO$_2$@SiO$_2$ NPs/PU cast on a PET matrix were dispersed to fabricate flexible VO$_2$@SiO$_2$/PU composite films. These films exhibited a high T$_{lum}$ (55.3%) and ΔT$_{sol}$ (7.5%). In addition to SiO$_2$, various types of oxides, such as ZnS [146], TiO$_2$ [11], and ZrO$_2$ [147], have been utilized for overcoating to prepare core–shell NPs. Saini et al. demonstrated an approach to improve the thermal stability and thermochromic properties of VO$_2$(M) NPs by overcoating with CeO$_2$ [148]. VO$_2$(M)@CeO$_2$ NPs were observed to be thermally stable for up to 320 °C in air, which confirmed the enhancement in stability after overcoating.

3. Perspectives

Flexible VO$_2$(M) films offer significant potential for the integration of energy-saving smart windows in existing buildings, as well as for application in novel flexile devices, such as sensors and actuators. Various methods for fabricating flexible thermochromic thin films based on the vacuum deposition and solution-based process have been reported; these methods are potentially suitable for commercialization. However, certain issues still remain to be resolved before VO$_2$-based smart windows can be utilized in practice. For example, flexible VO$_2$ films fabricated using vacuum deposition and film-transfer techniques show high T_{lum} and ΔT_{sol} values; however, these methods are still limited in terms of large-area and mass-production capabilities. In addition, deposition methods with uniform doping should be developed further to systematically reduce T_c while maintaining favorable phase-change optical properties. Conversely, the annealing-free, solution-based process offers advantages such as convenient, low-cost, large-area deposition of phase-change VO$_2$(M) on flexible substrates. Particularly, hydrothermal synthesis yields highly crystalline VO$_2$(M) NPs with colloidal stability and moderately useful phase-change behaviors. However, it is still challenging to prepare flexible VO$_2$(M) films with high T_{lum} and ΔT_{sol} values as well as a reduced T_c. The optical properties of the representative flexible VO$_2$ films fabricated using deposition and solution-based processes are summarized in Figure 13, which displays the opportunities for utilizing flexible VO$_2$(M) films in energy-saving smart windows. Therefore, large-scale, high-throughput, mass-production capabilities for the fabrication and commercialization of high-performance VO$_2$(M) films should be realized. Finally, certain limitations in terms of the intrinsic properties of VO$_2$(M) should be overcome to utilize flexible VO$_2$ films. First, phase-change VO$_2$ films show an inherent brown color, which is not desirable for window applications. Therefore, it is highly recommended to develop fabrication methods that can enable control of the apparent colors of VO$_2$(M) while ensuring a high T_{lum} and ΔT_{sol} and low T_c. Moreover, vanadium oxide has various stable phases and a stable stoichiometry; consequently, VO$_2$(M) films are easily oxidized into other phases under exposure in ambient conditions. Therefore, processes to prevent VO$_2$(M) from being oxidized, for example, overcoating of VO$_2$(M) films or using NPs with protective layers, should be developed to enable long-term usage of the films.

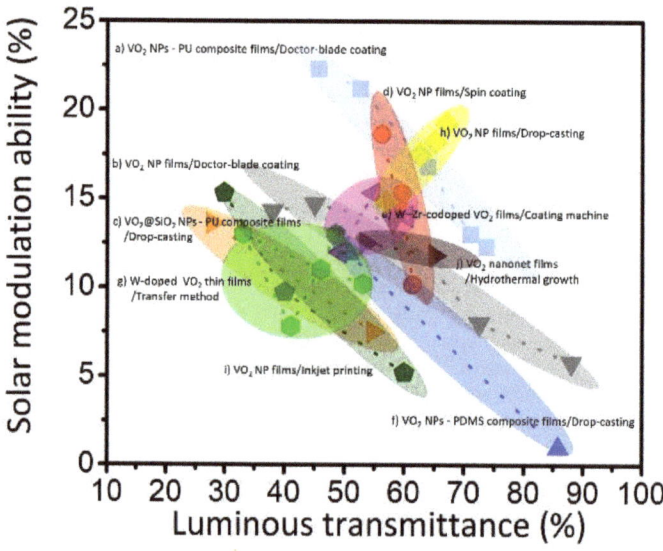

Figure 13. Comparison of luminous transmittance (T_{lum}) and solar modulation ability (ΔT_{sol}) between flexible VO$_2$ thin films fabricated via hydrothermal reaction: (a) [80], (b) [139], (c) [56], (d) [136], (e) [118], (f) [149], (g) [74], (h), [64], (i) [143], (j) [109].

Author Contributions: The manuscript was written through equal contributions from all authors. Investigation, J.K. and T.P.; writing—review and editing, J.K. and T.P.; supervision, T.P.; funding acquisition, T.P. All authors have read and agreed to the published version of the manuscript.

Funding: This research was supported by Creative Materials Discovery Program through the National Research Foundation of Korea (NRF) funded by the Ministry of Science and ICT (NRF-2018M3D1A1059001) and by the National Research Foundation of Korea (NRF) grant funded by the Korea government (MSIT) (NRF-2021R1A2C1013604). This research was also supported by the Chung-Ang University Graduate Research Scholarship in 2020.

Institutional Review Board Statement: Not applicable.

Informed Consent Statement: Not applicable.

Data Availability Statement: The data presented in this study are available on request from the corresponding author.

Conflicts of Interest: The authors declare no conflict of interest.

References

1. Gao, Y.; Luo, H.; Zhang, Z.; Kang, L.; Chen, Z.; Du, J.; Kanehira, M.; Cao, C. Nanoceramic VO2 thermochromic smart glass: A review on progress in solution processing. *Nano Energy* **2012**, *1*, 221–246. [CrossRef]
2. Granqvist, C.G. Transparent conductors as solar energy materials: A panoramic review. *Sol. Energy Mater. Sol. Cells* **2007**, *91*, 1529–1598. [CrossRef]
3. Ke, Y.; Zhou, C.; Zhou, Y.; Wang, S.; Chan, S.H.; Long, Y. Emerging thermal-responsive materials and integrated techniques targeting the energy-efficient smart window application. *Adv. Funct. Mater.* **2018**, *28*, 1800113. [CrossRef]
4. Chao, D.; Zhu, C.; Xia, X.; Liu, J.; Zhang, X.; Wang, J.; Liang, P.; Lin, J.; Zhang, H.; Shen, Z.X. Graphene quantum dots coated VO$_2$ arrays for highly durable electrodes for Li and Na ion batteries. *Nano Lett.* **2015**, *15*, 565–573. [CrossRef] [PubMed]
5. Morin, F. Oxides which show a metal-to-insulator transition at the Neel temperature. *Phys. Rev. Lett.* **1959**, *3*, 34. [CrossRef]
6. Tripathi, A.; John, J.; Kruk, S.; Zhang, Z.; Nguyen, H.S.; Berguiga, L.; Romeo, P.R.; Orobtchouk, R.; Ramanathan, S.; Kivshar, Y. Tunable Mie-Resonant Dielectric Metasurfaces Based on VO$_2$ Phase-Transition Materials. *ACS Photonics* **2021**, *8*, 1206–1213. [CrossRef]
7. Jorgenson, G.; Lee, J. Doped vanadium oxide for optical switching films. *Sol. Energy Mater.* **1986**, *14*, 205–214. [CrossRef]
8. Budai, J.D.; Hong, J.; Manley, M.E.; Specht, E.D.; Li, C.W.; Tischler, J.Z.; Abernathy, D.L.; Said, A.H.; Leu, B.M.; Boatner, L.A. Metallization of vanadium dioxide driven by large phonon entropy. *Nature* **2014**, *515*, 535–539. [CrossRef]
9. Aetukuri, N.B.; Gray, A.X.; Drouard, M.; Cossale, M.; Gao, L.; Reid, A.H.; Kukreja, R.; Ohldag, H.; Jenkins, C.A.; Arenholz, E. Control of the metal–insulator transition in vanadium dioxide by modifying orbital occupancy. *Nat. Phys.* **2013**, *9*, 661–666. [CrossRef]
10. Wu, C.; Feng, F.; Xie, Y. Design of vanadium oxide structures with controllable electrical properties for energy applications. *Chem Soc Rev* **2013**, *42*, 5157–5183. [CrossRef]
11. Li, Y.; Ji, S.; Gao, Y.; Luo, H.; Kanehira, M. Core-shell VO$_2$@TiO$_2$ nanorods that combine thermochromic and photocatalytic properties for application as energy-saving smart coatings. *Sci. Rep.* **2013**, *3*, 1–13. [CrossRef]
12. Goodenough, J.B. The two components of the crystallographic transition in VO$_2$. *J. Solid State Chem.* **1971**, *3*, 490–500. [CrossRef]
13. Whittaker, L.; Patridge, C.J.; Banerjee, S. Microscopic and nanoscale perspective of the metal− insulator phase transitions of VO$_2$: Some new twists to an old tale. *J. Phys. Chem. Lett.* **2011**, *2*, 745–758. [CrossRef]
14. Wu, B.; Zimmers, A.; Aubin, H.; Ghosh, R.; Liu, Y.; Lopez, R. Electric-field-driven phase transition in vanadium dioxide. *Phys. Rev. B* **2011**, *84*, 241410. [CrossRef]
15. Kikuzuki, T.; Lippmaa, M. Characterizing a strain-driven phase transition in VO$_2$. *Appl. Phys. Lett.* **2010**, *96*, 132107. [CrossRef]
16. Gea, L.A.; Boatner, L. Optical switching of coherent VO$_2$ precipitates formed in sapphire by ion implantation and annealing. *Appl. Phys. Lett.* **1996**, *68*, 3081–3083. [CrossRef]
17. Wu, C.; Feng, F.; Feng, J.; Dai, J.; Peng, L.; Zhao, J.; Yang, J.; Si, C.; Wu, Z.; Xie, Y. Hydrogen-incorporation stabilization of metallic VO$_2$ (R) phase to room temperature, displaying promising low-temperature thermoelectric effect. *J. Am. Chem. Soc.* **2011**, *133*, 13798–13801. [CrossRef]
18. Xie, J.; Wu, C.; Hu, S.; Dai, J.; Zhang, N.; Feng, J.; Yang, J.; Xie, Y. Ambient rutile VO$_2$(R) hollow hierarchitectures with rich grain boundaries from new-state nsutite-type VO$_2$, displaying enhanced hydrogen adsorption behavior. *Phys. Chem. Chem. Phys.* **2012**, *14*, 4810–4816. [CrossRef]
19. Yoon, H.; Choi, M.; Lim, T.-W.; Kwon, H.; Ihm, K.; Kim, J.K.; Choi, S.-Y.; Son, J. Reversible phase modulation and hydrogen storage in multivalent VO$_2$ epitaxial thin films. *Nat. Mater.* **2016**, *15*, 1113–1119. [CrossRef]
20. Hu, B.; Ding, Y.; Chen, W.; Kulkarni, D.; Shen, Y.; Tsukruk, V.V.; Wang, Z.L. External-strain induced insulating phase transition in VO$_2$ nanobeam and its application as flexible strain sensor. *Adv. Mater.* **2010**, *22*, 5134–5139. [CrossRef]

21. Liu, N.; Mesch, M.; Weiss, T.; Hentschel, M.; Giessen, H. Infrared perfect absorber and its application as plasmonic sensor. *Nano Lett.* **2010**, *10*, 2342–2348. [CrossRef] [PubMed]
22. Anker, J.N.; Hall, W.P.; Lyandres, O.; Shah, N.C.; Zhao, J.; Van Duyne, R.P. Biosensing with plasmonic nanosensors. *Nat. Mater.* **2008**, *7*, 442–453. [CrossRef] [PubMed]
23. Sengupta, S.; Wang, K.; Liu, K.; Bhat, A.K.; Dhara, S.; Wu, J.; Deshmukh, M.M. Field-effect modulation of conductance in VO_2 nanobeam transistors with HfO_2 as the gate dielectric. *Appl. Phys. Lett.* **2011**, *99*, 062114. [CrossRef]
24. Kim, H.-T.; Chae, B.-G.; Youn, D.-H.; Maeng, S.-L.; Kim, G.; Kang, K.-Y.; Lim, Y.-S. Mechanism and observation of Mott transition in VO_2-based two-and three-terminal devices. *N. J. Phys.* **2004**, *6*, 52. [CrossRef]
25. Zhou, Y.; Ramanathan, S. Relaxation dynamics of ionic liquid—VO_2 interfaces and influence in electric double-layer transistors. *J. Appl. Phys* **2012**, *111*, 084508. [CrossRef]
26. Liu, M.; Hwang, H.Y.; Tao, H.; Strikwerda, A.C.; Fan, K.; Keiser, G.R.; Sternbach, A.J.; West, K.G.; Kittiwatanakul, S.; Lu, J. Terahertz-field-induced insulator-to-metal transition in vanadium dioxide metamaterial. *Nature* **2012**, *487*, 345–348. [CrossRef]
27. Schurig, D.; Mock, J.J.; Justice, B.; Cummer, S.A.; Pendry, J.B.; Starr, A.F.; Smith, D.R. Metamaterial electromagnetic cloak at microwave frequencies. *Science* **2006**, *314*, 977–980. [CrossRef]
28. Lee, C.-W.; Choi, H.J.; Jeong, H. Tunable metasurfaces for visible and SWIR applications. *Nano Converg.* **2020**, *7*, 1–11. [CrossRef]
29. Lu, J.; Liu, H.; Deng, S.; Zheng, M.; Wang, Y.; van Kan, J.A.; Tang, S.H.; Zhang, X.; Sow, C.H.; Mhaisalkar, S.G. Highly sensitive and multispectral responsive phototransistor using tungsten-doped VO_2 nanowires. *Nanoscale* **2014**, *6*, 7619–7627. [CrossRef]
30. Babulanam, S.; Eriksson, T.; Niklasson, G.; Granqvist, C. Thermochromic VO_2 films for energy-efficient windows. *Sol. Energy Mater.* **1987**, *16*, 347–363. [CrossRef]
31. Kamalisarvestani, M.; Saidur, R.; Mekhilef, S.; Javadi, F. Performance, materials and coating technologies of thermochromic thin films on smart windows. *Renew. Sust. Energ. Rev.* **2013**, *26*, 353–364. [CrossRef]
32. Lampert, C.M. Large-area smart glass and integrated photovoltaics. *Sol. Energy Mater. Sol. Cells* **2003**, *76*, 489–499. [CrossRef]
33. Lin, S.; Bai, X.; Wang, H.; Wang, H.; Song, J.; Huang, K.; Wang, C.; Wang, N.; Li, B.; Lei, M. Roll-to-roll production of transparent silver-nanofiber-network electrodes for flexible electrochromic smart Windows. *Adv. Mater.* **2017**, *29*, 1703238. [CrossRef]
34. Kim, M.-J.; Sung, G.; Sun, J.-Y. Stretchable and reflective displays: Materials, technologies and strategies. *Nano Converg.* **2019**, *6*, 1–24. [CrossRef]
35. Chen, Y.; Ai, B.; Wong, Z.J. Soft optical metamaterials. *Nano Converg.* **2020**, *7*, 1–17. [CrossRef]
36. Hoffmann, S.; Lee, E.S.; Clavero, C. Examination of the technical potential of near-infrared switching thermochromic windows for commercial building applications. *Sol. Energy Mater. Sol. Cells* **2014**, *123*, 65–80. [CrossRef]
37. Chang, T.; Cao, X.; Long, Y.; Luo, H.; Jin, P. How to properly evaluate and compare the thermochromic performance of VO_2-based smart coatings. *J. Mater. Chem. A* **2019**, *7*, 24164–24172. [CrossRef]
38. Li, M.; Magdassi, S.; Gao, Y.; Long, Y. Hydrothermal synthesis of VO_2 polymorphs: Advantages, challenges and prospects for the application of energy efficient smart windows. *Small* **2017**, *13*, 1701147. [CrossRef] [PubMed]
39. Li, S.-Y.; Niklasson, G.A.; Granqvist, C.-G. Thermochromic fenestration with VO_2-based materials: Three challenges and how they can be met. *Thin Solid Films* **2012**, *520*, 3823–3828. [CrossRef]
40. Saeli, M.; Piccirillo, C.; Parkin, I.P.; Binions, R.; Ridley, I. Energy modelling studies of thermochromic glazing. *Energy Build.* **2010**, *42*, 1666–1673. [CrossRef]
41. Majid, S.; Sahu, S.; Ahad, A.; Dey, K.; Gautam, K.; Rahman, F.; Behera, P.; Deshpande, U.; Sathe, V.; Shukla, D. Role of VV dimerization in the insulator-metal transition and optical transmittance of pure and doped VO_2 thin films. *Phys. Rev. B* **2020**, *101*, 014108. [CrossRef]
42. Zhao, L.; Miao, L.; Liu, C.; Li, C.; Asaka, T.; Kang, Y.; Iwamoto, Y.; Tanemura, S.; Gu, H.; Su, H. Solution-processed VO_2-SiO_2 composite films with simultaneously enhanced luminous transmittance, solar modulation ability and anti-oxidation property. *Sci. Rep.* **2014**, *4*, 1–11. [CrossRef]
43. Whittaker, L.; Wu, T.-L.; Patridge, C.J.; Sambandamurthy, G.; Banerjee, S. Distinctive finite size effects on the phase diagram and metal–insulator transitions of tungsten-doped vanadium (iv) oxide. *J. Mater. Chem.* **2011**, *21*, 5580–5592. [CrossRef]
44. Xu, Y.; Huang, W.; Shi, Q.; Zhang, Y.; Song, L.; Zhang, Y. Synthesis and properties of Mo and W ions co-doped porous nano-structured VO_2 films by sol–gel process. *J. Solgel Sci. Technol.* **2012**, *64*, 493–499. [CrossRef]
45. Gao, Y.; Cao, C.; Dai, L.; Luo, H.; Kanehira, M.; Ding, Y.; Wang, Z.L. Phase and shape controlled VO_2 nanostructures by antimony doping. *Energy Environ. Sci.* **2012**, *5*, 8708–8715. [CrossRef]
46. Griffiths, C.; Eastwood, H. Influence of stoichiometry on the metal-semiconductor transition in vanadium dioxide. *J. Appl. Phys* **1974**, *45*, 2201–2206. [CrossRef]
47. Muraoka, Y.; Hiroi, Z. Metal–insulator transition of VO_2 thin films grown on TiO_2 (001) and (110) substrates. *Appl. Phys. Lett.* **2002**, *80*, 583–585. [CrossRef]
48. Dai, L.; Cao, C.; Gao, Y.; Luo, H. Synthesis and phase transition behavior of undoped VO_2 with a strong nano-size effect. *Sol. Energy Mater. Sol. Cells* **2011**, *95*, 712–715. [CrossRef]
49. Liang, S.; Shi, Q.; Zhu, H.; Peng, B.; Huang, W. One-step hydrothermal synthesis of W-doped VO_2 (M) nanorods with a tunable phase-transition temperature for infrared smart windows. *ACS Omega* **2016**, *1*, 1139–1148. [CrossRef] [PubMed]

50. Strelcov, E.; Tselev, A.; Ivanov, I.; Budai, J.D.; Zhang, J.; Tischler, J.Z.; Kravchenko, I.; Kalinin, S.V.; Kolmakov, A. Doping-based stabilization of the M_2 phase in free-standing VO_2 nanostructures at room temperature. *Nano Lett.* **2012**, *12*, 6198–6205. [CrossRef] [PubMed]
51. Panagopoulou, M.; Gagaoudakis, E.; Boukos, N.; Aperathitis, E.; Kiriakidis, G.; Tsoukalas, D.; Raptis, Y. Thermochromic performance of Mg-doped VO_2 thin films on functional substrates for glazing applications. *Sol. Energy Mater. Sol. Cells* **2016**, *157*, 1004–1010. [CrossRef]
52. Zhao, Z.; Liu, Y.; Wang, D.; Ling, C.; Chang, Q.; Li, J.; Zhao, Y.; Jin, H. Sn dopants improve the visible transmittance of VO_2 films achieving excellent thermochromic performance for smart window. *Sol. Energy Mater. Sol. Cells* **2020**, *209*, 110443. [CrossRef]
53. Patridge, C.J.; Whittaker, L.; Ravel, B.; Banerjee, S. Elucidating the Influence of Local Structure Perturbations on the Metal–Insulator Transitions of $V_{1-x}Mo_xO_2$ Nanowires: Mechanistic Insights from an X-ray Absorption Spectroscopy Study. *J. Phys. Chem. C* **2012**, *116*, 3728–3736. [CrossRef]
54. Li, D.; Li, M.; Pan, J.; Luo, Y.; Wu, H.; Zhang, Y.; Li, G. Hydrothermal synthesis of Mo-doped VO_2/TiO_2 composite nanocrystals with enhanced thermochromic performance. *ACS Appl. Mater. Interfaces* **2014**, *6*, 6555–6561. [CrossRef] [PubMed]
55. Dai, L.; Chen, S.; Liu, J.; Gao, Y.; Zhou, J.; Chen, Z.; Cao, C.; Luo, H.; Kanehira, M. F-doped VO_2 nanoparticles for thermochromic energy-saving foils with modified color and enhanced solar-heat shielding ability. *Phys. Chem. Chem. Phys.* **2013**, *15*, 11723–11729. [CrossRef] [PubMed]
56. Gao, Y.; Wang, S.; Luo, H.; Dai, L.; Cao, C.; Liu, Y.; Chen, Z.; Kanehira, M. Enhanced chemical stability of VO_2 nanoparticles by the formation of SiO_2/VO_2 core/shell structures and the application to transparent and flexible VO_2-based composite foils with excellent thermochromic properties for solar heat control. *Energy Environ. Sci.* **2012**, *5*, 6104–6110. [CrossRef]
57. Cui, Y.; Ke, Y.; Liu, C.; Chen, Z.; Wang, N.; Zhang, L.; Zhou, Y.; Wang, S.; Gao, Y.; Long, Y. Thermochromic VO_2 for energy-efficient smart windows. *Joule* **2018**, *2*, 1707–1746. [CrossRef]
58. Li, S.-Y.; Niklasson, G.A.; Granqvist, C.-G. Nanothermochromics: Calculations for VO_2 nanoparticles in dielectric hosts show much improved luminous transmittance and solar energy transmittance modulation. *J. Appl. Phys* **2010**, *108*, 063525. [CrossRef]
59. Ke, Y.; Balin, I.; Wang, N.; Lu, Q.; Tok, A.I.Y.; White, T.J.; Magdassi, S.; Abdulhalim, I.; Long, Y. Two-dimensional SiO_2/VO_2 photonic crystals with statically visible and dynamically infrared modulated for smart window deployment. *ACS Appl. Mater. Interfaces* **2016**, *8*, 33112–33120. [CrossRef]
60. Kang, L.; Gao, Y.; Luo, H.; Chen, Z.; Du, J.; Zhang, Z. Nanoporous thermochromic VO_2 films with low optical constants, enhanced luminous transmittance and thermochromic properties. *ACS Appl. Mater. Interfaces* **2011**, *3*, 135–138. [CrossRef]
61. Mlyuka, N.; Niklasson, G.A.; Granqvist, C.-G. Thermochromic multilayer films of VO_2 and TiO_2 with enhanced transmittance. *Sol. Energy Mater. Sol. Cells* **2009**, *93*, 1685–1687. [CrossRef]
62. Ke, Y.; Chen, J.; Lin, G.; Wang, S.; Zhou, Y.; Yin, J.; Lee, P.S.; Long, Y. Smart windows: Electro-, thermo-, mechano-, photochromics, and beyond. *Adv. Energy Mater.* **2019**, *9*, 1902066. [CrossRef]
63. Cao, X.; Chang, T.; Shao, Z.; Xu, F.; Luo, H.; Jin, P. Challenges and opportunities toward real application of VO_2-based smart glazing. *Matter* **2020**, *2*, 862–881. [CrossRef]
64. Chang, Q.; Wang, D.; Zhao, Z.; Ling, C.; Wang, C.; Jin, H.; Li, J. Size-Controllable M-Phase VO_2 Nanocrystals for Flexible Thermochromic Energy-Saving Windows. *ACS Appl. Nano Mater.* **2021**, *4*, 6778–6785. [CrossRef]
65. Gao, Y.; Wang, S.; Kang, L.; Chen, Z.; Du, J.; Liu, X.; Luo, H.; Kanehira, M. VO_2–Sb: SnO_2 composite thermochromic smart glass foil. *Energy Environ. Sci.* **2012**, *5*, 8234–8237. [CrossRef]
66. Choi, Y.; Sim, D.M.; Hur, Y.H.; Han, H.J.; Jung, Y.S. Synthesis of colloidal VO_2 nanoparticles for thermochromic applications. *Sol. Energy Mater. Sol. Cells* **2018**, *176*, 266–272. [CrossRef]
67. Manca, N.; Pellegrino, L.; Kanki, T.; Venstra, W.J.; Mattoni, G.; Higuchi, Y.; Tanaka, H.; Caviglia, A.D.; Marré, D. Selective high-frequency mechanical actuation driven by the VO_2 electronic instability. *Adv. Mater.* **2017**, *29*, 1701618. [CrossRef] [PubMed]
68. Warwick, M.E.; Ridley, I.; Binions, R. Thermochromic vanadium dioxide thin films prepared by electric field assisted atmospheric pressure chemical vapour deposition for intelligent glazing application and their energy demand reduction properties. *Sol. Energy Mater. Sol. Cells* **2016**, *157*, 686–694. [CrossRef]
69. Jiazhen, Y.; Yue, Z.; Wanxia, H.; Mingjin, T. Effect of Mo-W Co-doping on semiconductor-metal phase transition temperature of vanadium dioxide film. *Thin Solid Films* **2008**, *516*, 8554–8558. [CrossRef]
70. Warwick, M.E.; Binions, R. Thermochromic vanadium dioxide thin films from electric field assisted aerosol assisted chemical vapour deposition. *Sol. Energy Mater. Sol. Cells* **2015**, *143*, 592–600. [CrossRef]
71. Gagaoudakis, E.; Michail, G.; Aperathitis, E.; Kortidis, I.; Binas, V.; Panagopoulou, M.; Raptis, Y.S.; Tsoukalas, D.; Kiriakidis, G. Low temperature rf-sputtered thermochromic VO_2 films on flexible glass substrates. *Adv. Mater. Lett.* **2017**, *8*, 757–761. [CrossRef]
72. Bukhari, S.A.; Kumar, S.; Kumar, P.; Gumfekar, S.P.; Chung, H.-J.; Thundat, T.; Goswami, A. The effect of oxygen flow rate on metal–insulator transition (MIT) characteristics of vanadium dioxide (VO_2) thin films by pulsed laser deposition (PLD). *Appl. Surf. Sci* **2020**, *529*, 146995. [CrossRef]
73. Garry, G.; Durand, O.; Lordereau, A. Structural, electrical and optical properties of pulsed laser deposited VO2 thin films on R-and C-sapphire planes. *Thin Solid Films* **2004**, *453*, 427–430. [CrossRef]
74. Chae, J.-Y.; Lee, D.; Woo, H.-Y.; Kim, J.B.; Paik, T. Direct transfer of thermochromic tungsten-doped vanadium dioxide thin-films onto flexible polymeric substrates. *Appl. Surf. Sci* **2021**, *545*, 148937. [CrossRef]
75. Lee, W.S.; Jeon, S.; Oh, S.J. Wearable sensors based on colloidal nanocrystals. *Nano Converg.* **2019**, *6*, 1–13. [CrossRef] [PubMed]

76. Zhou, Y.; Huang, A.; Li, Y.; Ji, S.; Gao, Y.; Jin, P. Surface plasmon resonance induced excellent solar control for VO$_2$@SiO$_2$ nanorods-based thermochromic foils. *Nanoscale* **2013**, *5*, 9208–9213. [CrossRef]
77. Guo, D.; Ling, C.; Wang, C.; Wang, D.; Li, J.; Zhao, Z.; Wang, Z.; Zhao, Y.; Zhang, J.; Jin, H. Hydrothermal one-step synthesis of highly dispersed M-phase VO$_2$ nanocrystals and application to flexible thermochromic film. *ACS Appl. Mater. Interfaces* **2018**, *10*, 28627–28634. [CrossRef] [PubMed]
78. Cao, C.; Gao, Y.; Luo, H. Pure single-crystal rutile vanadium dioxide powders: Synthesis, mechanism and phase-transformation property. *J. Phys. Chem. C* **2008**, *112*, 18810–18814. [CrossRef]
79. Zhang, J.; He, H.; Xie, Y.; Pan, B. Theoretical study on the tungsten-induced reduction of transition temperature and the degradation of optical properties for VO$_2$. *J. Chem. Phys.* **2013**, *138*, 114705. [CrossRef]
80. Chen, Z.; Gao, Y.; Kang, L.; Cao, C.; Chen, S.; Luo, H. Fine crystalline VO$_2$ nanoparticles: Synthesis, abnormal phase transition temperatures and excellent optical properties of a derived VO$_2$ nanocomposite foil. *J. Mater. Chem. A* **2014**, *2*, 2718–2727. [CrossRef]
81. Parkin, I.P.; Manning, T.D. Intelligent thermochromic windows. *J. Chem. Educ.* **2006**, *83*, 393. [CrossRef]
82. Zhang, J.; Wang, J.; Yang, C.; Jia, H.; Cui, X.; Zhao, S.; Xu, Y. Mesoporous SiO$_2$/VO$_2$ double-layer thermochromic coating with improved visible transmittance for smart window. *Sol. Energy Mater. Sol. Cells* **2017**, *162*, 134–141. [CrossRef]
83. Howard, S.A.; Evlyukhin, E.; Páez Fajardo, G.; Paik, H.; Schlom, D.G.; Piper, L.F. Digital Tuning of the Transition Temperature of Epitaxial VO$_2$ Thin Films on MgF$_2$ Substrates by Strain Engineering. *Adv. Mater. Interfaces* **2021**, *8*, 2001790. [CrossRef]
84. Choi, Y.; Lee, D.; Song, S.; Kim, J.; Ju, T.S.; Kim, H.; Kim, J.; Yoon, S.; Kim, Y.; Phan, T.B. Correlation between Symmetry and Phase Transition Temperature of VO$_2$ Films Deposited on Al$_2$O$_3$ Substrates with Various Orientations. *Adv. Electron. Mater.* **2021**, *7*, 2000874. [CrossRef]
85. Yan, J.; Huang, W.; Zhang, Y.; Liu, X.; Tu, M. Characterization of preferred orientated vanadium dioxide film on muscovite (001) substrate. *Phys. Status Solidi A* **2008**, *205*, 2409–2412. [CrossRef]
86. Li, C.-I.; Lin, J.-C.; Liu, H.-J.; Chu, M.-W.; Chen, H.-W.; Ma, C.-H.; Tsai, C.-Y.; Huang, H.-W.; Lin, H.-J.; Liu, H.-L. Van der Waal epitaxy of flexible and transparent VO$_2$ film on muscovite. *Chem. Mater.* **2016**, *28*, 3914–3919. [CrossRef]
87. Geim, A.K.; Grigorieva, I.V. Van der Waals heterostructures. *Nature* **2013**, *499*, 419–425. [CrossRef]
88. Novoselov, K.; Mishchenko, O.A.; Carvalho, O.A.; Neto, A.C. 2D materials and van der Waals heterostructures. *Science* **2016**, *353*, aac9439. [CrossRef]
89. Liu, Y.; Huang, Y.; Duan, X. Van der Waals integration before and beyond two-dimensional materials. *Nature* **2019**, *567*, 323–333. [CrossRef]
90. Liang, W.; Jiang, Y.; Guo, J.; Li, N.; Qiu, W.; Yang, H.; Ji, Y.; Luo, S.N. Van der Waals heteroepitaxial VO$_2$/mica films with extremely low optical trigger threshold and large THz field modulation depth. *Adv. Opt. Mater.* **2019**, *7*, 1900647. [CrossRef]
91. Wang, J.N.; Xiong, B.; Peng, R.W.; Li, C.Y.; Hou, B.Q.; Chen, C.W.; Liu, Y.; Wang, M. Flexible Phase Change Materials for Electrically-Tuned Active Absorbers. *Small* **2021**, *17*, 2101282. [CrossRef]
92. Chen, Y.; Fan, L.; Fang, Q.; Xu, W.; Chen, S.; Zan, G.; Ren, H.; Song, L.; Zou, C. Free-standing SWNTs/VO$_2$/Mica hierarchical films for high-performance thermochromic devices. *Nano Energy* **2017**, *31*, 144–151. [CrossRef]
93. Xiao, L.; Ma, H.; Liu, J.; Zhao, W.; Jia, Y.; Zhao, Q.; Liu, K.; Wu, Y.; Wei, Y.; Fan, S. Fast adaptive thermal camouflage based on flexible VO$_2$/graphene/CNT thin films. *Nano Lett.* **2015**, *15*, 8365–8370. [CrossRef] [PubMed]
94. Chang, T.; Zhu, Y.; Huang, J.; Luo, H.; Jin, P.; Cao, X. Flexible VO$_2$ thermochromic films with narrow hysteresis loops. *Sol. Energy Mater. Sol. Cells* **2021**, *219*, 110799. [CrossRef]
95. Chang, T.; Cao, X.; Li, N.; Long, S.; Gao, X.; Dedon, L.R.; Sun, G.; Luo, H.; Jin, P. Facile and low-temperature fabrication of thermochromic Cr$_2$O$_3$/VO$_2$ smart coatings: Enhanced solar modulation ability, high luminous transmittance and UV-shielding function. *ACS Appl. Mater. Interfaces* **2017**, *9*, 26029–26037. [CrossRef]
96. Martins, L.G.; Song, Y.; Zeng, T.; Dresselhaus, M.S.; Kong, J.; Araujo, P.T. Direct transfer of graphene onto flexible substrates. *Proc. Natl. Acad. Sci. USA* **2013**, *110*, 17762–17767. [CrossRef]
97. Malarde, D.; Powell, M.J.; Quesada-Cabrera, R.; Wilson, R.L.; Carmalt, C.J.; Sankar, G.; Parkin, I.P.; Palgrave, R.G. Optimized atmospheric-pressure chemical vapor deposition thermochromic VO$_2$ thin films for intelligent window applications. *ACS Omega* **2017**, *2*, 1040–1046. [CrossRef]
98. Kim, H.; Kim, Y.; Kim, K.S.; Jeong, H.Y.; Jang, A.-R.; Han, S.H.; Yoon, D.H.; Suh, K.S.; Shin, H.S.; Kim, T. Flexible thermochromic window based on hybridized VO$_2$/graphene. *ACS Nano* **2013**, *7*, 5769–5776. [CrossRef]
99. Paik, T.; Hong, S.-H.; Gaulding, E.A.; Caglayan, H.; Gordon, T.R.; Engheta, N.; Kagan, C.R.; Murray, C.B. Solution-processed phase-change VO$_2$ metamaterials from colloidal vanadium oxide (VO$_x$) nanocrystals. *ACS Nano* **2014**, *8*, 797–806. [CrossRef] [PubMed]
100. Parameswaran, C.; Gupta, D. Large area flexible pressure/strain sensors and arrays using nanomaterials and printing techniques. *Nano Converg.* **2019**, *6*, 1–23. [CrossRef]
101. Wang, S.; Liu, M.; Kong, L.; Long, Y.; Jiang, X.; Yu, A. Recent progress in VO$_2$ smart coatings: Strategies to improve the thermochromic properties. *Prog. Mater. Sci.* **2016**, *81*, 1–54. [CrossRef]
102. Chae, B.-G.; Kim, H.-T.; Yun, S.-J.; Kim, B.-J.; Lee, Y.-W.; Youn, D.-H.; Kang, K.-Y. Highly oriented VO$_2$ thin films prepared by sol-gel deposition. *Electrochem. Solid-State Lett.* **2005**, *9*, C12. [CrossRef]

103. Speck, K.; Hu, H.-W.; Sherwin, M.; Potember, R. Vanadium dioxide films grown from vanadium tetra-isopropoxide by the sol-gel process. *Thin Solid Films* **1988**, *165*, 317–322. [CrossRef]
104. Cao, X.; Wang, N.; Law, J.Y.; Loo, S.C.J.; Magdassi, S.; Long, Y. Nanoporous thermochromic VO_2 (M) thin films: Controlled porosity, largely enhanced luminous transmittance and solar modulating ability. *Langmuir* **2014**, *30*, 1710–1715. [CrossRef] [PubMed]
105. Jo, Y.-R.; Lee, W.-J.; Yoon, M.-H.; Kim, B.-J. In Situ Tracking of Low-Temperature VO_2 Crystallization via Photocombustion and Characterization of Phase-Transition Reliability on Large-Area Flexible Substrates. *Chem. Mater.* **2020**, *32*, 4013–4023. [CrossRef]
106. Zhang, J.; Jin, H.; Chen, Z.; Cao, M.; Chen, P.; Dou, Y.; Zhao, Y.; Li, J. Self-assembling VO_2 nanonet with high switching performance at wafer-scale. *Chem. Mater.* **2015**, *27*, 7419–7424. [CrossRef]
107. Zhong, L.; Luo, Y.; Li, M.; Han, Y.; Wang, H.; Xu, S.; Li, G. TiO_2 seed-assisted growth of VO_2 (M) films and thermochromic performance. *CrystEngComm* **2016**, *18*, 7140–7146. [CrossRef]
108. Makarevich, A.; Makarevich, O.; Ivanov, A.; Sharovarov, D.; Eliseev, A.; Amelichev, V.; Boytsova, O.; Gorodetsky, A.; Navarro-Cia, M.; Kaul, A. Hydrothermal epitaxy growth of self-organized vanadium dioxide 3D structures with metal–insulator transition and THz transmission switch properties. *CrystEngComm* **2020**, *22*, 2612–2620. [CrossRef]
109. Guo, D.; Zhao, Z.; Li, J.; Zhang, J.; Zhang, R.; Wang, Z.; Chen, P.; Zhao, Y.; Chen, Z.; Jin, H. Symmetric confined growth of superstructured vanadium dioxide nanonet with a regular geometrical pattern by a solution approach. *Cryst. Growth Des.* **2017**, *17*, 5838–5844. [CrossRef]
110. Melnik, V.; Khatsevych, I.; Kladko, V.; Kuchuk, A.; Nikirin, V.; Romanyuk, B. Low-temperature method for thermochromic high ordered VO2 phase formation. *Mater. Lett.* **2012**, *68*, 215–217. [CrossRef]
111. Ivanov, A.V.; Makarevich, O.N.; Boytsova, O.V.; Tsymbarenko, D.M.; Eliseev, A.A.; Amelichev, V.A.; Makarevich, A.M. Citrate-assisted hydrothermal synthesis of vanadium dioxide textured films with metal-insulator transition and infrared thermochromic properties. *Ceram. Int.* **2020**, *46*, 19919–19927. [CrossRef]
112. Zhao, Z.; Liu, Y.; Yu, Z.; Ling, C.; Li, J.; Zhao, Y.; Jin, H. Sn–W Co-doping Improves Thermochromic Performance of VO_2 Films for Smart Windows. *ACS Appl. Energy Mater* **2020**, *3*, 9972–9979. [CrossRef]
113. Théobald, F. Étude hydrothermale du système VO_2-VO_2, 5-H_2O. *J. Less-Common Met.* **1977**, *53*, 55–71. [CrossRef]
114. Shi, J.; Zhou, S.; You, B.; Wu, L. Preparation and thermochromic property of tungsten-doped vanadium dioxide particles. *Sol. Energy Mater. Sol. Cells* **2007**, *91*, 1856–1862. [CrossRef]
115. Liu, P.; Zhu, K.; Gao, Y.; Wu, Q.; Liu, J.; Qiu, J.; Gu, Q.; Zheng, H. Ultra-long VO_2 (A) nanorods using the high-temperature mixing method under hydrothermal conditions: Synthesis, evolution and thermochromic properties. *CrystEngComm* **2013**, *15*, 2753–2760. [CrossRef]
116. Liu, L.; Cao, F.; Yao, T.; Xu, Y.; Zhou, M.; Qu, B.; Pan, B.; Wu, C.; Wei, S.; Xie, Y. New-phase VO_2 micro/nanostructures: Investigation of phase transformation and magnetic property. *New J. Chem.* **2012**, *36*, 619–625. [CrossRef]
117. Wu, C.; Hu, Z.; Wang, W.; Zhang, M.; Yang, J.; Xie, Y. Synthetic paramontroseite VO_2 with good aqueous lithium–ion battery performance. *Chem. Commun.* **2008**, 3891–3893. [CrossRef]
118. Shen, N.; Chen, S.; Chen, Z.; Liu, X.; Cao, C.; Dong, B.; Luo, H.; Liu, J.; Gao, Y. The synthesis and performance of Zr-doped and W–Zr-codoped VO_2 nanoparticles and derived flexible foils. *J. Mater. Chem. A* **2014**, *2*, 15087–15093. [CrossRef]
119. Popuri, S.R.; Miclau, M.; Artemenko, A.; Labrugere, C.; Villesuzanne, A.; Pollet, M. Rapid hydrothermal synthesis of VO_2 (B) and its conversion to thermochromic VO_2 (M_1). *Inorg. Chem.* **2013**, *52*, 4780–4785. [CrossRef]
120. Corr, S.A.; Grossman, M.; Shi, Y.; Heier, K.R.; Stucky, G.D.; Seshadri, R. VO_2 (B) nanorods: Solvothermal preparation, electrical properties, and conversion to rutile VO_2 and V_2O_3. *J. Mater. Chem.* **2009**, *19*, 4362–4367. [CrossRef]
121. Sun, Y.; Jiang, S.; Bi, W.; Long, R.; Tan, X.; Wu, C.; Wei, S.; Xie, Y. New aspects of size-dependent metal-insulator transition in synthetic single-domain monoclinic vanadium dioxide nanocrystals. *Nanoscale* **2011**, *3*, 4394–4401. [CrossRef] [PubMed]
122. Zhong, L.; Li, M.; Wang, H.; Luo, Y.; Pan, J.; Li, G. Star-shaped VO_2 (M) nanoparticle films with high thermochromic performance. *CrystEngComm* **2015**, *17*, 5614–5619. [CrossRef]
123. Song, Z.; Zhang, L.; Xia, F.; Webster, N.A.; Song, J.; Liu, B.; Luo, H.; Gao, Y. Controllable synthesis of VO_2 (D) and their conversion to VO_2 (M) nanostructures with thermochromic phase transition properties. *Inorg. Chem. Front.* **2016**, *3*, 1035–1042. [CrossRef]
124. Li, M.; Ji, S.; Pan, J.; Wu, H.; Zhong, L.; Wang, Q.; Li, F.; Li, G. Infrared response of self-heating VO_2 nanoparticles film based on Ag nanowires heater. *J. Mater. Chem. A* **2014**, *2*, 20470–20473. [CrossRef]
125. Li, M.; Wu, X.; Li, L.; Wang, Y.; Li, D.; Pan, J.; Li, S.; Sun, L.; Li, G. Defect-mediated phase transition temperature of VO 2 (M) nanoparticles with excellent thermochromic performance and low threshold voltage. *J. Mater. Chem. A* **2014**, *2*, 4520–4523. [CrossRef]
126. Chen, R.; Miao, L.; Liu, C.; Zhou, J.; Cheng, H.; Asaka, T.; Iwamoto, Y.; Tanemura, S. Shape-controlled synthesis and influence of W doping and oxygen nonstoichiometry on the phase transition of VO_2. *Sci. Rep.* **2015**, *5*, 1–12. [CrossRef]
127. Narayan, J.; Bhosle, V. Phase transition and critical issues in structure-property correlations of vanadium oxide. *J. Appl. Phys* **2006**, *100*, 103524. [CrossRef]
128. Zeng, W.; Chen, N.; Xie, W. Research progress on the preparation methods for VO_2 nanoparticles and their application in smart windows. *CrystEngComm* **2020**, *22*, 851–869. [CrossRef]
129. Lopez, R.; Feldman, L.C.; Haglund Jr, R.F. Size-Dependent Optical Properties of VO_2 Nanoparticle Arrays. *Phys. Rev. Lett.* **2004**, *93*, 177403. [CrossRef]

130. Liu, Y.; Liu, J.; Li, Y.; Wang, D.; Ren, L.; Zou, K. Effect of annealing temperature on the structure and properties of vanadium oxide films. *Opt. Mater. Express* **2016**, *6*, 1552–1560. [CrossRef]
131. Zhang, H.; Wu, Z.; Wu, X.; Yang, W.; Jiang, Y. Transversal grain size effect on the phase-transition hysteresis width of vanadium dioxide films comprising spheroidal nanoparticles. *Vacuum* **2014**, *104*, 47–50. [CrossRef]
132. Alie, D.; Gedvilas, L.; Wang, Z.; Tenent, R.; Engtrakul, C.; Yan, Y.; Shaheen, S.E.; Dillon, A.C.; Ban, C. Direct synthesis of thermochromic VO_2 through hydrothermal reaction. *J. Solid State Chem.* **2014**, *212*, 237–241. [CrossRef]
133. Li, W.; Ji, S.; Sun, G.; Ma, Y.; Guo, H.; Jin, P. Novel VO_2 (M)–ZnO heterostructured dandelions with combined thermochromic and photocatalytic properties for application in smart coatings. *New J. Chem.* **2016**, *40*, 2592–2600. [CrossRef]
134. Ji, S.; Zhang, F.; Jin, P. Preparation of high performance pure single phase VO_2 nanopowder by hydrothermally reducing the V_2O_5 gel. *Sol. Energy Mater. Sol. Cells* **2011**, *95*, 3520–3526. [CrossRef]
135. Chen, R.; Miao, L.; Cheng, H.; Nishibori, E.; Liu, C.; Asaka, T.; Iwamoto, Y.; Takata, M.; Tanemura, S. One-step hydrothermal synthesis of $V_{1-x}W_xO_2$ (M/R) nanorods with superior doping efficiency and thermochromic properties. *J. Mater. Chem. A* **2015**, *3*, 3726–3738. [CrossRef]
136. Kim, J.B.; Lee, D.; Yeo, I.H.; Woo, H.Y.; Kim, D.W.; Chae, J.-Y.; Han, S.H.; Paik, T. Hydrothermal synthesis of monoclinic vanadium dioxide nanocrystals using phase-pure vanadium precursors for high-performance smart windows. *Sol. Energy Mater. Sol. Cells* **2021**, *226*, 111055. [CrossRef]
137. Dahiya, A.S.; Shakthivel, D.; Kumaresan, Y.; Zumeit, A.; Christou, A.; Dahiya, R. High-performance printed electronics based on inorganic semiconducting nano to chip scale structures. *Nano Converg.* **2020**, *7*, 1–25. [CrossRef] [PubMed]
138. Li, S.-Y.; Niklasson, G.A.; Granqvist, C.-G. Nanothermochromics with VO_2-based core-shell structures: Calculated luminous and solar optical properties. *J. Appl. Phys* **2011**, *109*, 113515. [CrossRef]
139. Shen, N.; Chen, S.; Wang, W.; Shi, R.; Chen, P.; Kong, D.; Liang, Y.; Amini, A.; Wang, J.; Cheng, C. Joule heating driven infrared switching in flexible VO_2 nanoparticle films with reduced energy consumption for smart windows. *J. Mater. Chem. A* **2019**, *7*, 4516–4524. [CrossRef]
140. Zhou, J.; Gao, Y.; Liu, X.; Chen, Z.; Dai, L.; Cao, C.; Luo, H.; Kanahira, M.; Sun, C.; Yan, L. Mg-doped VO_2 nanoparticles: Hydrothermal synthesis, enhanced visible transmittance and decreased metal–insulator transition temperature. *Phys. Chem. Chem. Phys.* **2013**, *15*, 7505–7511. [CrossRef]
141. Nam, V.B.; Giang, T.T.; Koo, S.; Rho, J.; Lee, D. Laser digital patterning of conductive electrodes using metal oxide nanomaterials. *Nano Converg.* **2020**, *7*, 1–17. [CrossRef]
142. Yang, S.; Vaseem, M.; Shamim, A. Fully inkjet-printed VO_2-based radio-frequency switches for flexible reconfigurable components. *Adv. Mater. Technol.* **2019**, *4*, 1800276. [CrossRef]
143. Ji, H.; Liu, D.; Cheng, H.; Zhang, C. Inkjet printing of vanadium dioxide nanoparticles for smart windows. *J. Mater. Chem. C* **2018**, *6*, 2424–2429. [CrossRef]
144. Ji, H.; Liu, D.; Cheng, H.; Tao, Y. Large area infrared thermochromic VO_2 nanoparticle films prepared by inkjet printing technology. *Sol. Energy Mater. Sol. Cells* **2019**, *194*, 235–243. [CrossRef]
145. Wang, Y.; Zhao, F.; Wang, J.; Khan, A.R.; Shi, Y.; Chen, Z.; Zhang, K.; Li, L.; Gao, Y.; Guo, X. VO_2@SiO_2/Poly(N-isopropylacrylamide) hybrid nanothermochromic microgels for smart window. *Ind. Eng. Chem. Res.* **2018**, *57*, 12801–12808. [CrossRef]
146. Ji, H.; Liu, D.; Zhang, C.; Cheng, H. VO_2/ZnS core-shell nanoparticle for the adaptive infrared camouflage application with modified color and enhanced oxidation resistance. *Sol. Energy Mater. Sol. Cells* **2018**, *176*, 1–8. [CrossRef]
147. Wen, Z.; Ke, Y.; Feng, C.; Fang, S.; Sun, M.; Liu, X.; Long, Y. Mg-Doped VO_2@ZrO_2 Core—Shell Nanoflakes for Thermochromic Smart Windows with Enhanced Performance. *Adv. Mater. Interfaces* **2021**, *8*, 2001606. [CrossRef]
148. Saini, M.; Dehiya, B.S.; Umar, A. VO_2 (M)@CeO_2 core-shell nanospheres for thermochromic smart windows and photocatalytic applications. *Ceram. Int.* **2020**, *46*, 986–995. [CrossRef]
149. Moot, T.; Palin, C.; Mitran, S.; Cahoon, J.F.; Lopez, R. Designing Plasmon-Enhanced Thermochromic Films Using a Vanadium Dioxide Nanoparticle Elastomeric Composite. *Adv. Opt. Mater.* **2016**, *4*, 578–583. [CrossRef]

MDPI
St. Alban-Anlage 66
4052 Basel
Switzerland
Tel. +41 61 683 77 34
Fax +41 61 302 89 18
www.mdpi.com

Nanomaterials Editorial Office
E-mail: nanomaterials@mdpi.com
www.mdpi.com/journal/nanomaterials

www.ingramcontent.com/pod-product-compliance
Lightning Source LLC
LaVergne TN
LVHW070742100526
838202LV00013B/1284